Towards a Sustainable Management of Mine Wastes

Towards a Sustainable Management of Mine Wastes

Reprocessing, Reuse, Revalorization and Repository

Special Issue Editors

Mostafa Benzaazoua
Yassine Taha

MDPI • Basel • Beijing • Wuhan • Barcelona • Belgrade

Special Issue Editors
Mostafa Benzaazoua
University of Quebec
Canada

Yassine Taha
Mohammed VI Polytechnic University
Morocco

Editorial Office
MDPI
St. Alban-Anlage 66
4052 Basel, Switzerland

This is a reprint of articles from the Special Issue published online in the open access journal *Minerals* (ISSN 2075-163X) from 2018 to 2019 (available at: https://www.mdpi.com/journal/minerals/special_issues/industrial_mining_wastes).

For citation purposes, cite each article independently as indicated on the article page online and as indicated below:

LastName, A.A.; LastName, B.B.; LastName, C.C. Article Title. *Journal Name* **Year**, *Article Number*, Page Range.

ISBN 978-3-03928-174-9 (Pbk)
ISBN 978-3-03928-175-6 (PDF)

© 2020 by the authors. Articles in this book are Open Access and distributed under the Creative Commons Attribution (CC BY) license, which allows users to download, copy and build upon published articles, as long as the author and publisher are properly credited, which ensures maximum dissemination and a wider impact of our publications.

The book as a whole is distributed by MDPI under the terms and conditions of the Creative Commons license CC BY-NC-ND.

Contents

About the Special Issue Editors vii

Preface to "Towards a Sustainable Management of Mine Wastes" ix

Yassine Taha and Mostafa Benzaazoua
Editorial for Special Issue "Towards a Sustainable Management of Mine Wastes: Reprocessing, Reuse, Revalorization, and Repository"
Reprinted from: *Minerals* **2020**, *10*, 21, doi:10.3390/min10010021 - 1

José A. Aznar-Sánchez, José J. García-Gómez, Juan F. Velasco-Muñoz and Anselmo Carretero-Gómez
Mining Waste and Its Sustainable Management: Advances in Worldwide Research
Reprinted from: *Minerals* **2018**, *8*, 284, doi:10.3390/min8070284 5

Maedeh Tayebi-Khorami, Mansour Edraki, Glen Corder and Artem Golev
Re-Thinking Mining Waste through an Integrative Approach Led by Circular Economy Aspirations
Reprinted from: *Minerals* **2019**, *9*, 286, doi:10.3390/min9050286 32

Aurélie Chopard, Philippe Marion, Raphaël Mermillod-Blondin, Benoît Plante and Mostafa Benzaazoua
Environmental Impact of Mine Exploitation: An Early Predictive Methodology Based on Ore Mineralogy and Contaminant Speciation
Reprinted from: *Minerals* **2019**, *9*, 397, doi:10.3390/min9070397 45

Abdellatif Elghali, Mostafa Benzaazoua, Bruno Bussière and Thomas Genty
Spatial Mapping of Acidity and Geochemical Properties of Oxidized Tailings within the Former Eagle/Telbel Mine Site
Reprinted from: *Minerals* **2019**, *9*, 180, doi:10.3390/min9030180 76

Abdelkrim Nadeif, Yassine Taha, Hassan Bouzahzah, Rachid Hakkou and Mostafa Benzaazoua
Desulfurization of the Old Tailings at the Au-Ag-Cu Tiouit Mine (Anti-Atlas Morocco)
Reprinted from: *Minerals* **2019**, *9*, 401, doi:10.3390/min9070401 99

Boujemaa Drif, Yassine Taha, Rachid Hakkou and Mostafa Benzaazoua
Recovery of Residual Silver-Bearing Minerals from Low-Grade Tailings by Froth Flotation: The Case of Zgounder Mine, Morocco
Reprinted from: *Minerals* **2018**, *8*, 273, doi:10.3390/min8070273 114

Richard A. Crane and Devin J. Sapsford
Towards Greener Lixiviants in Value Recovery from Mine Wastes: Efficacy of Organic Acids for the Dissolution of Copper and Arsenic from Legacy Mine Tailings
Reprinted from: *Minerals* **2018**, *8*, 383, doi:10.3390/min8090383 131

Bingxin Zhou, Shaotao Cao, Fangfang Chen, Fangfang Zhang and Yi Zhang
Recovery of Alkali from Bayer Red Mud Using CaO and/or MgO
Reprinted from: *Minerals* **2019**, *9*, 269, doi:10.3390/min9050269 147

Chang Tang, Keqing Li, Wen Ni and Duncheng Fan
Recovering Iron from Iron Ore Tailings and Preparing Concrete Composite Admixtures
Reprinted from: *Minerals* **2019**, *9*, 232, doi:10.3390/min9040232 162

Wilson Mugera Gitari, Rendani Thobakgale and Segun Ajayi Akinyemi
Mobility and Attenuation Dynamics of Potentially Toxic Chemical Species at an Abandoned Copper Mine Tailings Dump
Reprinted from: *Minerals* **2018**, *8*, 64, doi:10.3390/min8020064 176

Abdellatif Elghali, Mostafa Benzaazoua, Bruno Bussière and Thomas Genty
In Situ Effectiveness of Alkaline and Cementitious Amendments to Stabilize Oxidized Acid-Generating Tailings
Reprinted from: *Minerals* **2019**, *9*, 314, doi:10.3390/min9050314 194

Liuhua Yang, Hongjiang Wang, Hong Li and Xu Zhou
Effect of High Mixing Intensity on Rheological Properties of Cemented Paste Backfill
Reprinted from: *Minerals* **2019**, *9*, 240, doi:10.3390/min9040240 214

About the Special Issue Editors

Mostafa Benzaazoua joined the University of Quebec (UQAT) in 1996, as postdoctoral fellow. He became professor in June 1997 in the same university and held a Canada Research Chair (CRC 2003–2011) in "mine waste integrated management". In 2009, he co-chairs an International Research Chair funded by the International Development Research Centre (IDRC 2009–2014) jointly with the CRC program jointly with UCA Marrakech University (Morocco). He is presently Affiliated Professor at the University Mohamed IV Polytechnique in Morocco (UM6P) leading the program of Mining Environment and Circular Economy. He is also Visiting Professor at the Wuhan Institute of technology in China (WIT). He has worked on large number of government-funded and industry-sponsored projects, dealing with applied mineralogy and geochemistry for mine pollution control, waste management and valorization, mine site rehabilitation, and mineral processing. At the end of 2010, he took a secondment and joined the National Institute of Applied Sciences at Lyon in France as a University Professor, which allow him to diversify his research themes working on other industrial mineral waste management, environmental evaluation, treatment and reuse (dredged sediments and sewage sludges, incineration by-products, etc.). In 2012, he returned to UQAT, joining its recent Research Institute of Mining and Environment. M. Benzaazoua is familiar with various spectroscopic and microbeam-based mineralogical techniques, most mineral processing techniques, as well as lab and field physical modeling applied to study the pollution generation potential from mining and industrial wastes before and after treatment/stabilization/valorization. Presently, he is working on the adaptation of Circular Economy, Industrial Ecology and Geometallurgy to mine industry.

Yassine Taha is an Assistant Professor at Mohammed VI Polytechnic University (UM6P), Morocco. His research interests revolve around phosphate sustainable mining, mine waste management and valorization, development of high-value added materials based on industrial solid wastes, resource recovery and low-grade ore beneficiation, low carbon footprint materials for construction, life cycle assessment and circular economy.

Preface to "Towards a Sustainable Management of Mine Wastes"

The mining industry could be the synonym for wealth creation, goods production, services and infrastructure insurance and life quality improvement. However, it could also be the origin of many harmful potential impacts on the surrounding environments, causing water, soil and air pollution. Mine operations continuously produce huge volumes of mine waste, more remarkably in the case of open pit mining. In this Special Issue, an emphasis was given to the main challenges facing the mining industry and the need to manage mine waste in sustainable ways. It also presents many lessons that can be learned from mining's liability and the numerous abandoned mine sites in many places around the world. Different approaches and management strategies have been developed by researchers from all around the five continents. The papers consider the social, economic, environmental, and technical aspects of sustainable management of mine waste. The developed solutions are generally very site-specific and are difficult to apply from site to site but could inspire further mining development.

Site remediation, waste reuse in situ and out of mine sites, and the evaluation of the potential alternative uses for mining waste are the most currently covered topics, since they allow the mine industry to subscribe to the circular economy framework and to have a reduced ecological footprint. The sustainable management of mining waste and the contribution of mine waste management to sustainability is an emerging and widely explored research field. This research area has considerable potential due to the urgent need for more sustainable practices. Waste management and mine closure planning could be realized using the best available practices, particularly if earlier detection is conducted, possibly even during the exploration step of the mine lifecycle. In this way, if the environmental impacts of mine waste are well planned during the mine development phase, and well controlled during the operation step, the mine closure can be completed more easily and sustainably.

Fourteen research papers in the form of reviews, articles, technical notes, and case studies about the sustainable management of mine waste were published in this Special Issue. The United States, Canada, Spain, Australia, and China are the countries that produce the most papers in mine waste management. We are thankful to all of the authors for sharing their research activities and to all reviewers for improving the quality of the submitted manuscripts. The research in the field helps legislation be move to a more strict framing by concerned governments, which can finally make the mine industry more environmental friendly
.

Mostafa Benzaazoua, Yassine Taha
Special Issue Editors

Editorial

Editorial for Special Issue "Towards a Sustainable Management of Mine Wastes: Reprocessing, Reuse, Revalorization, and Repository"

Yassine Taha [1,*] and Mostafa Benzaazoua [1,2,*]

[1] Materials Science and Nano-Engineering Department, Mohammed VI Polytechnic University, Lot 660, Hay Moulay Rachid, Ben Guerir 43150, Morocco
[2] Institut de Recherche en Mines et Environnement (IRME), Université du Québec en Abitibi-Témiscamingue (UQAT), 445 Boul de l'Université, Rouyn-Noranda, QC J9X 5E4, Canada
* Correspondence: Yassine.TAHA@um6p.ma (Y.T.); Mostafa.Benzaazoua@uqat.ca (M.B.)

Received: 20 December 2019; Accepted: 24 December 2019; Published: 25 December 2019

The mining industry continues to face many challenges due to its potential environmental impacts. These challenges are becoming harder to overcome with growing social awareness, increasing governmental pressure on disposal alternatives, and the scarcity of available spaces for waste disposal [1]. The need to manage mine wastes in a sustainable way also continues to grow. In this Special Issue, research papers in the form of reviews, articles, technical notes, and case studies about the sustainable management of mine wastes were selected. Different approaches and management strategies have been developed by researchers from all around the world. The papers presented in this Special Issue consider the social, economic, environmental, and technical aspects of sustainable management of mine wastes. Many different types of mine wastes can be produced during a mine's life cycle (Figure 1), and the choice of methods for managing them depends on many factors. Therefore, the developed solutions are generally very site-specific and difficult to apply in other cases.

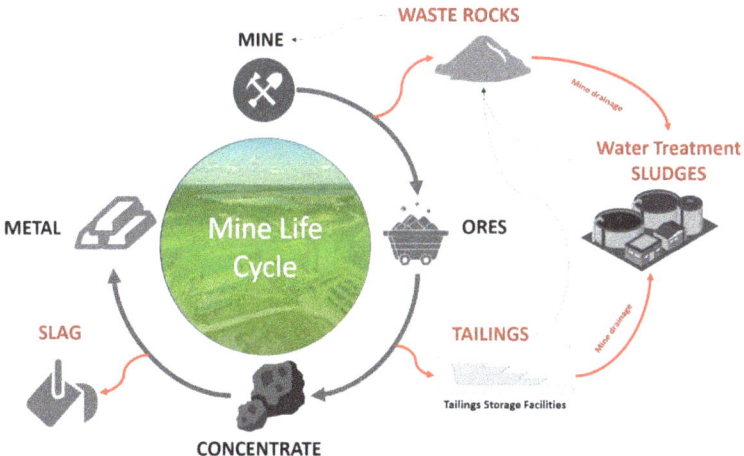

Figure 1. A mine's life cycle and types of mine wastes.

A systematic review paper from the University of Almeria (Spain), based on a bibliometric analysis of a sample of 3577 articles from 1988 to 2017, analyzed the dynamics of research focused on mining wastes and their sustainable management [2]. The results indicated an increase in the number of published papers about mining wastes due to an increase in attention to, and the growing social awareness about, mining wastes. It was revealed also that the United States, followed by Canada, Spain, Australia, and China are the countries that produce the most papers in mine waste management. Remediation, reuse, and evaluation of the potential alternative uses of mining wastes are the most currently studied aspects since they allow for the mine industry to subscribe to the circular economy, and allow for the mining industry to have a smaller ecological footprint. The sustainable management of mining wastes and the contribution of mine waste management to sustainability is an emerging field of research that has considerable potential due to the increased demand for natural resources and for sustainable practices.

Another review paper from the University of Queensland (Australia) investigated current knowledge about mining waste management and the possible opportunities related to the application of circular economy thinking to mine wastes [3]. Five key areas were explored (Figure 2): social dimensions, geoenvironmental aspects, geometallurgy specifications, economic drivers, and legal implications. Much research work remains necessary to identify efficient and effective solutions in each key area, as these kinds of approaches require cross-disciplinary skills to transform waste materials into more valuable resources.

Figure 2. Key areas of an integrative approach to mining waste management (adapted from [3]).

The evaluation of the environmental impacts of mine wastes has been also included in this Special Issue through different articles. Acid mine drainage (AMD) from sulfidic and oxidized ores and the related impacts on the geochemical behavior of mine wastes and pollutant mobility were assessed. A new predictive methodology based on ore mineralogy and contaminant speciation was developed by Chopard et al. [4]. It was concluded in this study that it is possible to improve waste management practices and mine closure planning, while saving time and money thanks to an earlier detection of the environmental impacts of mine wastes from the beginning of the mine development phase. In the case of closed mines, geochemical properties in an oxidized tailings storage facility (TSF) were evaluated by Elghali et al. [5]. The spatial variability of the geochemical properties of tailings in the TSF is mainly governed by the formation of hardpans at the subsurface of the tailings. The goal of the study was also to identify the most appropriate reclamation scenario for each area based on a geographic information system (GIS)-based approach and multi-technique characterization.

The environmental desulfurization of sulfidic tailings was used to depollute tailings by removing the problematic residual sulfides in different metallic and nonmetallic tailings [6,7]. This technique was applied by Nadeif et al. [8] to reprocess tailings from an abandoned mine to avoid acid mine drainage and to recover the residual metals (Au, Ag, and Cu). The same approach was adopted by Drif et al. [9], who succeeded in recovering silver-bearing minerals, mainly in the form of sulfides, from low-grade tailings by froth flotation processing. By doing so, the authors proved that the cleaned tailings are inert and do not present any leaching risk according to a toxicity characteristic leaching procedure test (TCLP). Another method was developed by Crane and Sapsford [10], who assessed the performance of a leaching-based remediation approach using greener lixiviants for the recovery of value from legacy mine wastes and tailings. The results of this study demonstrated that the use of organic acids could provide similar As and Cu recovery efficiencies while preserving the environment. Moreover, this Special Issue also presents information about a hydrothermal method that was used to recover alkali from Bayer red mud using CaO and/or MgO for resource utilization and environmental protection [11]. As an example of mine waste that could have economic potential, historic tailings from the gravity separation process in an old Bergwerkswohlfahrt mine waste dump in Germany were investigated. The spatial distribution of valuable metals (Pb, Zn, Cu, Ag, and Sb) was assessed using a laser-induced breakdown spectroscopy (LIBS) core scanner coupled with other textural, geochemical, and mineralogical characterization techniques. The results of this study demonstrated that a total tonnage close to 8000 t of lead and 610,000 ounces of silver could be recovered from these tailings. Finally, in the framework of circular economy objectives, a technical note has focused on the integrated management of iron ore tailings by recovering the residual iron and preparing concrete composites from high silica residues and blast furnace slag [12].

As is the case in many developing countries where environmental policies and practices remain underdeveloped, it was reported that, due to many years of unregulated and unsustainable mining activities, large volumes of disposed mine tailings could present various environmental risks related to the potential mobility of toxic chemical species into surface and groundwater systems [13]. Gitari et al. [13] present the case of copper mine wastes from Musina in South Africa. After a characterization of tailings and leachates, it was concluded that the mobility of potentially toxic chemical species to the aqueous phase was very low. This was explained mainly by the adsorption of metals to the surface of precipitated hematite.

In the contrary, in developed countries such as Canada, local regulation obliges every closed mine site to be rehabilitated. At least some mitigating techniques can be applied successfully. This was found to be the case at the Joutel mine site, where stabilization of acidic tailings was investigated to allow for improvement in the effluents' quality before final mine reclamation [14]. In this context, Elghali et al. [14] assessed the effectiveness of alkaline and cementitious additives (limestones, ordinary Portland cement, and fly ash) for the neutralization and stabilization of acid-generating tailings from a closed gold mine in Quebec, Canada. The results highlighted the efficiency of the used additives in increasing pH to circumneutral values and decreasing the release of metals/metalloids.

Cemented paste backfill was also investigated as another option for the disposal of tailings. This option is widely used due to its environmental and economic benefits. However, many parameters could affect the quality of the cemented backfill, including the quality of the tailings and the mixing performance. Therefore, the effect of high mixing intensity on the rheological properties of cemented paste backfill was investigated by Yang et al. [15]. It was demonstrated that the mixing intensity affects the rheological behavior of the paste backfill.

Conflicts of Interest: The authors declare no conflict of interest.

References

1. Taha, Y.; Benzaazoua, M.; Hakkou, R.; Mansori, M. Natural clay substitution by calamine processing wastes to manufacture fired bricks. *J. Clean. Prod.* **2016**, *135*, 847–858. [CrossRef]
2. Aznar-Sánchez, J.A.; García-Gómez, J.J.; Velasco-Muñoz, J.F.; Carretero-Gómez, A. Mining waste and its sustainable management: Advances in worldwide research. *Minerals* **2018**, *8*, 284. [CrossRef]
3. Tayebi-Khorami, M.; Edraki, M.; Corder, G.; Golev, A. Re-Thinking Mining Waste through an Integrative Approach Led by Circular Economy Aspirations. *Minerals* **2019**, *9*, 286. [CrossRef]
4. Chopard, A.; Marion, P.; Mermillod-Blondin, R.; Plante, B.; Benzaazoua, M. Environmental Impact of Mine Exploitation: An Early Predictive Methodology Based on Ore Mineralogy and Contaminant Speciation. *Minerals* **2019**, *9*, 397. [CrossRef]

5. Elghali, A.; Benzaazoua, M.; Bussière, B.; Genty, T. Spatial Mapping of Acidity and Geochemical Properties of Oxidized Tailings within the Former Eagle/Telbel Mine Site. *Minerals* **2019**, *9*, 180. [CrossRef]
6. Benzaazoua, M.; Bussière, B.; Demers, I.; Aubertin, M.; Fried, É.; Blier, A. Integrated mine tailings management by combining environmental desulphurization and cemented paste backfill: Application to mine Doyon, Quebec, Canada. *Miner. Eng.* **2008**, *21*, 330–340. [CrossRef]
7. Çelik, M.; Yildirim, I. A new physical process for desulfurization of low-rank coals. *Fuel* **2000**, *79*, 1665–1669. [CrossRef]
8. Nadeif, A.; Taha, Y.; Bouzahzah, H.; Hakkou, R.; Benzaazoua, M. Desulfurization of the Old Tailings at the Au-Ag-Cu Tiouit Mine (Anti-Atlas Morocco). *Minerals* **2019**, *9*, 401. [CrossRef]
9. Drif, B.; Taha, Y.; Hakkou, R.; Benzaazoua, M. Recovery of Residual Silver-Bearing Minerals from Low-Grade Tailings by Froth Flotation: The Case of Zgounder Mine, Morocco. *Minerals* **2018**, *8*, 273. [CrossRef]
10. Crane, R.; Sapsford, D. Towards Greener Lixiviants in Value Recovery from Mine Wastes: Efficacy of Organic Acids for the Dissolution of Copper and Arsenic from Legacy Mine Tailings. *Minerals* **2018**, *8*, 383. [CrossRef]
11. Zhou, B.; Cao, S.; Chen, F.; Zhang, F.; Zhang, Y. Recovery of Alkali from Bayer Red Mud Using CaO and/or MgO. *Minerals* **2019**, *9*, 269. [CrossRef]
12. Tang, C.; Li, K.; Ni, W.; Fan, D. Recovering Iron from Iron Ore Tailings and Preparing Concrete Composite Admixtures. *Minerals* **2019**, *9*, 232. [CrossRef]
13. Gitari, W.; Thobakgale, R.; Akinyemi, S. Mobility and Attenuation Dynamics of Potentially Toxic Chemical Species at an Abandoned Copper Mine Tailings Dump. *Minerals* **2018**, *8*, 64. [CrossRef]
14. Elghali, A.; Benzaazoua, M.; Bussière, B.; Genty, T. In Situ Effectiveness of Alkaline and Cementitious Amendments to Stabilize Oxidized Acid-Generating Tailings. *Minerals* **2019**, *9*, 314. [CrossRef]
15. Yang, L.; Wang, H.; Li, H.; Zhou, X. Effect of High Mixing Intensity on Rheological Properties of Cemented Paste Backfill. *Minerals* **2019**, *9*, 240. [CrossRef]

© 2019 by the authors. Licensee MDPI, Basel, Switzerland. This article is an open access article distributed under the terms and conditions of the Creative Commons Attribution (CC BY) license (http://creativecommons.org/licenses/by/4.0/).

Review

Mining Waste and Its Sustainable Management: Advances in Worldwide Research

José A. Aznar-Sánchez *, José J. García-Gómez, Juan F. Velasco-Muñoz and Anselmo Carretero-Gómez

Department of Economics and Business, Research Centre CAESCG and CIAIMBITAL, University of Almería, 04120 Almería, Spain; josejgg@ual.es (J.J.G.-G.); jfvelasco@ual.es (J.F.V.-M.); acarrete@ual.es (A.C.-G.)
* Correspondence: jaznar@ual.es; Tel.: +34-950-015-192

Received: 23 May 2018; Accepted: 30 June 2018; Published: 2 July 2018

Abstract: Growing social awareness of the need to adequately treat mining waste in order to protect the environment has led to an increase in the research in this field. The aim of this study was to analyze the dynamics of the research focused on mining waste and its sustainable management on a worldwide scale from 1988 to 2017. A systematic review and a bibliometric analysis of 3577 articles were completed. The results show that research into mining waste has increased, with studies focusing on waste management accounting for almost 40% of the total. The most productive journals in this field were Applied Geochemistry and Science of the Total Environment. The five most productive countries were the United States, Canada, Spain, Australia, and China. Works on the sustainable management of mining waste were in the minority, but it is an area of research that has considerable potential given the growing social awareness of the environmental repercussions of mining activities and the demands for increasingly sustainable practices. The findings of this study could prove useful for studies into mine waste, as they depict a global view of this line of research.

Keywords: mine waste; management; sustainability; bibliometric analysis; systematic review

---

1. Introduction

Mining activity has considerably increased due to notable population growth and worldwide demand for mineral resources [1]. This increase coincides with a new awareness in which environmental concerns have become a growing challenge for all of the agents within the sector [2,3]. The social demand has increased for the sustainable development of all of the activities related to mining, particularly the adequate management of waste products during each phase of the mining process, including prospection and exploration, development, extraction, transport and treatment of product obtained, etc. [4]. The mining process generates a large quantity of residues that must be strategically treated and managed to combine economic efficiency with demands for environmental sustainability. Energy requirements, environmental and human health risks, demands on water resources, and the required technology must all be taken into account [5].

The waste generated by mineral extraction may be solid, tailings, or slurry, with the most common being tailings, waste rock, slag, and tail ends, although in certain circumstances, the vegetation and overburden may also be considered waste [6,7]. To avoid negative effects on the environment, waste is maintained in tailing ponds, dams, or tips, in accordance with the local legislation on waste control treatment that is applicable to each mining area, and on recycling where technically possible [8,9]. In turn, each of these structures may be considered inert when they present no danger to human health or the environment, or dangerous when they cause negative effects to the soil, ground and surface water, vegetation, and even the local fauna and population [10,11]. Danger occurs due to the toxicity of the waste (acute, chronic, or extrinsic), flammability, reactivity, corrosivity, etc. In these cases, waste management activities that minimize or annul the dangers are required [12].

Mining activities lead to many negative environmental and socio-economic impacts. Many changes take place in the territory and society, such as: alterations of soil use, ecosystem variations, pollution, water shortages and disturbance of groundwater flows, modifications in the infrastructure networks, unbalanced industrial development, forced resettlement, and changes in the economic structure and local population, among others [13]. In the last few years, some studies have focused on the analysis of mining impacts on soil. First, the ground must be tested for contaminants, and the average levels of these elements must be measured in the various soil levels and sediment to establish the margins of safety [14]. Next, studies must be undertaken to establish the concentration of these trace elements in the mine, understand their capacity to produce acid mine drainage, identify the primary and secondary minerals in the waste, and estimate the mobility of the dangerous elements. Sequential extraction techniques are usually used to determine the environmental risks posed by these trace elements. This indicates the degree of adherence of these elements to the soil, and subsequently how easily they may contaminate the air, water, and food chain [15,16].

Analyses were undertaken of the dispersion of contaminating elements in residues and the structural stability of the deposits of these elements. Studies of human bioaccessibility [17], estimates of enrichment factor (EF), and geoaccumulation index (Igeo) are also common [18,19]. Sediment quality guidelines (SQGs) [20] have been introduced, and ecotoxicological risks were evaluated. The risks of the mobility of trace elements to surface and ground water were also evaluated through sediments [21].

Mineral deposits have traditionally been sealed off, although the traditionally used techniques have not been environmentally optimal [22]. Mine waste management systems recommend a geographic description of the residue and its mobility, a revision of the biogenetic and mineral dismantling of sulfide-based residue, a study of jarosite formation and soluble iron sulfates, monitoring the weathering of slag, an analysis of oxidation on the marine floor, the use of wetlands to immobilize trace elements, and the use of microorganisms to reduce the reactivity of mine residues [23].

Although mineral waste management has traditionally been based on the linear economy, the current challenge is to apply the possibilities presented by a circular economy to this problem, so that society changes its fundamentally negative perception of the sector. Recently, concentrations of graphite have been used to reduce tin mine foundry slag. Cement filling processes using superfine tailings have also been used to control sink holes in underground mining. New methods of transporting cement to fill tubes have been introduced, as has the reuse of residues in different geo-engineering applications. The recycling of leaching residue and new tailings procedures have been aimed at deep sea mining [24].

Mining waste management includes the characterization and remediation of residues. The state-of-the-art proposes new methods such as the use of mapping to determine the extent of wastes, the use of hyperspectral instruments [25], the mobility of sediments containing toxic residues [26], the mitigation of toxic metals spread in redox areas [27], the use of biochemical and mineral dissolution processes in sulfurous tailings [28], the dilution of tailings products, and the geochemical and mineral elimination of submarine tailings [29]. Other remediation systems include the use of heat to volatilize toxic components, and the use of microorganisms to reduce the reactivity and toxicity [30].

The potential environmental threat of waste generated by mining, along with an increasing societal awareness of the need to adequately treat mining waste, have led to the increased importance of this line of research. Nevertheless, no analysis of the developments in this research area has been completed as of late. This study aimed to fill that gap by analyzing the dynamics of the research into mining waste and its sustainable management since 1988 on a worldwide scale. The results may prove interesting for researchers of mining waste by offering a global view of the dynamics of this line of research.

2. Methodology

We analyzed two parameters to achieve our intended aim: a quantitative analysis using bibliometry, and a systematic, qualitative revision.

2.1. Bibliometric Method

The bibliometric analysis was first introduced by Garfield in the mid-20th century [31]. The main objective of this methodology is to identify, organize, and analyze the main components within a specific research field [32,33]. Since then, the method has been applied to areas such as engineering, biology, energy, medicine, and administration [34]. Over the last few decades, it has contributed to the review of scientific knowledge. Bibliometry is used to study the evolution and research trends of a topic. Through some statistical and mathematic analyzing tools, the publication relevance within a specific field can be assessed [35]. It also enables identifying the most productive authors, institutions, and countries so that the main researchers of a field can be stated [36].

Moreover, the use of mapping tools enables the identification of collaboration areas between some actors [37]. Thanks to these tools, the bibliographic information of a database can be shown, as well as main research trends [38,39]. Links between authors of different subject areas, institutions and countries can be viewed graphically. This application has been very useful and relevant in areas where international collaboration is essential, as is the case for the mineral sector [40].

Traditionally, co-occurrence analysis, co-quotation, and bibliographic coupling have been the main bibliometric approaches. They have been applied to database metadata according to the year of document publication, theme categories of classification, and obtained quotations of the works and keywords [41]. Currently, the traditional bibliometric methods have given way to new applications: viewing tools and information through text extraction techniques and data mining [42]; techniques of overlaying maps and variable associations [43]; the development of analysis frames to assess innovation [44]; tool developments based on routine types of automatized software [31,41]; and methods to identify and view evolution ways of scientific topics within a time segment [45]. In order to achieve the main goals of our work, a traditional approach based on co-occurrence has been considered best suited, since a general character is pursued. Furthermore, current processing and mapping tools have been applied due to their reliability [43].

Durieux and Gevenois defined three types of indicators when applying bibliometric analysis [46]. These are divided up into: (i) quantity indicators referring to productivity and counting; (ii) qualitative indicators regarding publication impact; and (iii) structural indicators that measure established links between agents. These three types of indicators have been taken into account in this paper. In this way, counting has been used to measure the productivity of authors, journals, institutions, and countries. The quotation number, H-index, and Scopus Journal Ranking (SJR) impact factor, have been used to measure impact. Network maps have been generated to view international links between different actors and analyze hotspots trends in this study field [47].

Obtained results through this type of analysis are very useful for many users. The evolution representation of the main research lines, identification of the most productive and relevant agents, and recognition of future trends are very helpful for junior and senior researchers of a specific research field [42]. Information compilations on new technologies and innovations in rapidly evolving areas mean new investment options for analysts and business agents [44]. Moreover, data on the hottest topics, such as material management, environmental protection, natural resources, and climate change are of particular interest in the decision-making processes from a business and administration point of view.

2.2. Data and Processing

The majority of bibliometric analyses use the Scopus data base, as it is considered the largest repository of peer-reviewed literature, it is easily accessible, provides different tools for viewing and analyzing publications, and allows data to be downloaded in different formats for processing by software [48].

Currently, there is a debate about the comparability and stability of the gained statistical data based on the two main databases: Web of Science (WoS) and Scopus [49]. Some studies have attempted to answer the question of which database is more adequate to be used for a bibliometric work. It has been proven that Scopus has more indexed journals than WoS [50]. It has also been demonstrated that only the 54% of the indexed publications in Scopus are comprised in WoS, while 84% of the WoS titles are indexed in Scopus [51]. The Google Scholar database has not been taken into account, since it has some limitations. Some studies compared the utility of diverse databases to Google Scholar when reviewing literature. Borrett et al. [52] pointed out that Google Scholar includes a greater quantity of non-relevant variables such as help files. Therefore, cleaning the data up requires more effort. Wildgaard [53] argued that Google Scholar includes a great number of non-peer reviewed articles, which generally implies publications with a low quality level. Ştirbu et al. [54] concluded that result processing and classification require a higher effort when using Google Scholar due to its total data amount and limited functioning. For these reasons, Scopus was chosen to perform the bibliometric analysis in this study. Many recent publications have used Scopus to perform bibliometric studies: Judd [55]; Feng, Zhu, and Lai [56]; Mugomeri et al. [57]; Mateo-Sanguino [58], and Kokol, Blazun-Vosner, and Zeleznik [59].

To study the various topics in our research, including mining waste, mining waste management, and the sustainable management of mining waste, we performed a descending search. This kind of search means first selecting a sample of a wider general topic, and subsequently, more restricted searches of the sample are conducted until a specific topic is defined. The main reason for this procedure is to compare the relevance of a specific topic with a broader research field. Initially, a search was performed using the parameters [TITLE-ABS-KEY ("mine waste")], with the aim of covering all of the works related to mining waste. The time scale of the sample was established between 1988–2017. As non-original publications undergo a less rigorous peer-review process, are less available, and may present duplicate information, these were excluded from our sample [60]. The resulting final sample totaled 3577 articles and reviews. A second search using the parameters: [TITLE-ABS-KEY ("mine waste") AND TITLE-ABS-KEY ("reprocessing" OR "reuse" OR "revalorization" OR "reposition" OR "re-using" OR "recycling" OR "remediation" OR "treatment" OR "stabilization" OR "valorization" OR "integrated management")] was undertaken to study the research into mining waste management. Different search terms were tested before reaching the final sample. Various parameters were included, and some of them were finally removed, such as "management", since they introduced a high noise level in the article sample. We applied the same restrictions as the initial search, obtaining a sample of 1092 articles. Finally, to analyze works on the sustainable management of mining waste, a third search was completed that included the following parameters: [TITLE-ABS-KEY ("mine waste") AND TITLE-ABS-KEY ("reprocessing" OR "reuse" OR "revalorization" OR "reposition" OR "re-using" OR "recycling" OR "remediation" OR "treatment" OR "stabilization" OR "valorization" OR "integrated management") AND TITLE-ABS-KEY ("sustainable" OR "sustainability")]. The result was a sample of 59 published articles.

To study the characteristics of the research in these three areas, the selected variables were: the year of publication, publishing journals, authors, institutions, and countries of author affiliation, and keywords. Once data were downloaded in two formats (RIS and csv), the first task to be undertaken was the depuration of information. Later on, data analysis and processing took place. Excel (version 2016) and SciMAT (v1.1.04) were the used software tools. VOSviewer was applied to analyze the links between different authors and keywords, as well as create the corresponding network maps. This software was chosen due to its suitability and frequent use in these kinds of works. Finally, the study of keywords was used to analyze the evolution of research trends and identify future ones. Figure 1 summarizes the followed methodology.

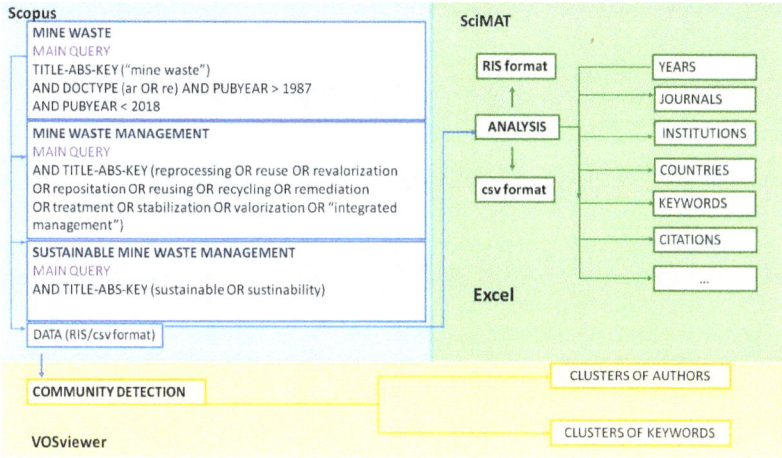

Figure 1. Methodology flow diagram.

3. Results and Discussion

3.1. Mine Waste

Table 1 shows the evolution of the main variables in research into mine waste (MW) on a global scale from 1988 to 2017. The number of articles published on this subject (A) increased notably from 14 in 1988 to 279 in 2017. This trend indicates that research into MW has increased in importance, culminating in the maximum number of articles published in 2017. The comparison between the total growth of articles on MW and the total growth of articles within all of the disciplines would be highly interesting, but unfortunately, it was not possible to achieve these data under the current research framework. The remaining variables in Table 1 show a similar growing trend. The number of authors in this field (AU) grew from 28 in 1988 to 1221 in 2017. The number of references (NR) increased exponentially from 86 to 11,323. The number of journals (J) also increased during this period, from 13 in 1988 to 153 in 2017. The internationalization of the field is reflected in the number of countries (C), which increased from four in 1988 to 51 in 2017. The total number of cited articles (TC) on MW was three in 1989, increasing to 7413 citations in 2017. The number of citations per article increased from 0.10 in 1989 to 16.89 in 2017.

Table 1. Major characteristics of the articles published on mine waste (MW).

Year	A	AU	NR	J	C	TC	CTC/CA
1988	14	28	86	13	4	0	0.00
1989	15	31	217	12	7	3	0.10
1990	18	40	248	13	4	10	0.28
1991	21	45	465	19	7	18	0.46
1992	8	30	134	7	4	24	0.72
1993	26	60	343	24	6	35	0.88
1994	22	59	652	19	9	52	1.15
1995	32	79	478	30	10	80	1.42
1996	38	99	1012	30	14	122	1.77
1997	42	105	905	32	14	166	2.16
1998	79	193	1752	55	26	204	2.27
1999	98	293	2387	54	19	322	2.51
2000	108	307	2596	65	25	370	2.70
2001	107	304	2709	68	30	463	2.98

Table 1. Cont.

Year	A	AU	NR	J	C	TC	CTC/CA
2002	126	395	2983	73	33	664	3.36
2003	129	387	3039	82	35	965	3.96
2004	137	382	4292	77	32	1020	4.43
2005	159	521	4796	91	38	1465	5.07
2006	127	436	4194	83	38	1853	6.00
2007	165	559	5147	97	41	2197	6.82
2008	161	513	4835	101	44	2678	7.79
2009	172	572	5633	97	43	3162	8.80
2010	168	578	5901	99	44	3536	9.84
2011	205	714	6944	105	47	4249	10.87
2012	177	648	6166	103	50	4467	11.95
2013	233	797	9007	127	50	5183	12.88
2014	230	863	9384	129	48	6087	13.98
2015	226	872	9086	122	52	6352	15.03
2016	255	1007	10,779	126	51	7245	16.07
2017	279	1221	11,323	153	51	7413	16.89

A: annual number of articles; AU: annual number of authors; NR: total number of references for all of the articles; J: annual number of journals; C: annual number of countries; TC: annual number of citations for all articles; CTC/CA: annual total citations per cumulative article.

Figure 2 shows the evolution of the principle subjects under which Scopus classifies articles on MW. Note that one article may be simultaneously included in more than one category. During the period studied, 58.6% of published articles were classified under the Environmental Sciences category, 47.2% were in Earth and Planetary Sciences, 16.4% were in Agricultural and Biological Sciences, 8% were in Engineering, and 7% were in Chemistry. These were followed by the categories of Materials Science, Medicine, Social Sciences, Pharmacology, Toxicology, Pharmaceutics, and Energy, but none accounted for 4% of the total of articles. Since the beginning of the analyzed period, Environmental Science and Earth and Planetary Sciences have been the principle categories. However, since 2006, Environmental Science has become the leading discipline in this area, which indicates that MW research is being principally studied from an environmental perspective.

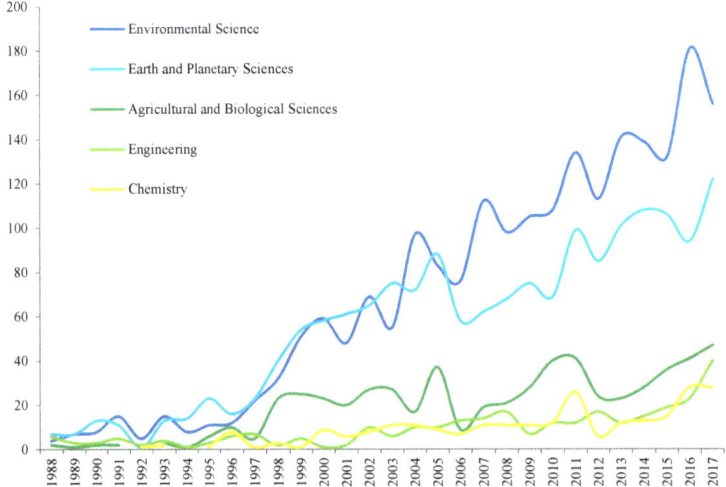

Figure 2. Trend in the subject categories of MW articles published from 1988 to 2017.

Table 2 shows the 10 journals with the most publications on MW. This group is entirely made up of European journals, specifically British, Dutch, and German journals, with the exception of one Iranian publication. These journals publish 28% of all of the articles in the field, indicating no great concentration of publication in this area. In the first column, we can see the total number of articles published by each journal along the whole period. Moreover, the evolution of the article number per journal is shown during the three 10-year periods, into which the studied time was divided up. Applied Geochemistry was the most productive journal on this subject from 1988 to 2017, with 155 articles, followed by Science of the Total Environment with 130 articles, the Journal of Geochemical Exploration with 109 articles, and Environmental Earth Sciences with 88 articles. Environmental Earth Sciences was established in 2009 under that name; however, it was previously published under the name of Environmental Geology. This journal occupied the first position in terms of the number of articles published from 1997 to 2009, the year in which it changed its name. From this date, Applied Geochemistry took the first position. During the sub-period of 2008–2017, Environmental Earth Sciences established itself once again in first position. Both appear separately in the fourth and fifth position of the most productive journals, but if the publications were totaled, this journal would take first position with 172 articles and a total of 2875 citations.

Journals with a greater SJR index were: Environmental Pollution with 1.786, the Journal of Hazardous Materials with 1.727, and Science of the Total Environment with 1.621. Applied Geochemistry was the most cited journal, followed by Science of the Total Environment, Environmental Science and Technology, and Chemosphere. However, considering the average number of citations per article, Environmental Science and Technology was the journal with the greatest impact, with a total of 48.8 citations per article. Chemosphere took second position with 44.2 citations per article, and Environmental Pollution was in third place with 37.1 citations per article. This journal had the greatest record within the top 10, since it first published an article on this subject in 1989. Notably, the journals in the top 10 are of the highest quality; they all appear in the first two quartiles of the Scopus classification.

Table 3 shows the 10 most productive countries in the publication of articles on MW. The United States led the group, followed by Canada, Spain, Australia, and China. The number of articles published per million inhabitants (APC) is also shown in this table. This variable is led by Canada with 13.75 articles per capita, followed by Australia with 12.47, Portugal with 10.07, and Spain with 7.29. The United States placed first in the total number of citations, followed by Canada, Spain, and the United Kingdom. However, considering the average number of citations per article, the United Kingdom placed first with 29 citations per article, followed by the United States with 22.3, Spain with 21.4, Portugal with 19.2, and Canada with 18.1. Figure 3 shows the elevated correlation existing between the H index and number of articles published by each country.

Table 2. Top 10 most productive journals for MW research.

Journal	A	SJR	H Index	C	TC	TC/A	1st A	R (A) 1988–1997	R (A) 1998–2007	R (A) 2008–2017
Applied Geochemistry	155	1.019 (Q1)	41	UK	4840	31.2	1991	12 (3)	2 (62)	1 (90)
Science of the Total Environment	130	1.621 (Q1)	37	Netherlands	4289	33.0	1997	3 (6)	3 (44)	3 (84)
Journal of Geochemical Exploration	109	1.047 (Q1)	28	Netherlands	2635	24.2	1995	5 (5)	5 (31)	4 (73)
Environmental Earth Sciences	88	0.574 (Q2)	14	Germany	544	6.2	2009	0	0	2 (88)
Environmental Geology	84	ND	31	Germany	2331	27.8	1993	1 (7)	1 (67)	33 (10)
Water Air and Soil Pollution	69	0.578 (Q2)	20	Netherlands	1120	16.2	1991	12 (3)	9 (27)	9 (39)
Journal of Hazardous Materials	68	1.727 (Q1)	25	Netherlands	1717	25.3	1995	37 (1)	16 (12)	5 (55)
Chemosphere	67	1.417 (Q1)	29	UK	2962	44.2	2000	0	11 (20)	7 (47)
Environmental Science and Technology	67	0.575 (Q2)	31	Iran	3270	48.8	1992	3 (6)	4 (32)	12 (29)
Environmental Pollution	57	1.786 (Q1)	27	UK	2111	37.0	1989	1 (7)	7 (28)	17 (22)
Environmental Earth Sciences*	172	0.574 (Q2)	31	Germany	2875	16.7	1993	1 (7)	1 (67)	1 (98)

A: annual number of total articles; SJR: Scopus Journal Ranking; C: country; TC: annual number of citations for all articles; TC/A: number of citations by article; 1stA: first article of MW research by journal; R: ranking position; UK: United Kingdom.

Table 3. Most productive countries in MW research.

Country	A	APC	TC	TC/A	R (A) 1988–1997	R (A) 1998–2007	R (A) 2008–2017
United States	613	1.897	13,646	22.3	1 (62)	1 (248)	1 (303)
Canada	499	13.752	9050	18.1	3 (21)	2 (205)	2 (273)
Spain	339	7.299	7247	21.4	8 (3)	5 (99)	3 (237)
Australia	301	12.476	4113	13.7	4 (17)	5 (99)	5 (185)
China	280	0.203	4364	15.6	14 (1)	7 (62)	4 (217)
United Kingdom	241	3.672	6977	29.0	2 (32)	3 (113)	7 (96)
Germany	187	2.262	3191	17.1	6 (9)	4 (105)	10 (73)
India	170	0.128	1492	8.8	8 (3)	8 (61)	6 (106)
Portugal	104	10.073	1996	19.2	0	21 (16)	8 (88)
Poland	103	2.714	935	9.1	8 (3)	13 (22)	9 (78)

A: annual number of total articles; APC: number of articles per 1 mill. inhabitants; TC: annual number of citations for all articles; TC/A: number of citations by article; R: ranking position.

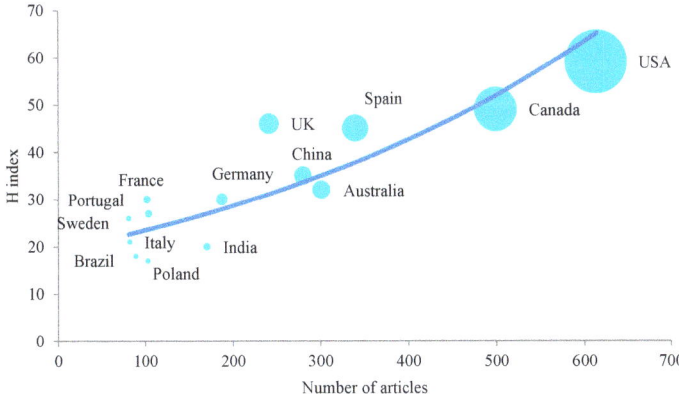

Figure 3. Correlation between H index and the number of articles by country in MW research.

Table 4 indicates the different variables related to the international collaboration between the group of the 10 most productive countries on the subject of MW. The United Kingdom had the largest percentage of articles produced in collaboration with other countries, with 46.9% of the total. The United States, Australia, Spain, Canada, and Germany were its main collaborators. These were followed by Germany with 42.8%, Portugal with 40.4%, and Australia with 38.9% of the total. The United Kingdom was also the country with the greatest number of collaborators, with 46 associates, followed by the United States with 45, and Australia with 38. The United States stands out as the principle collaborator among the remaining top 10 countries, being the foremost collaborator of these five countries: Canada, China, the United Kingdom (UK), Germany, and India. This table also shows the average number of citations (TC/A) per article produced in international collaboration (IC), and those produced without collaboration (NIC), for each country. The number of average citations per article, in every country, was greater with international collaboration, except in the case of the United States, Spain, and Portugal.

Table 4. International collaboration between the most productive countries in MW research.

Country	IC (%)	NC	Main Collaborators	TC/A	
				IC	NIC
United States	31.81	45	Canada, United Kingdom, China, Spain, Australia	21.4	22.7
Canada	24.05	34	United States, Australia, Morocco, United Kingdom, Germany	19.8	17.6
Spain	37.46	30	Portugal, United States, United Kingdom, Germany, Netherlands	20.3	22.0
Australia	38.87	38	Canada, United Kingdom, China, United States, Germany	18.7	10.5
China	26.07	29	United States, Australia, Canada, United Kingdom, Norway	25.4	12.1
United Kingdom	46.89	46	United States, Australia, Spain, Canada, Germany	31.3	26.9
Germany	42.78	37	United States, Spain, Australia, Canada, France	18.7	15.8
India	18.82	22	United States, Australia, United Kingdom, China, Russian Federation	11.2	8.2
Portugal	40.38	19	Spain, Australia, Brazil, Tunisia, United States	17.3	20.5
Poland	16.50	24	Czech Republic, Netherlands, United States, China, Germany	13.1	8.3

IC: international collaborations; NC: total number of international collaborators; TC/A: total citations per article; NIC: no international collaborations.

The principle characteristics of the institutions with the largest number of publications on MW are displayed in Table 5. Half of these were found in Canada, with the remainder in Spain, China, Australia, and the United States. Canada's University of British Columbia was the institution with the greatest number of articles published, followed by the Chinese Academy of Sciences, the University of Queensland, the United States Geological Survey, and Western University in Canada. The University of Waterloo (Canada) had the largest number of cited publications, followed by the University of British Columbia, the Chinese Academy of Sciences, and the United States Geological Survey. The University of Waterloo also took first position in terms of the average number of citations per article with 37.7, followed by Spain's National Research Council with 27.2, the United States Geological Survey with 24.5, and Western University with 19.3. Spanish institutions were those with the largest percentage of research completed with international collaboration.

Table 5. Most productive institutions in MW research.

Institution	C	A	TC	TC/A	H Index *	IC (%)	TC/A	
							IC	NIC
The University of British Columbia	Canada	79	1417	17.9	23	25.32	11.1	20.3
Chinese Academy of Sciences	China	75	1412	18.8	20	29.33	22.2	17.4
University of Queensland	Australia	55	495	9.0	12	25.45	14.1	7.2
United States Geological Survey	USA	54	1325	24.5	20	22.22	21.1	25.5
Western University	Canada	51	985	19.3	17	11.76	19.8	19.2
Universidad Politecnica de Cartagena	Spain	47	867	18.4	17	42.55	24.6	13.9
Consejo Superior de Investigaciones Científicas	Spain	46	1250	27.2	19	41.30	19.3	32.7
University of Saskatchewan	Canada	45	731	16.2	15	22.22	16.5	16.2
Universite du Quebec en Abitibi-Temiscamingue	Canada	43	455	10.6	12	34.88	11.7	10.0
University of Waterloo	Canada	41	1546	37.7	21	26.83	28.8	41.0

* Only sample items. C: country; A: annual number of total articles; TC: annual number of citations in total articles; TC/A: number of citations by article; IC: international collaborations; NIC: no international collaborations.

Table 6 shows the authors with the largest number of MW articles. The four most prolific authors were affiliated with Canadian institutions. David Blowes of the University of Waterloo was the most seasoned of the ranking with a paper from 1994. He was the most cited, with a total of 1394 citations and the highest H index (20). Ernest K. Yanful of Western University had 481 citations and an H index of 14. Following this were Mostafa Benzaazoua and Bruno Bussière of the Université du Quebec. The most recent author to join the ranks was R. Hakkou of the University Cadi Ayyad Marrakech of Morocco, with the first paper published in 2008. Even so, Hakkou managed to place ninth. Karen A. Hudson-Edwards of the University of Exeter was the author with the largest average number of citations per article with a total of 48.7. Figure 4 shows a network map illustrating the collaborative relationships of co-authorship between the different authors of MW articles. The size of the circle indicates the number of articles, whereas the thickness of the line indicates the number of collaborations between authors. The formation of different clusters can be observed through the colored representation. The group made up by Blowes, Smith, Ptacek, and Jambor stands out. Yanful leads a cluster that includes Simms, Hendry, Morris, and Song, among others. In the Benzaazoua group, we also find Bussière and Hakkou, whereas Craw, Lottermoser, and Schippers create another cluster. Next to Öhlander we can find Nason, Mäkitalo, Alakangas, and Maurice. Conesa shares the group with Jiménez-Cárceles, Robinson, Schulin, Álvarez-Rogel, and Elbaz-Poulichet. Hudson-Edwards builds a group together with Macklin, Bird, and Kossoff, among others.

Table 6. Most productive authors in MW research.

Author	A	TC	TC/A	H Index *	C	Affiliation	1st A	Last A
Blowes, David W.	33	1394	42.2	20	Canada	University of Waterloo	1994	2017
Yanful, Ernest Kwesi	33	481	14.6	14	Canada	Western University	1997	2013
Benzaazoua, Mostafa	28	378	13.5	11	Canada	Universite du Quebec en Abitibi-Temiscamingue	2004	2017
Bussière, Bruno	28	404	14.4	11	Canada	Universite du Quebec en Abitibi-Temiscamingue	2004	2017
Craw, David	24	501	20.9	14	New Zealand	University of Otago	1999	2017
Öhlander, Björn	23	311	13.5	10	Sweden	Lulea tekniska Universitet	1999	2016
Conesa, Héctor Miguel	21	573	27.3	11	Spain	Universidad Politecnica de Cartagena	2006	2017
Lottermoser, Bernd G.	18	419	23.3	12	Germany	Rheinisch-Westfalische Technische Hochschule Aachen	1999	2016
Hakkou, R.	17	223	13.1	8	Morocco	University Cadi Ayyad Marrakech	2008	2017
Hudson-Edwards, Karen A.	17	828	48.7	11	UK	University of Exeter	1996	2017
Schippers, Axel	17	489	28.8	13	Germany	Bundesanstalt fur Geowissenschaften und Rohstoffe	1995	2014

* Only sample items. A: annual number of total articles; TC: annual number of citations in total articles; TC/A: number of citations by article; C: country.

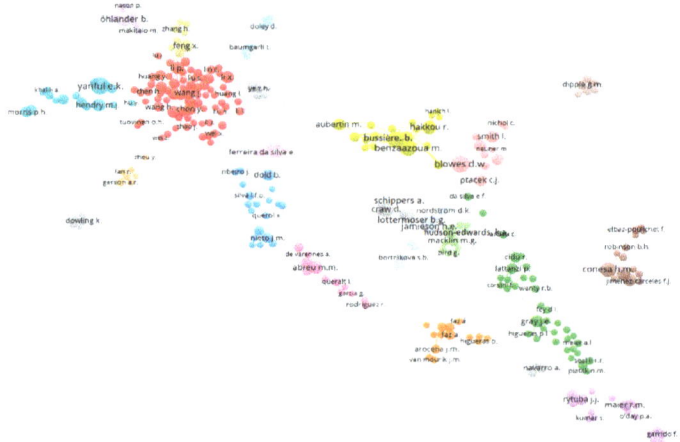

Figure 4. Cooperation based on co-authorship between authors.

We analyzed keywords to identify trends in MW research, which was necessary in order to previously remove duplicities. This pre-treatment of keywords has been undertaken with the SciMAT software. Words such as "article" and "priority journal" were excluded from this process, as they were irrelevant for our purposes. Table 7 shows the 20 most frequently used keywords in articles during the period of 1988 to 2017. This table also shows the evolution of these words through the three different 10-year sub-periods, into which the complete period may be divided. The values refer to the number of articles in which each keyword appears (A), the position the word occupies in relation to the others in terms of the number of repetitions (R), and the percentage of appearances with respect to the total number of articles analyzed in the period (%). Among the most often-used keywords were mining products (zinc, lead, copper, metals, heavy metal, iron, and arsenic), different terms relating to the processes and elements of mining (tailings, acid mine drainage, concentration, industrial waste, oxidation, and environmental monitoring), and soil contamination (soils, soil pollutants, pH, and soil pollution).

Table 7. Most frequently used keywords in MW research.

Keywords	1988–2017		1988–1997		1998–2007		2008–2017	
	A	%	R (A)	%	R (A)	%	R (A)	%
Mining	1315	36.8	1 (49)	20.8	1 (435)	35.2	1 (831)	39.5
Tailings	895	25.0	30 (8)	3.4	2 (319)	25.8	2 (568)	27.0
Heavy Metal	686	19.2	3 (30)	12.7	3 (247)	20.0	3 (409)	19.4
Lead	571	16.0	6 (20)	8.5	8 (145)	11.7	4 (406)	19.3
Zinc	561	15.7	8 (19)	8.1	6 (165)	13.4	7 (377)	17.9
Soil Pollution	547	15.3	8 (19)	8.1	9 (137)	11.1	5 (391)	18.6
Soils	537	15.0	5 (22)	9.3	10 (131)	10.6	6 (384)	18.2
Copper	527	14.7	4 (23)	9.7	5 (172)	13.9	9 (332)	15.8
Acid Mine Drainage	490	13.7	19 (13)	5.9	4 (185)	15.0	13 (291)	13.8
Arsenic	460	12.9	23 (11)	4.7	7 (153)	12.4	11 (296)	14.1
Metals	420	11.7	23 (11)	4.7	24 (92)	7.4	10 (317)	15.1
Mine Tailings	407	11.4	15 (14)	5.9	24 (92)	7.4	11 (296)	14.1
pH	385	10.8	20 (12)	5.1	20 (108)	8.7	14 (265)	12.6

Table 7. Cont.

Keywords	1988–2017		1988–1997		1998–2007		2008–2017	
	A	%	R (A)	%	R (A)	%	R (A)	%
Concentration (Composition)	366	10.2	0	0.0	291 (13)	1.1	8 (353)	16.8
Environmental Monitoring	356	10.0	72 (4)	1.7	17 (114)	9.2	15 (238)	11.3
Industrial Waste	349	9.8	13 (15)	6.4	12 (122)	9.9	19 (212)	10.1
Non-human	347	9.7	15 (14)	5.9	13 (119)	9.6	18 (214)	10.2
Iron	327	9.1	30 (8)	3.4	18 (109)	8.8	20 (210)	10.0
Oxidation	318	8.9	15 (14)	5.9	11 (128)	10.4	27 (176)	8.4
Soil Pollutants	306	8.6	100 (3)	1.3	33 (72)	5.8	16 (231)	11.0

R: ranking position; A: annual number of total articles.

As expected, the term most used during the entire study period was mining. The rest of the keywords varied their positions in accordance with the research preferences of each period. Although the words in the table were the most used, their importance oscillated over time. From 1988 to 1997, the most common keywords were mining, contamination, heavy metal, copper, soil, lead, water pollution, soil pollution, zinc, and environmental impact. During this time, the materials that were most studied were heavy metals, copper, lead, zinc, cadmium, and uranium. Attention was focused on both soil and water contamination (contamination, environmental impact, industrial wastes, waste disposal, sediment). The most frequently named countries in keywords were Canada, the United States, and Australia.

The most relevant keywords during the 1998–2007 sub-period, apart from mining, were: "tailings", "heavy metal", "acid mine drainage", "copper", "zinc", "arsenic", "lead", "soil pollution", and "soils". The principle elements that were analyzed were copper, zinc, lead, and arsenic, with the latter attracting more attention in this period compared with the previous. Acid drainage received particular attention, moving from position 19 in the previous period to fourth. Notably, the amount of attention paid to tailings in this period rose from 30th position to second place. However, the use of monitoring to study the environment experienced the greatest boost in this period, entering the list of 20 principle themes, from position number 72 during the 1988–1997 period. Conversely, studies on water contamination were no longer among the most numerous. In terms of geographic location, the regions with most studies on MW were Eurasia and Europe, and the countries were the United States, Spain, and Canada. The term "world" appears for the first time, indicating the gaining global significance of the research in this field.

The largest number of articles was published from 2008 to 2017; therefore, the greatest number of keyword repetitions were concentrated in this period. This conditioned the current framework of keywords. The principle keywords during this time were "mining", "tailings", "heavy metal", "lead", "soil pollution", "soils", "zinc", "concentration", "copper", and "metals". The two things of note during this period were the consolidation of a preference for studies of the ground rather than water, and the emergence of the term "concentration" in MW articles. From no presence at all in the previous periods, "concentration" became the eighth most common keyword. Studies of abandoned mines began to appear more frequently. Geographical reference takes 52nd place among keywords, with the United States closely followed by Spain and China.

Figure 5 shows a network map of the co-occurrence of the main keywords. The size of the circle represents the number of repetitions, and the color shows the different clusters in which the words are grouped according to the number of ties between the different words. Three main groups were found. The first (green) is titled "Contamination and public health". In this cluster, elements such as potassium, arsenic, aluminum, antimony, cobalt, copper, zinc, and lead are analyzed. The cluster includes terminology related to health, both human and animal, such as "health risk", "health hazard", "public health", "drinking water", "animals", "fish", "human", "pollution exposure", etc. The principle methodology terms gathered here are: "multivariate analysis", "principal component analysis", and "risk assessment". The main countries in this line were the

United States, the United Kingdom, Spain, and Portugal. The second group (red) is called "Waste management", and includes the terms: "metal recovery", "heavy metal removal", "neutralization", "waste management", and "waste treatment". This group presents a perspective from the fields of Hydrology and Hydrogeology. The most significant methodology terms were "analytic method", "analytical geochemistry", "chemical analysis", "computer simulation", and "experimental study". Brazil, Canada, Germany, South Africa, and Sweden appear in this group. The last cluster (blue), called "Ecological restoration", includes China, Australia, and India as the foremost countries with an environmental orientation. Terminology relating to the ground appears in this cluster, including: soil composition, microbiological activity, and revegetation (ecology, plant restoration, revegetation, ecosystem restoration, soils, soil microbiology, soil conservation, soil analysis, soil remediation, microbiology, microbial activity, etc.). The outstanding methodology terms are: "microbial analysis", "controlled study", and "comparative study".

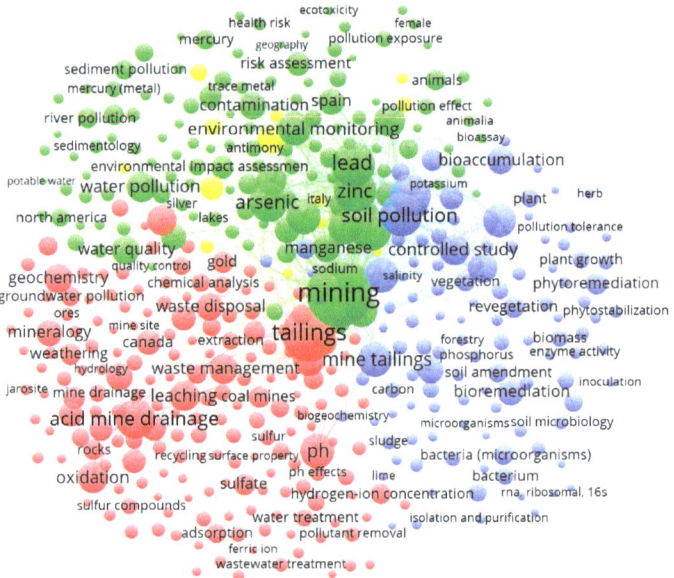

Figure 5. Main keywords' co-occurrence network in MW research.

3.2. Management of Mining Waste

In this section, we present the main analysis results of the evolution of worldwide research into the management of mining waste (MMW) during the period of 1988 to 2017. Table 8 shows the evolution of the principle indicators of research in this field. The number of articles on MMW (A) increased from one in 1988 to 100 in 2017. To contextualize the increase in articles in this line of research, Figure 6 displays the growing trend in the number of research articles on mining, mine waste (MW), and the management of mine waste (MMW). To facilitate the comparison and homogenize variables, logarithms were applied to them, and the annual average accumulated growth rate was calculated. As a result, although articles on mining increased by 7.54% on average per year, those on MW increased by 10.87%, and those on MMW increased by 17.52%. Figure 6 shows a great variability regarding the growth trends of the published articles on these three fields of research until the last decade of the 20th century. As far as the growth trend of mining articles (in green) is concerned, it did not start to be positive until 1998. Therefore, in the field of mining research, MW and particularly MMW have become increasingly important.

To analyze the contribution of research into MMW to MW, Table 8 shows a variable indicating the percentage of MW articles corresponding to research on MMW (AMW). Research into MMW gained importance within the field of MW in terms of the number of articles. In 1988, articles on management represented only 7.1% of the total; in 2017, they represented 38.7%.

Table 8. Major characteristics of the articles published on MMW.

Year	A	AU	NR	J	C	TC	CTC/CA	AMW	TCMW
1988	1	1	ND	1	1	0	0.0	7.1	0.0
1989	2	5	7	2	2	0	0.0	13.3	0.0
1990	6	9	42	5	3	0	0.0	33.3	0.0
1991	3	11	82	3	2	1	0.1	14.3	5.6
1992	1	5	ND	1	3	1	0.2	12.5	4.2
1993	7	13	106	7	4	2	0.2	26.9	5.7
1994	3	8	58	3	2	2	0.3	13.6	3.8
1995	2	6	44	2	2	12	0.7	6.3	15.0
1996	13	27	426	11	6	13	0.8	34.2	10.7
1997	6	13	53	6	3	25	1.3	14.3	15.1
1998	20	58	416	18	14	26	1.3	25.3	12.7
1999	27	86	712	21	9	64	1.6	27.6	19.9
2000	34	100	877	25	12	76	1.8	31.5	20.5
2001	30	95	709	24	13	102	2.1	28.0	22.0
2002	30	101	744	25	18	168	2.7	23.8	25.3
2003	36	136	805	28	17	258	3.4	27.9	26.7
2004	53	152	1707	39	24	264	3.7	38.7	25.9
2005	42	142	1434	31	21	349	4.3	26.4	23.8
2006	39	129	1314	23	21	472	5.2	30.7	25.5
2007	41	137	1144	32	24	613	6.2	24.8	27.9
2008	41	135	1304	31	20	810	7.5	25.5	30.2
2009	57	194	1864	42	26	914	8.4	33.1	28.9
2010	54	205	2204	33	30	1062	9.6	32.1	30.0
2011	59	194	2305	34	25	1304	10.8	28.8	30.7
2012	61	224	2219	36	24	1393	11.9	34.5	31.2
2013	73	234	3091	51	28	1589	12.8	31.3	30.7
2014	72	320	2995	48	33	1995	14.2	31.3	32.8
2015	72	280	2784	52	31	2061	15.3	31.9	32.4
2016	100	418	4562	60	37	2457	16.3	39.2	33.9
2017	108	504	4791	71	36	2677	17.1	38.7	36.1

A: annual number of articles; AU: annual number of authors; NR: total number of references for all articles; J: annual number of journals; C: annual number of countries; TC: annual number of citations for all articles; CTC/CA: annual total citations per cumulative article; AMW: percentage of annual contribution of MMW to MW (annual number of articles of MMW/annual number of articles of MW); TCMW: percentage of annual contribution of MMW citation to MW citation (annual number of citations of MMW/annual number of citations of MW).

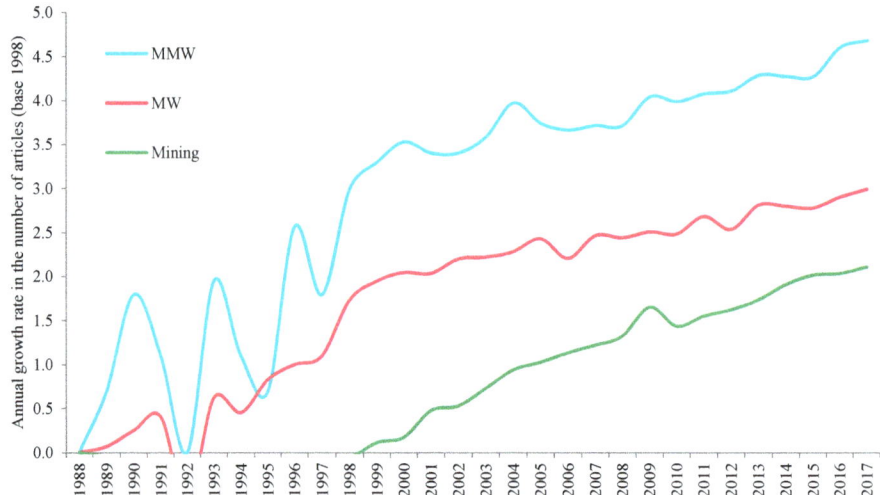

Figure 6. Comparative trends in the number of articles of mining, MW, and the management of mining waste (MMW) research.

Table 9 shows the main variables of the most productive journals on MMW. If we compare this group of journals to Table 2, we find a group of journals that published both on MW and MMW, their position in the ranking of most productive journals changed. In both cases, Applied Geochemistry was the most productive journal. The articles on MMW make up 29% of the total number of articles on MW. The Journal of Hazardous Materials and Chemosphere occupied the first and fourth positions, respectively, and are the only publications that improved their positions with respect to research on MW. For the former, articles on MMW accounted for 66.2% of the total number of articles on MW, whereas the latter accounted for 55.2% of the total articles. The Journal of Environmental Management, Environmental Science and Pollution Research, and Ecological Engineering were placed seventh, eighth, and 10th, respectively. These three publications did not appear in the most productive group on MW. Similarly, Environmental Geology, Environmental Science and Technology, and Environmental Pollution were not among the most productive journals on MMW research.

Table 9. Top 10 most productive journals for MMW research.

Journal	A	SJR	H Index	C	TC	TC/A	1st A	R (A) 1988–1997	R (A) 1998–2007	R (A) 2008–2017
Applied Geochemistry	45	1.019 (Q1)	22	UK	1312	29.2	1991	5 (1)	2 (13)	3 (31)
Journal of Hazardous Materials	45	1.727 (Q1)	20	Netherlands	1205	26.8	2002	0	6 (8)	1 (37)
Science of the Total Environment	43	1.621 (Q1)	20	Netherlands	1887	43.9	1999	0	1 (16)	5 (27)
Chemosphere	37	1.417 (Q1)	16	UK	1293	34.9	2003	0	11 (7)	4 (30)
Environmental Earth Sciences	33	0.574 (Q2)	10	Germany	268	8.1	2009	0	0	2 (33)
Journal of Geochemical Exploration	29	1.047 (Q1)	15	Netherlands	631	21.8	1998	0	11 (7)	7 (22)
Journal of Environmental Management	26	1.141 (Q1)	11	USA	346	13.3	2005	0	35 (2)	6 (24)
Environmental Science and Pollution Research	22	0.813 (Q2)	8	Germany	150	6.8	2010	0	0	7 (22)
Water Air and Soil Pollution	22	0.578 (Q2)	13	Netherlands	479	21.8	2002	0	6 (8)	10 (14)
Ecological Engineering	20	1.053 (Q1)	9	Netherlands	219	11.0	2002	0	24 (3)	9 (17)

A: annual number of total articles; SJR: Scopus Journal Ranking; C: country; TC: annual number of citations for all articles; TC/A: number of citations by article; 1st A: first article of MMW research by journal; R: ranking position.

The group of the 10 most productive journals on MMW included 29% of the total, indicating a wide distribution of publications on this theme. All of the publications included were among the first or second quartile in SJR ranking. The journal with the most citations was Science of the Total Environment with 1887; it also had the greatest average number of citations per article, with 43.9. The Journal of Hazardous Materials was the publication with the greatest SJR index (1.727).

Table 10 shows the list of the 10 most productive countries publishing articles on MMW. Once again, the United States was the country with the most articles, followed by Canada, Spain, China, and Australia. This means that the five most productive countries on MMW coincided with those on MW. In terms of the number of citations, the United States was the most important, followed by Canada, Spain, and the United Kingdom. As with research on MW, taking into account the average number of citations per article, the United Kingdom was placed first, with 27.5 citations per article, followed by Portugal with 22, Sweden with 21.5, and the United States with 21. The table also includes the percentage of articles on MMW compared with the number of articles on MW (AMW) of each of the nine most productive countries in both research fields. The country with the largest percentage of articles on management of the total works on MW was India with 40%, followed by Portugal with 36.5%, the United States with 33.4%, and Spain with 32.4%.

Table 10. Most productive countries in MMW research.

Country	A	TC	TC/A	AMW	R (A) 1988–1997	R (A) 1998–2007	R (A) 2008–2017
United States	205	4310	21.0	33.4	1 (15)	1 (82)	1 (108)
Canada	128	2333	18.2	25.7	3 (3)	2 (51)	3 (74)
Spain	110	2229	20.3	32.4	0	6 (22)	2 (88)
China	90	1589	17.7	32.1	0	7 (19)	4 (71)
Australia	79	1063	13.5	26.2	6 (1)	7 (19)	5 (59)
United Kingdom	75	2065	27.5	31.1	2 (7)	4 (30)	7 (38)
India	68	711	10.5	40.0	0	5 (24)	6 (44)
Germany	56	850	15.2	29.9	3 (3)	3 (33)	13 (20)
Portugal	38	837	22.0	36.5	0	13 (7)	8 (31)
South Africa	32	323	10.1	ND	6 (1)	11 (8)	11 (23)
Sweden	32	688	21.5	ND	5 (2)	9 (15)	17 (15)

A: annual number of total articles; TC: annual number of citations for all articles; TC/A: number of citations by article; AMW: percentage of contribution of MMW to MW (number of articles of MMW/number of articles of MW); R: ranking position.

Table 11 shows the main characteristics of the institutions with the greatest number of articles on MMW. Sweden's Lulea Tekniska Universitet had the most publications. This institution does not stand out for its production of articles on MW; however, it is a reference for MMW. Spain's Universidad Politécnica de Cartagena had the same number of publications, and was included among the group of most important producers of MW articles. However, its first-place ranking in MMW research means this is one of its most important areas of research. Other institutions that gained ground with respect to Table 5 are the United States Geological Survey, which was ranked third, and Spain's National Research Council in fifth place. Other institutions that did not appear among the most productive in MW, but did for MMW, were the Universidade de Lisboa (Portugal), the United States Environmental Protection Agency, the Universidad Politécnica de Cataluña (Spain), and the Universidade de Aveiro (Portugal). The United States Geological Survey is the institution with the greatest number of citations, followed by the National Research Council of Spain, the United States Environmental Protection Agency, and the Universidad Politécnica de Cartagena (Spain). In the ranking of average citations per article, these institutions remained in the same order. Those with the largest percentage of articles produced in collaboration were: the Universidade de Lisboa and the Université du Quebec in Abitibi-Temiscamingue with 50% of the total. These were followed by Spain's National Research Council with 43.8%, and Sweden's Lulea Tekniska Universitet with 38.9%.

Table 11. Most productive institutions in MMW research.

Institution	C	A	TC	TC/A	H Index *	IC (%)	TC/A IC	TC/A NIC
Lulea tekniska Universitet	Sweden	18	248	13.8	9	38.89	9.1	16.7
Universidad Politecnica de Cartagena	Spain	18	422	23.4	9	33.33	49.7	10.3
United States Geological Survey	USA	17	658	38.7	10	11.76	8.0	42.8
Consejo Superior de Investigaciones Científicas	Spain	16	538	33.6	10	43.75	20.7	43.7
Chinese Academy of Sciences	China	16	339	21.2	10	25.00	27.3	19.2
The University of British Columbia	Canada	15	142	9.5	6	26.67	8.3	9.9
Universidade de Lisboa	Portugal	14	154	11.0	6	50.00	3.3	18.7
United States Environmental Protection Agency	USA	13	433	33.3	8	15.38	5.5	38.4
University of Queensland	Australia	13	73	5.6	6	15.38	0.5	6.5
Universite du Quebec en Abitibi-Temiscamingue	Canada	12	61	5.1	5	50.00	8.2	2.0
Universitat Politecnica de Catalunya	Spain	12	201	16.8	8	33.33	12.0	19.1
Universidade de Aveiro	Portugal	12	128	10.7	7	25.00	6.3	12.1

* Only sample items. C: country; A: annual number of total articles; TC: annual number of citations in total articles; TC/A: number of citations by article; IC: international collaborations; NIC: no international collaborations.

3.3. Sustainable Management of Mining Waste

3.3.1. Quantitative Analysis

This section shows the main results obtained from the quantitative analysis of the article sample on sustainable management of mine waste (SMMW). The definitive sample was made up of 59 articles. Due to this small number of results, documents can be shown in an abbreviated manner.

The first found article dates from 1992. From then on, the number of articles published on this topic is highly irregular. Until 1998, there was no continuity in the publication of articles. In 2017, a total amount of 10 articles can be found. The main journals regarding the publication number on this field are Ecological Engineering, Minerals, Environmental Earth Sciences, Environmental Science and Pollution Research, the Journal of Environmental Management, the Journal of Geochemical Exploration, and the Journal of Hazardous Materials. When the groups of the main journals, according to the number of articles on the three main studied topics, are compared (MW, MMW, SMMW), some different publication trends are observed. Applied Geochemistry is the most productive journal in the two main topics. It is the leader publication on waste management studies. However, this is not the case within the sustainability field. It also occurs with Science of the Total Environment, Chemosphere, and Water, Air, and Soil Pollution. Differently, we could find three journals ranking at the first positions within the three research fields. They are the Journal of Hazardous Materials, Environmental Earth Sciences, and the Journal of Geochemical Exploration. These journals embrace the whole spectrum of mine waste management, including sustainability. The Journal of Environmental Management, Environmental Science and Pollution Research, and Ecological Engineering also stand out in the general management and sustainable management areas. The latter is also the most productive one on SMMW. Finally, a journal that was not outstanding in the two first topics occupies the second position on SMMW. We refer to the journal Minerals, which has placed itself as a leading specialist and a reference journal on the sustainable management of mine waste.

Authors with the highest number of publications were Banning, N.C.; Fan, R.; Gerson, A.R.; Huang, L.; Kawashima, N.; Li, J.; Li, X.; Lottermoser, B.G.; Qian, G.; Schumann, R.C.; Short, M.D.; and Smart, R.S.C. As far as authors are concerned, only six of the 10 most important ones on WM were to be found publishing on MMW. Öhlander and Conesa share the first position; followed by Blowes, Benzaazoua, Hudson-Edwards, and Lottermoser. The last author is also one of the most prolific ones on SMMW. This can suggest the great diversity of fields within the study of mine waste. The most relevant institutions came from Australia: the University of Queensland, the University of Western Australia, and the University of South Australia. All of the institutions in Table 5 are also active in Table 11, except for the Western University, the University of Saskatchewan, and the University of Waterloo. It can be stated that the three Australian institutions are the only institutions that stand out in the three analyzed fields of research. The country with the highest number of articles is Australia, followed by Canada, the United States, China, India, Spain, and the United Kingdom. These countries configure the most relevant group regarding the publications on the three studied topics.

The Environmental Science category includes a greater number of articles, resulting in 59.3% out of the total, followed by Earth and Planetary Sciences with 38.9%, Agricultural and Biological Sciences with 15.3%, Engineering with 15.3%, and Materials Science with 6.8%. In contrast, the Economics, Econometrics and Finance, and Social Sciences categories have only two documents, and the multidisciplinary category only has one. This implies an absolute predominance of Environmental Sciences and Engineering regarding these studies. This result can be also obtained within the other two analyzed topics (MW and MMW).

3.3.2. Qualitative Analysis

In the qualitative analysis, works on SMMW were divided up according to the two main aspects on which the articles focus: scope of sustainability and focus of action. Table 12 comprises the 59-article sample grouped by these two features.

Table 12. Articles published on the sustainable management of mine waste (SMMW).

Scope of Sustainability	Focus of Action	Article
Environmental	Stabilization and waste treatment	Li, X., Huang, L. [61]; Li, X. et al. [62]; Qian et al. [63]; Ogbughalu et al. [64]
	Evaluation of environmental impacts	Ellis, D.V. [65]; Miler, M.; Gosar, M. [66]; Méndez-Ramírez, M.; Hernández, M.A.A. [67]; Popovic et al. [13]; Bansch, C.; Topp, W. [68]; Gatzweiler et al. [69]; Ghose, M.K. [70]; Elshorbagy et al. [71]; Bowen et al. [72]; Van Deventer, P.W.; Bloem, A.A.; Hattingh, J.M. [73]; Maddocks, G.; Lin, C.; McConchie, D. [74]; Lottermoser, B.G.; Glass, H.J.; Page, C.N. [75]; Wu et al. [76]; Pepper et al. [77]; Melgar-Ramírez et al. [78]
	Remediation and recovery of soils and landscape	Valente et al. [79]; Courtney, R. [80]; Bigot et al. [81]; Adams, A.; Raman, A.; Hodgkins, D. [82]; Sjöberg et al. [83]; Naeth, M.A.; Wilkinson, S.R. [84]; Banning et al. [85]; Johansson et al. [86]; Li, J.J.; Yan, J.X.; Li, H.J. [87]; Anawar, H.M. [88]; Santini, T.C.; Banning, N.C. [89]; Nirola et al. [90]; Párraga-Aguado et al. [91]; Nancucheo et al. [92]; Plaza et al. [93]; Sözen et al. [94]; Mwandira, W.; Nakashima, K.; Kawasaki, S. [95]
	Reuse and recycling of materials	Dold, B. [2]; Bartke, K. [96]
	Treatment and remediation of contaminated water	Kalin, M. [97]; Azam, S. [98]; Younger, P.L. [99]; Macklin et al. [100]
	Cooperation in the development of I + D + I	Meech et al. [101]
Environmental Economic	Recycling of materials and rehabilitation of landscape	Careddu et al. [102]
	Remediation and recovery of soils and landscape	Shukla, M.K.; Lal, R. [103]; Wajima, T.; Ikegami, Y. [104]; Hwang, T.; Neculita, C.M.; Han, J.-I. [105]
	Reuse and recycling of materials	Emery et al. [106]; Venkatarama Reddy, B.V. [107]; Gabzdyl, W.; Hanak, B. [108]; Arrigo et al. [109]; Haibin, L.; Zhenling, L. [110]; Lottermoser, B.G. [111]; Cadierno et al. [112]; Kundu et al. [113]; Yang et al. [114]; Rana, A.; Kalla, P.; Csetenyi, L.J. [115]; Taha et al. [116]; Gorakhki, M.H.; Bareither, C.A. [117]

The Sustainability concept includes three fields: the economic, the environmental, and the social one [48]. According to this, the sustainable development of any activity should assure an economical use, the integrity of ecological systems, and a contribution to social welfare for current and future generations [13]. Mining activities, and especially its waste management, raise conflicts between these three fields. Waste management has an economic impact. Wastes of the mining activities are one of the main polluting agents for soil, water, and air. As far as the social aspect is concerned, mining raises interest conflicts between the main stakeholders. The welfare of the population living within the mining influence areas depends to a large extent on the appropriate management of mine waste. We can mention health hazard as an example. As the table shows, all of the analyzed articles speak of aspects relevant to the environment. The main concerns regarding sustainable waste management focus on environmental impacts. They handle pollution prevention or decontamination treatments.

Only 28.8% of the articles include an economic perspective of waste management. No article of our selected sample was found where the impact of mine waste management on human welfare is analyzed, which is further beyond the economic and environmental impacts.

As far as the focus of action is concerned, the articles have been grouped regarding the two main sets of activities detected during the review. Sustainable waste management concentrates on two activities: material reuse and waste depollution. On the one hand, a set of articles is devoted to the incorporation of waste materials in the production process. This can be achieved through recycling, reuse, and recuperation, among other processes. On the other hand, a further set of articles focus on the treatment processes of waste and cleanup, such as bioremediation and phytoremediation. In works with an exclusive focus on environmental sustainability, 95.2% of the articles concentrate on different depollution aspects, while 66.7% of the articles treat the remediation of soils polluted by wastes. The impacts on ecosystems and biodiversity, the stabilization of polluting agents, and the remediation of polluted water are considered by 9.5% of the publications. Material reuse can only be found in 4% of this article group.

Regarding the works focusing on economic and environmental sustainability, the reuse of waste materials stands out within 70.6% of the total publications, while only 17.6% of the articles analyzed profitable processes for the recuperation of polluted soils. One work is jointly devoted to material reutilization and depollution. A further study aims at the study of interinstitutional cooperation for the development of waste management projects that contribute to improve profitability and reduce environmental impacts.

4. Conclusions

This study analyzed the dynamics of global research into mining waste analysis and its sustainable management from 1988 to 2017. A systematic and bibliometric analysis was completed on a sample of 3577 articles. The results indicated a rapid increase in the number of published articles each year, growing from 14 in 1988 to 279 in 2017. This increase has occurred particularly since 2008, with 63% of the overall total. This increase in mining waste articles and journals, authors, institutions, and countries indicated that this line of research is receiving growing worldwide attention. This is due to several factors, including concerns over environmental threats, a greater social awareness of environmental issues, and new and more restrictive regulations in developed countries. We demonstrated that mining waste and mining waste management are two fields of research with a marked differential growth rate within the field of mining research worldwide.

Applied Geochemistry, Science of the Total Environment, and the Journal of Geochemical Exploration were the journals with the largest number of articles published on mining waste. Along with the first two, the Journal of Hazardous Materials was one of the journals that published the most articles on mining waste. The United States was the country with the largest number of articles published on mining waste management, followed by Canada, Spain, Australia, and China. These are also the most prolific countries in terms of articles on managing mining waste. If you consider the average number of citations per article on mining waste, the order changes to: UK, the United States, Spain, Portugal, and Canada. Considering the population of each country, Canada was placed first. In the list of the 10 most published authors on mining waste, the top four were Canadian: Blowes, Yanful, Benzaazoua, and Bussière. The University of British Columbia, the Chinese Academy of Sciences, and the University of Queensland were the three institutions with the largest number of published papers on mining waste, whereas Sweden's Lulea Tekniska Universitet, Spain's Universidad Politécnica de Cartagena, and the United States Geological Survey were those with most articles on mining waste management.

The keywords analysis that was used in the articles studied showed that various mining products were among the most frequently used words, including: zinc, lead, copper, metals, heavy metals, iron, and arsenic. The most common terms related to processes and mining elements were: "tailings", "acid mine drainage", "concentration", "industrial waste", "oxidation", and "environmental monitoring". The most common terms related to soil contamination were: "soils", "soil pollutants", "pH", and "soil pollution". The network mapping of co-occurrence of keywords revealed three different clusters focused on contamination and public health, waste management, and environmental restoration.

Regarding the sustainable management of mining waste, it has been proven that this is a recent field of study. Only 59 articles were found in the sample. Although studies on sustainable mining waste management are of secondary importance, it is a field of research that shows great potential given the increasing social awareness about the environmental repercussions of mining and the increasing demands for sustainable production methods. Our analysis shows a twofold action in order to achieve the sustainable management of mining waste. On the one side, efforts to depollute mining waste are in progress. This action embraces air, water, and soil. Nevertheless, the last one has attracted the most attention to date. On the other side, the recycling of mining wastes is being developed. It enables reductions in energy consumption, the emission of greenhouse gases, and waste generation. Moreover, it also results in cost reduction and higher profitability.

Currently, the treatment of mining waste focuses on remediation, reuse, and evaluating the mined area for alternative use. Lines of research are oriented toward the application of biotechnology, the use of microbes, and bioremediation with algae, and phytoremediation. To resolve water contamination issues, the use of nutrient-enriched sediments has been proposed to reduce metal acidity and increase pH, in addition to applying engineering systems for storage following ecological principles. Concerning the reuse of residues, it is proposed to use the link between mining and construction to convert waste into building materials. Another area of research involves investigating the use of slag and gases to generate electricity.

A relevant issue that has arisen during this research work refers to the contribution of mine waste management to sustainability. In the studies on mining waste, the term sustainability is commonly associated with environmental protection, since most works focus on it. Fewer articles have analyzed mine waste management from an economic point of view. No articles have been found where waste management contributes to social welfare, apart from those comments on health hazards. We can therefore state that there is a relevant gap in this research field. The approach to sustainability analysis should be based on multidisciplinary frameworks where technical and socio-economic methods are taken into account. This can provide relevant information for all of the involved stakeholders in the decision-making processes regarding the management of material and natural resources.

Author Contributions: The four authors have equally contributed to this paper. All authors have revised and approved the final manuscript.

Acknowledgments: This work has been partially supported by the Spanish Ministry of Economy and Competitiveness and the European Regional Development Fund by means of the research projects ECO2017-82347-P and HAR2014-56428-C3-2, and by the Research Plan of the University of Almería through a Predoctoral Contract to Juan F. Velasco Muñoz. This paper was developed during the research stay by José A. Aznar-Sánchez at the Humboldt-Universität zu Berlin.

Conflicts of Interest: The authors declare no conflict of interest.

References

1. Reichl, C.; Schatz, M.; Zsak, G. World-Mining-Data. In *Minerals Production*; International Organizing Committee for the World Mining Congresses: Vienna, Austria, 2016; Volume 31.
2. Dold, B. Sustainability in metal mining: From exploration, over processing to mine waste management. *Rev. Environ. Sci. Bio/Technol.* **2008**, *7*, 275–285. [CrossRef]
3. Gómez Ros, J.M.; García, G.; Peñas, J.M. Assessment of restoration success of former metal mining areas after 30 years in a highly polluted Mediterranean mining area: Cartagena-La Union. *Ecol. Eng.* **2013**, *57*, 393–402. [CrossRef]
4. Bakken, G.M. Montana, Anaconda, and the Price of Pollution. *Historian* **2007**, *69*, 36–48. Available online: http://www.jstor.org/stable/24453910 (accessed on 14 April 2018). [CrossRef]

5. Durucan, S.; Korre, A.; Muñoz-Melendez, G. Mining life cycle modelling: A cradle-to-gate approach to environmental management in the minerals industry. *J. Clean. Prod.* **2006**, *14*, 1057–1070. [CrossRef]
6. Alloway, B.J. *Heavy Metals in Soils*; Blackie: Glasgow, UK, 1995; ISBN 0751401986.
7. Pérez Cebada, J.D. Mining corporations and air pollution science before the Age of Ecology. *Ecol. Econ.* **2016**, *123*, 77–83. [CrossRef]
8. Pasariello, B.; Giuliano, V.; Quaresima, S.; Barbaro, M.; Caroli, S.; Forte, G.; Carelli, G.; Iavicoli, I. Evaluation of the environmental contamination at abandoned mining site. *Microchem. J.* **2002**, *73*, 245–250. [CrossRef]
9. Hudson-Edwards, K.A.; Dold, B. Mine Waste Characterization, Management and Remediation. *Minerals* **2015**, *5*, 82–85. [CrossRef]
10. Fetter, C.W. *Contaminant Hydrogeology*; Prentice Hall: Upper Saddle River, NJ, USA, 1999; ISBN 13 978-1577665830.
11. Iribar, V.; Izco, F.; Tames, P.; Antigüedad, I.; da Silva, A. Water contamination and remedial measures at the Troya abandoned Pb-Zn mine (The Basque Country, Northern Spain). *Environ. Geol.* **2000**, *39*, 800–806. [CrossRef]
12. Alberruche del Campo, E.; Arranz-González, J.C.; Rodríguez-Pacheco, R.; Vadillo-Fernández, L.; Rodríguez-Gómez, V.; Fernández-Naranjo, F.J. *Manual para la Evaluación de Riesgos de Instalaciones de Residuos de Industrias Extractivas Cerradas o Abandonadas*; Instituto Geológico y Minero de España-Ministerio de Agricultura, Alimentación y Medio Ambiente: Madrid, Spain, 2014; ISBN 978-84-7840-934-1.
13. Popovic, V.; Miljkovic, J.Ž.; Subic, J.; Jean-Vasile, A.; Adrian, N.; Nicolaescu, E. Sustainable land management in mining areas in Serbia and Romania. *Sustainability* **2015**, *7*, 11857–11877. [CrossRef]
14. Salomons, W.; Förstner, U.; Mader, P. (Eds.) *Heavy Metals. Problems and Solutions*; Springer: Berlin, Germany, 1995; ISBN 978-3-642-79316-5.
15. Pérez-Santana, S.; Pomares, A.M.; Villanueva, T.M.; Peña-Icart, M.; Brunori, C.; Morabito, R. Total and partial digestion of sediments for the evaluation of trace element environmental pollution. *Chemosphere* **2007**, *66*, 1545–1553. [CrossRef] [PubMed]
16. Zhang, L.; Liao, Q.; Shao, S.; Zhang, N.; Shen, Q.; Liu, C. Heavy Metal Pollution, Fractionation, and Potential Ecological Risks in Sediments from Lake Chaohu (Eastern China) and the Surrounding Rivers. *Int. J. Environ. Res. Public Health* **2015**, *12*, 14115–14131. [CrossRef] [PubMed]
17. Paustenbach, D.J. The practice of exposure assessment: A state of the art review. *J. Toxicol. Environ. Health B Crit. Rev.* **2000**, *3*, 179–291. [CrossRef] [PubMed]
18. Christophoridis, C.; Dedepsidis, D.; Fytianos, K. Occurrence and distribution of selected heavy metals in the surface sediments of Thermaikos Gulf, N. of Greece. Assessment ussing pollution indicators. *J. Hazard. Mater.* **2009**, *168*, 1082–1091. [CrossRef] [PubMed]
19. Khan, M.Z.H.; Hasan, M.R.; Khan, M.; Aktar, S.; Fatema, K. Distribution of Heavy Metals in Surface Sediments of the Bay of Bengal Coast. *J. Toxicol.* **2017**, 9235764. [CrossRef] [PubMed]
20. Borja, A.; Heinrich, H. Implementing the European Water Framework: The debate continues. *Mar. Pollut. Bull.* **2005**, *50*, 486–488. [CrossRef] [PubMed]
21. Díaz de Alba, M.; Galindo-Riaño, M.D.; Casanueva-Marenco, M.J.; García-Vargas, M.; Kosore, C.M. Assessment of the metal pollution, potential toxicity and speciation of sediments from Algeciras Bay (South Spain) using chemometric tools. *J. Hazard. Mater.* **2011**, *190*, 177–187. [CrossRef] [PubMed]
22. Zhou, H.; GuoSoil, X. Soil Heavy Metal Pollution Evaluation around Mine Area with Traditional and Ecological Assessment Methods. *J. Geosci. Environ. Prot.* **2015**, *3*, 28–33. [CrossRef]
23. Pellegrini, S.; García, G.; Peñas-Castejón, J.M.; Vignozzi, N.; Constantini, E.A.C. Pedogenesis in mine tails affects macroporisity, hydrological properties and pollutant flow. *Catena* **2016**, *136*, 3–16. [CrossRef]
24. Chen, X.; Zhou, J.; Chen, Q.; Shi, X.; Gou, Y. CFD Simulation of Pipeline Transport Properties of Mine Tailings Three-Phase Foam Slurry Backfill. *Minerals* **2017**, *7*, 149. [CrossRef]
25. Buzzi, J.; Riaza, A.; García-Meléndez, E.; Weide, S.; Bachmann, M. Mapping changes in a recovering mine site with hyper spectral airborne HyMap imagery (Sotiel, SW Spain). *Minerals* **2014**, *4*, 313–329. [CrossRef]
26. Pattelli, G.; Rimondi, V.; Benvenuti, M.; Chiarantini, L.; Colica, A.; Costagliola, P.; Di Benedetto, F.; Lattanzi, P.; Paolieri, M.; Rinaldi, M. Effects of the November 2012 flood event on the mobilization of Hg from the Mount Amiata mining district to the sediments of the Paglia River Basin. *Minerals* **2014**, *4*, 241–256. [CrossRef]
27. Lynch, S.F.L.; Batty, L.C.; Byrne, P. Environmental risk of metal mining contaminated river bank sediment at redox-transitional zones. *Minerals* **2014**, *4*, 52–73. [CrossRef]

28. Nordstrom, D.K. Mine waters: Acidic to circumneutral. *Elements* **2011**, *7*, 393–398. [CrossRef]
29. Dold, B. Submarine tailings disposal (STD)—A review. *Minerals* **2014**, *4*, 642–666. [CrossRef]
30. Johnson, D.B. Recent developments in microbiological approaches for securing mine wastes and for recovering metals from mine waters. *Minerals* **2014**, *4*, 279–292. [CrossRef]
31. Huang, L.; Zhang, Y.; Guo, Y.; Zhu, D.; Porter, A.L. Four dimensional science and technology planning: A new approach based on bibliometrics and technology roadmapping. *Technol. Forecast. Soc. Chang.* **2014**, *81*, 39–48. [CrossRef]
32. Zhang, Y.; Zhang, Y.; Shi, K.; Yao, X. Research development, current hotspots, and future directions of water research based on MODIS images: A critical review with a bibliometric analysis. *Environ. Sci. Pollut. Res. Int.* **2017**, *24*, 15226–15239. [CrossRef] [PubMed]
33. Rodrigues-Vaz, C.; Shoeninger-Rauen, T.R.; Rojas-Lezana, A.G. Sustainability and innovation in the automotive sector: A structured content analysis. *Sustainability* **2017**, *9*, 880. [CrossRef]
34. Gusmão-Caiado, R.G.; de Freitas-Dias, R.; Veiga-Mattos, L.; Gonçalves-Quelhas, O.L.; Leal-Filho, W. Towards sustainable development through the perspective of eco-efficiency—A systematic literature review. *J. Clean. Prod.* **2017**, *165*, 890–904. [CrossRef]
35. Zhong, S.; Geng, Y.; Liu, W.; Gao, C.; Chen, W. A bibliometric review on natural resource accounting during 1995–2014. *J. Clean. Prod.* **2016**, *139*, 122–132. [CrossRef]
36. Li, W.; Zhao, Y. Bibliometric analysis of global environmental assessment research in a 20-year period. *Environ. Impact Assess. Rev.* **2015**, *50*, 158–166. [CrossRef]
37. Waltman, L.; van Eck, N.J.; Noyons, E.C. A unified approach to mapping and clustering of bibliometric networks. *J. Informetr.* **2010**, *4*, 629–635. [CrossRef]
38. Garfield, E. Citation Indexes for Science. *Science* **1955**, *122*, 108–111. [CrossRef] [PubMed]
39. Zhou, X.; Zhang, Y.; Porter, A.L.; Guo, Y.; Zhu, D. A patent analysis method to trace technology evolutionary pathways. *Scientometrics* **2014**, *100*, 705–721. [CrossRef]
40. Lee, S.; Lee, S.; Seol, H.; Park, Y. Using patent information for designing new product and technology: Keyword based technology roadmapping. *R D Manag.* **2008**, *38*, 169–188. [CrossRef]
41. Suominen, A.; Toivanen, H. Map of science with topic modeling: Comparison of unsupervised learning and human-assigned subject classification. *J. Assoc. Inf. Sci. Technol.* **2016**, *67*, 2464–2476. [CrossRef]
42. Zhang, Y.; Chen, H.; Lu, J.; Zhang, G. Detecting and predicting the topic change of Knowledge-based Systems: A topic-based bibliometric analysis from 1991 to 2016. *Knowl.-Based Syst.* **2017**, *133*, 255–268. [CrossRef]
43. Rafols, I.; Porter, A.L.; Leydesdorff, L. Science overlay maps: A new tool for research policy and library management. *J. Am. Soc. Inf. Sci. Technol.* **2010**, *61*, 1871–1887. [CrossRef]
44. Robinson, D.K.; Huang, L.; Guo, Y.; Porter, A.L. Forecasting Innovation Pathways (FIP) for new and emerging science and technologies. *Technol. Forecast. Soc. Chang.* **2013**, *80*, 267–285. [CrossRef]
45. Zhang, Y.; Zhang, G.; Zhu, D.; Lu, J. Science evolutionary pathways: Identifying and visualizing relationships for scientific topics. *J. Assoc. Inf. Sci. Technol.* **2017**, *68*, 1925–1939. [CrossRef]
46. Durieux, V.; Gevenois, P.A. Bibliometric Indicators: Quality Measurements of Scientific Publication. *Radiology* **2010**, *255*, 342. [CrossRef] [PubMed]
47. Garrido-Cárdenas, J.A.; Manzano-Agugliaro, F. The metagenomics worldwide research. *Curr. Genet.* **2017**, *63*, 819–829. [CrossRef] [PubMed]
48. Velasco-Muñoz, J.V.; Aznar-Sánchez, J.A.; Belmonte-Ureña, L.J.; Román-Sánchez, I.M. Sustainable water use in agriculture: A review of worldwide research. *Sustainability* **2018**, *10*, 1084. [CrossRef]
49. Salmerón-Manzano, E.; Manzano-Agugliaro, F. Worldwide scientific production indexed by Scopus on Labour Relations. *Publications* **2017**, *5*, 25. [CrossRef]
50. Mongeon, P.; Paul-Hus, A. The journal coverage of Web of Science and Scopus: A comparative analysis. *Scientometrics* **2016**, *106*, 213–228. [CrossRef]
51. Gavel, Y.; Iselid, L. Web of Science and Scopus: A journal title overlap study. *Online Inf. Rev.* **2008**, *32*, 8–21. [CrossRef]
52. Borrett, S.R.; Sheble, L.; Moody, J.; Anway, E.C. Bibliometric review of ecological network analysis: 2010–2016. *Ecol. Model.* **2018**, *382*, 63–82. [CrossRef]
53. Wildgaard, L. A comparison of 17 author-level bibliometric indicators for researchers in Astronomy, Environmental Science, Philosophy and Public Health in Web of Science and Google Scholar. *Scientometrics* **2015**, *104*, 873. [CrossRef]

54. Ştirbu, S.; Thirion, P.; Schmitz, S.; Haesbroeck, G.; Greco, N. The Utility of Google Scholar When Searching Geographical Literature: Comparison With Three Commercial Bibliographic Databases. *J. Acad. Librariansh.* **2015**, *41*, 322–329. [CrossRef]
55. Judd, S.J. Membrane technology costs and me. *Water Res.* **2017**, *122*, 1–9. [CrossRef] [PubMed]
56. Feng, Y.; Zhu, Q.; Lai, K.H. Corporate social responsibility for supply chain management: A literature review and bibliometric analysis. *J. Clean. Prod.* **2017**, *158*, 296–307. [CrossRef]
57. Mugomeri, E.; Bekele, B.S.; Mafaesa, M.; Maibvise, C.H.; Tarirai, C.; Aiyuk, S.E. A 30-year bibliometric analysis of research coverage on HIV and AIDS in Lesotho. *Health Res. Policy Syst.* **2017**, *15*, 1–9. [CrossRef] [PubMed]
58. Mateo-Sanguino, T.J. 50 years of rovers for planetary exploration: A retrospective review for future directions. *Robot. Auton. Syst.* **2017**, *94*, 172–185. [CrossRef]
59. Kokol, P.; Blazun-Vosner, E.; Zeleznik, D. Clinical simulation in nursing: A bibliometric analysis after its tenth anniversary. *Clin. Simul. Nurs.* **2017**, *13*, 161–167. [CrossRef]
60. Velasco-Muñoz, J.V.; Aznar-Sánchez, J.A.; Belmonte-Ureña, L.J.; López-Serrano, M.J. Advances in water use efficiency in agriculture: A bibliometric analysis. *Water* **2018**, *10*, 377. [CrossRef]
61. Li, X.; Huang, L. Toward a new paradigm for tailings phytostabilization—Nature of the substrates, amendment options, and anthropogenic pedogenesis. *Crit. Rev. Environ. Sci. Technol.* **2015**, *45*, 813–839. [CrossRef]
62. Li, X.; You, F.; Bond, P.L.; Huang, L. Establishing microbial diversity and functions in weathered and neutral Cu-Pb-Zn tailings with native soil addition. *Geoderma* **2015**, *247–248*, 108–116. [CrossRef]
63. Qian, G.; Schumann, R.C.; Li, J.; Short, M.D.; Fan, R.; Li, Y.; Kawashina, N.; Zhou, Y.; Smart, R.S.C.; Gerson, A.R. Strategies for reduced acid and metalliferous drainage by pyrite surface passivation. *Minerals* **2017**, *7*, 42. [CrossRef]
64. Ogbughalu, O.T.; Gerson, A.R.; Qian, G.; Smart, R.S.C.; Schumann, R.C.; Kawashima, N.; Fan, R.; Li, J.; Short, M.D. Heterotrophic microbial stimulation through biosolids addition for enhanced acid mine drainage control. *Minerals* **2017**, *7*, 105. [CrossRef]
65. Ellis, D.V. Effect of mine tailings on the biodiversity of the sea bed: Example of the Island Copper Mine, Canada. *Seas Millennium Environ. Eval.* **2000**, *3*, 235–246.
66. Miler, M.; Gosar, M. Characteristics and potential environmental influences of mine waste in the area of the closed Mežica Pb-Zn mine (Slovenia). *J. Geochem. Explor.* **2012**, *112*, 152–160. [CrossRef]
67. Méndez-Ramírez, M.; Hernández, M.A.A. Distribución de Fe, Zn, Pb, Cu, Cd y As originada por residuos mineros y aguas residuales en un transecto del Río Taxco en Guerrero, México. *Rev. Mex. Cienc. Geol.* **2012**, *29*, 450–462.
68. Bansch, C.; Topp, W. Woodland soil in a reclaimed lignite open-cast mine: A sustainable improvement of soil quality? *Verh. Ges. Okol.* **1998**, *29*, 511–518.
69. Gatzweiler, R.; Jahn, S.; Neubert, G.; Paul, M. Cover design for radioactive and AMD-producing mine waste in the Ronneburg area, Eastern Thuringia. *Waste Manag.* **2001**, *21*, 175–184. [CrossRef]
70. Ghose, M.K. Restoration and revegetation strategies for degraded mine land for sustainable mine closure. *Land Contam. Reclam.* **2004**, *12*, 363–378. [CrossRef]
71. Elshorbagy, A.; Jutla, A.; Barbour, L.; Kells, J. System dynamics approach to assess the sustainability of reclamation of disturbed watersheds. *Can. J. Civ. Eng.* **2005**, *32*, 144–158. [CrossRef]
72. Bowen, C.K.; Schuman, G.E.; Olson, R.A.; Ingram, L.J. Influence of topsoil depth on plant and soil attributes of 24-year old reclaimed mined lands. *Arid Land Res. Manag.* **2005**, *19*, 267–284. [CrossRef]
73. Van Deventer, P.W.; Bloem, A.A.; Hattingh, J.M. Soil quality as a key success factors in sustainable rehabilitation of kimberlite mine waste. *J. S. Afr. Inst. Min. Metall.* **2008**, *108*, 131–137. Available online: http://www.scielo.org.za/scielo.php?pid=S2225-62532008000300001&script=sci_arttext&tlng=en (accessed on 14 April 2018).
74. Maddocks, G.; Lin, C.; McConchie, D. Field scale remediation of mine wastes at an abandoned gold mine, Australia II: Effects on plant growth and groundwater. *Environ. Geol.* **2009**, *57*, 987. [CrossRef]
75. Lottermoser, B.G.; Glass, H.J.; Page, C.N. Sustainable natural remediation of abandoned tailings by metal-excluding heather (*Calluna vulgaris*) and gorse (*Ulex europaeus*), Carnon Valley, Cornwall, UK. *Ecol. Eng.* **2011**, *37*, 1249–1253. [CrossRef]

76. Wu, X.; Chen, Y.; Hu, J.; Yang, J.; Zhang, G. Current status and remediation measures for the solid mine ecological environment in Beijing, China. *Environ. Earth. Sci.* **2011**, *64*, 1555. [CrossRef]
77. Pepper, I.L.; Zerzghi, H.G.; Bengson, S.A.; Iker, B.C.; Banerjee, M.J.; Brooks, J.P. Bacterial populations within copper mine tailings: Long-term effects of amendment with Class A biosolids. *J. Appl. Microbiol.* **2012**, *113*, 569–577. [CrossRef] [PubMed]
78. Melgar-Ramírez, R.; González, V.; Sánchez, J.A.; García, I. Effects of application of organic and inorganic wastes for restoration of sulphur-mine soil. *Water Air Soil Pollut.* **2012**, *223*, 6123–6131. [CrossRef]
79. Valente, T.; Gomes, P.; Pamplona, J.; de la Torre, M.L. Natural stabilization of mine waste-dumps—Evolution of the vegetation cover in distinctive geochemical and mineralogical environments. *J. Geochem. Explor.* **2012**, *123*, 152–161. [CrossRef]
80. Courtney, R. Mine tailings composition in a historic site: Implications for ecological restoration. *Environ. Geochem. Health* **2013**, *35*, 79–88. [CrossRef] [PubMed]
81. Bigot, M.; Guterres, J.; Rossato, L.; Pudmenzky, A.; Doley, D.; Whittaker, M.; Pillai-McGarry, U.; Schmidt, S. Metal-binding hydrogel particles alleviate soil toxicity and facilitate healthy plant establishment of the native metallophyte grass *Astrebla lappacea* in mine waste rock and tailings. *J. Hazard. Mater.* **2013**, *248–249*, 424–434. [CrossRef] [PubMed]
82. Adams, A.; Raman, A.; Hodgkins, D. How do the plants used in phytoremediation in constructed wetlands, a sustainable remediation strategy, perform in heavy-metal-contaminated mine sites? *Water Environ. J.* **2013**, *27*, 373–386. [CrossRef]
83. Sjöberg, V.; Karlsson, S.; Grandin, A.; Allard, B. Conditioning sulfidic mine waste for growth of Agrostis capillaris-impact on solution chemistry. *Environ. Sci. Pollut. Res.* **2014**, *21*, 6888. [CrossRef] [PubMed]
84. Naeth, M.A.; Wilkinson, S.R. Establishment of restoration trajectories for upland tundra communities on diamond mine wastes in the Canadian arctic. *Restor. Ecol.* **2014**, *22*, 534–543. [CrossRef]
85. Banning, N.C.; Sawada, Y.; Phillips, I.R.; Murphy, D.V. Amendment of bauxite residue sand can alleviate constraints to plant establishment and nutrient cycling capacity in a water-limited environment. *Ecol. Eng.* **2014**, *62*, 179–187. [CrossRef]
86. Johansson, C.L.; Paul, N.A.; de Nys, R.; Roberts, D.A. The complexity of biosorption treatments for oxyanions in a multi-element mine effluent. *J. Environ. Manag.* **2015**, *151*, 386–392. [CrossRef] [PubMed]
87. Li, J.J.; Yan, J.X.; Li, H.J. Effects of different reclaimed measures on soil carbon mineralization and enzyme actives in mining areas. *Acta Ecol. Sin.* **2015**, *35*, 4178–4185. [CrossRef]
88. Anawar, H.M. Sustainable rehabilitation of mining waste and acid mine drainage using geochemistry, mine type, mineralogy, texture, ore extraction and climate knowledge. *J. Environ. Manag.* **2015**, *158*, 111–121. [CrossRef] [PubMed]
89. Santini, T.C.; Banning, N.C. Alkaline tailings as novel soil forming substrates: Reframing perspectives on mining and refining wastes. *Hydrometallurgy* **2016**, *164*, 38–47. [CrossRef]
90. Nirola, R.; Megharaj, M.; Beecham, S.; Aryal, R.; Thavamani, P.; Vankateswarlu, K.; Saint, C. Remediation of metalliferous mines, revegetation challenges and emerging prospects in semi-arid and arid conditions. *Environ. Sci. Pollut. Res.* **2016**, *23*, 20131–20150. [CrossRef] [PubMed]
91. Párraga-Aguado, I.; González-Alcaraz, M.N.; López-Orenes, A.; Ferrer-Ayala, M.A.; Conesa, H.M. Evaluation of the environmental plasticity in the xerohalophyte *Zygophyllum fabago* L. for the phytomanagement of mine tailings in semiarid areas. *Chemosphere* **2016**, *161*, 259–265. [CrossRef] [PubMed]
92. Nancucheo, I.; Bitencourt, J.A.P.; Sahoo, P.K.; Oliveira-Alves, J.; Siqueira, J.O.; Oliveira, G. Recent Developments for Remediating Acidic Mine Waters Using Sulfidogenic Bacteria. *BioMed Res. Int.* **2017**, *2017*, 7256582. [CrossRef] [PubMed]
93. Plaza, F.; Wen, Y.; Perone, H.; Xu, Y.; Liang, X. Acid rock drainage passive remediation: Potential use of alkaline clay, optimal mixing ratio and long-term impacts. *Sci. Total Environ.* **2017**, *576*, 572–585. [CrossRef] [PubMed]
94. Sözen, S.; Orhon, D.; Dinçer, H.; Ateşok, G.; Baştürkçü, H.; Yalçın, T.; Öznesil, H.; Karaca, C.; Allı, B.; Dulkadiroğlu, H.; et al. Resource recovery as a sustainable perspective for the remediation of mining wastes: Rehabilitation of the CMC mining waste site in Northern Cyprus. *Bull. Eng. Geol. Environ.* **2017**, *76*, 1535–1547. [CrossRef]

95. Mwandira, W.; Nakashima, K.; Kawasaki, S. Bioremediation of lead-contaminated mine waste by *Pararhodobacter* sp. based on the microbially induced calcium carbonate precipitation technique and its effects on strength of coarse and fine grained sand. *Ecol. Eng.* **2017**, *109*, 57–64. [CrossRef]
96. Bartke, K. Waste management in the mineral recovery industry according to EU guideline 2006/21/EC. *Wasser Abfall* **2009**, *11*, 40–44.
97. Kalin, M. Passive mine water treatment: The correct approach? *Ecol. Eng.* **2004**, *22*, 299–304. [CrossRef]
98. Azam, S. Thickening of mine waste slurries. *Geotech. News* **2004**, *22*, 40–43.
99. Younger, P.L. Environmental impacts of coal mining and associated wastes: A geochemical perspective. *Geol. Soc. Lond.* **2004**, *236*, 169–209. [CrossRef]
100. Macklin, M.G.; Brewer, P.A.; Hudson-Edwards, K.A.; Bird, G.; Coulthard, T.J.; Dennis, I.A.; Lechler, P.J.; Miller, J.R.; Turner, J.N. A geomorphological approach to the management of rivers contaminated by metal mining. *Geomorphology* **2006**, *79*, 423–447. [CrossRef]
101. Meech, J.A.; Scoble, M.; Wilson, W.; Lang, B.; Klein, B.; Veiga, M.M.; Hall, R.; Ghomshei, M.; Baldwin, S.; Lavkulich, L.M.; et al. CERM3 and its contribution to providing sustainable research for the mining industry. *CIM Bull.* **2003**, *96*, 72–81.
102. Careddu, N.; Siotto, G.; Siotto, R.; Tilocca, C. From landfill to water, land and life: The creation of the Centre for stone materials aimed at secondary processing. *Resour. Policy* **2013**, *38*, 258–265. [CrossRef]
103. Shukla, M.K.; Lal, R. Soil organic carbon stock for reclaimed minesoils in northeastern Ohio. *Land Degrad. Dev.* **2005**, *16*, 377–386. [CrossRef]
104. Wajima, T.; Ikegami, Y. Stabilization of mine waste using paper sludge ash under laboratory condition. *J. Jpn. Inst. Met.* **2008**, *72*, 903–910. [CrossRef]
105. Hwang, T.; Neculita, C.M.; Han, J.I. Biosulfides precipitation in weathered tailings amended with food waste-based compost and zeolite. *J. Environ. Qual.* **2012**, *41*, 1857–1864. [CrossRef] [PubMed]
106. Emery, J.J.; MacKay, M.H.; Umar, P.A.; Vanderveer, D.G.; Pichette, R.J. Use of wastes and byproducts as pavement construction materials. In Proceedings of the Canadian Geotechnical Conference, Toronto, ON, Canada, 25–28 October 1992; pp. 45/1–45/10.
107. Venkatarama-Reddy, B.V. Sustainable building technologies. *Curr. Sci.* **2004**, *87*, 899–907.
108. Gabzdyl, W.; Hanak, B. Raw materials from the Upper Silesia Coal Basin and from the adjacent areas. *Prz. Geol.* **2005**, *53*, 726–733.
109. Arrigo, I.; Catalfamo, P.; Cavallari, L.; Di Pasquale, S. Use of zeolitized pumice waste as a water softening agent. *J. Hazard. Mater.* **2007**, *147*, 513–517. [CrossRef] [PubMed]
110. Liu, H.; Liu, Z. Recycling utilization patterns of coal mining waste in China. *Resour. Conserv. Recycl.* **2010**, *54*, 1331–1340. [CrossRef]
111. Lottermoser, B.G. Recycling, reuse and rehabilitation of mine wastes. *Elements* **2011**, *7*, 405–410. [CrossRef]
112. Cadierno, J.F.; Romero, M.I.G.; Valdés, A.J.; Morán del Pozo, J.M.; García-González, J.; Robles, D.R.; Espinosa, J.V. Characterization of Colliery Spoils in León: Potential Uses in Rural Infrastructures. *Geotech. Geol. Eng.* **2014**, *32*, 439–452. [CrossRef]
113. Kundu, S.; Aggarwal, A.; Mazumdar, S.; Dutt, K.B. Stabilization characteristics of copper mine tailings through its utilization as a partial substitute for cement in concrete: Preliminary investigations. *Environ. Earth Sci.* **2016**, *75*, 227. [CrossRef]
114. Yang, Y.; Chen, T.; Morrison, L.; Gerrity, S.; Collins, G.; Porca, E.; Li, R.; Zhan, X. Nanostructured pyrrhotite supports autotrophic denitrification for simultaneous nitrogen and phosphorus removal from secondary effluents. *Chem. Eng. J.* **2017**, *328*, 511–518. [CrossRef]
115. Rana, A.; Kalla, P.; Csetenyi, L.J. Recycling of dimension limestone industry waste in concrete. *Int. J. Min. Reclam. Environ.* **2017**, *31*, 231–250. [CrossRef]
116. Taha, Y.; Benzaazoua, M.; Hakkou, R.; Mansori, M. Coal mine wastes recycling for coal recovery and eco-friendly bricks production. *Miner. Eng.* **2017**, *107*, 123–138. [CrossRef]
117. Gorakhki, M.H.; Bareither, C.A. Sustainable reuse of mine tailings and waste rock as water-balance covers. *Minerals* **2017**, *7*, 128. [CrossRef]

 © 2018 by the authors. Licensee MDPI, Basel, Switzerland. This article is an open access article distributed under the terms and conditions of the Creative Commons Attribution (CC BY) license (http://creativecommons.org/licenses/by/4.0/).

Review

Re-Thinking Mining Waste through an Integrative Approach Led by Circular Economy Aspirations

Maedeh Tayebi-Khorami, Mansour Edraki, Glen Corder and Artem Golev *

Centre for Mined Land Rehabilitation, Sustainable Minerals Institute, The University of Queensland, St Lucia QLD 4072, Australia; m.tayebikhorami@uq.edu.au (M.T.-K.); m.edraki@cmlr.uq.edu.au (M.E.); g.corder@smi.uq.edu.au (G.C.)
* Correspondence: a.golev@uq.edu.au

Received: 5 April 2019; Accepted: 9 May 2019; Published: 10 May 2019

Abstract: Mining wastes, particularly in the form of waste rocks and tailings, can have major social and environmental impacts. There is a need for comprehensive long-term strategies for transforming the mining industry to move toward zero environmental footprint. "How can the mining industry create new economic value, minimise its social and environmental impacts and diminish liability from mining waste?" This would require cross-disciplinary skills, across the social, environmental, technical, legal, regulatory, and economic domains, to produce innovative solutions. The aim of this paper is to review the current knowledge across these domains and integrate them in a new approach for exploiting or "re-thinking" mining wastes. This approach includes five key areas of social dimensions, geoenvironmental aspects, geometallurgy specifications, economic drivers and legal implications for improved environmental outcomes, and circular economy aspirations, which are aligned with the 10 principles of the International Council on Mining and Metals (ICMM). Applying circular economy thinking to mining waste presents a major opportunity to reduce the liability and increase the value of waste materials arising from mining and processing operations.

Keywords: sustainable development; tailings management; industrial ecology; sustainable resource management; mining waste; circular economy

1. Introduction

Each year, mining operations generate large volumes of mining waste. According to the Mining, Minerals, and Sustainable Development Project (MMSD), there are approximately 3500 active mining waste facilities worldwide, consisting of waste rock dumps and tailing dams. The estimated worldwide generation of solid wastes from the primary production of mineral and metal commodities is over 100 billion tonnes per year and can range from several times the mass of the valuable element, such as iron and aluminium ores, up to millions of times for some scarce elements such as gold ore [1]. This large amount of mining waste can have major environmental impacts and require appropriate management strategies both in the short- and long-term [2]. In addition, increasing demand for essential metals is leading to the extraction and development of complex finer-grained low-grade orebodies. The mining of lower grade and/or complex ores will produce greater quantities of tailings per ton of product and will also increase the fines content of the tailings [3]. If not managed properly, mining waste can generate significant pollution, both through air pathways (dust and gas emissions) and water leaching (acid mine drainage) [4]. Moreover, failure to manage can result in costly catastrophic consequences. Recent events, such as the Brumadinho tailings dam failure in Brazil, have heightened the attention of the industry and society to the catastrophic impacts of mining waste when failure of tailings management systems occurs. One way to better manage mine waste is to reduce its generation from mining, such as transitioning from open-pit surface operations to underground mining by the development of more powerful and efficient underground equipment [5].

The environmental, economic, and social impacts of mining waste indicate that the mining industry needs to re-think waste management. This re-thinking will require expertise from many disciplines to make a transformational change to how the mining industry and its stakeholders manage and utilise mining waste in current operations and future projects. Potential solutions should draw on existing approaches, such as cleaner production, by-products from waste, re-engineering of processes, closed-loop systems, and product stewardship [1].

This paper aims to assess current management frameworks and the application of mining waste management strategies, and to determine current gaps and challenges. It also highlights the benefits of an integrative approach for exploiting mining waste across the social, environmental, technical, legal, regulatory, and economic domains.

2. Sustainable Development of Mining Waste

Over the last few decades, several approaches have emerged that aim to include sustainability aspects across the mining life cycle [6]. Several industry organisations and companies have implemented sustainability principles and strategies for establishing commitment to resource development in a socially and environmentally responsible manner. One such organisation, the International Council on Mining and Metals (ICMM), has 10 principles for effective sustainable development, presented in Table 1 [7].

Table 1. Sustainable development principles of ICMM.

1	Ethical business and sound governance
2	Integrate sustainable development in decision-making
3	Respect for human rights
4	Effective risk management strategies
5	Health and safety performance
6	Environmental performance
7	Conservation of biodiversity and land-use planning
8	Responsible design, use, reuse, recycling, and disposal of materials
9	Social contribution
10	Engagement, communication and independently-verified reporting

These principles are similar to international sustainable management standards, including the Rio Declaration [8], Global Reporting Initiative [9], World Bank Operational Guidelines, the International Labour Organization (ILO) Conventions, and the Voluntary Principles on Security and Human Rights [7]. These 10 principles cover many topics associated with sustainability, although there is no uniform approach that allows the industry to better integrate sustainability into the design process. Ideally, a sustainability framework should connect corporate and operational level activities and engage with technical professionals [10].

In this regard, we have identified five key areas, which are aligned with the 10 ICMM principles, that will cover the most important aspects for improved environmental outcomes, and circular economy aspirations. We believe that a dynamic interaction across these five areas will drive the mindset change for re-thinking mining wastes. It includes social dimensions, geoenvironmental aspects, geometallurgy specifications, economic drivers, and legal implications (Figure 1). At its core is the overarching question: "How can the mining industry create new economic value, minimise its social and environmental impacts and diminish liability from mining waste?" To find potential solutions to this question, the following guiding questions should be addressed in the following order:

- Social dimensions: What are the local, regional, and global societal dimensions related to managing mining waste?

- Geoenvironmental aspects: What are the spatial and temporal geoenvironmental impacts resulting from mining waste, and how can potential liabilities be prevented or substantially mitigated?
- Geometallurgy specifications: What are the geometallurgical properties to create additional value and improve environmental outcomes in waste from mining and mineral processing?
- Economic drivers and legal implications: How and what economic drivers should lead the changes in regulatory systems, to transform business approaches for creating value, diminishing risk and drastically mitigating liabilities from mining waste?
- Circular economy aspirations: How can the mining industry assess and quantify their contribution to the circular economy?

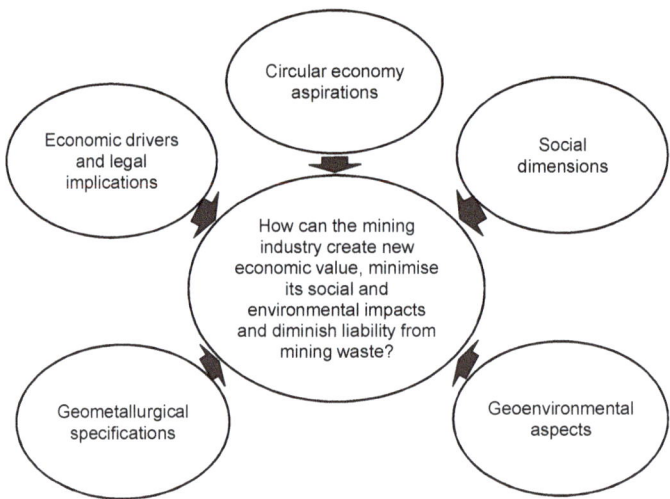

Figure 1. Key areas of an integrative approach for mining waste.

By taking this approach there will be a far greater chance that social dimensions and expectations will be achieved within the technical, regulatory, and economic constraints, while ensuring potential outcomes meet the aims and desires of the circular economy. If all of this is achieved, then the social expectations, which are closely aligned with the aspirations of the circular economy, should be broadly met.

3. Five Areas of Integrative Approach

In this section, the key aspects of practice in the five areas of an integrative approach are reviewed. These reviews provide a basis for developing the connections across the five areas.

3.1. Social Dimensions

Mining companies have had a long reputation for social responsibility [11], however, the costs of community conflict remain poorly understood. A recent study investigated the cost of conflict between the community and companies in the extractive sector, and found that the majority of mining and gas companies do not identify the full range of costs of conflict with local communities [12]. Conflict between an extractive company and the community is often triggered by the environmental impacts of company operations [12]. Conflict therefore acts as a means by which environmental and social risks are translated into actual business cost [13].

Esteves et al. [14] noted the importance of social impact assessment (SIA) in managing issues associated with mining and development, and described a mechanism for identifying, assessing, and managing the social impacts of mining in a systematic manner. SIAs can be used not just for understanding and resolving existing conflicts, but also for preventing conflicts [15], particularly by tracking changes in economic and social conditions over time [16].

Kemp and Vanclay [17] examined the inclusion of human rights in impact assessments and noted that the discourse surrounding a company's human rights related responsibilities is framed around avoiding or minimising harm, and does not extend to the development of benefits or shared value. Similarly, within the mining industry, the concept of "social license to operate" is viewed and applied as a way of reducing opposition to industry, rather than an opportunity for a company to collaborate with community to create an agenda that would satisfy the community's desire for development and the company's desire for a profitable and sustainable business [18].

One area that poses a substantial risk to the environment, human rights, and a company's social licence to operate is the disposal of mining and mineral processing wastes. Franks et al. [19] developed a set of seven principles based on sustainable development to guide decision making in regard to mining waste disposal, including: (1) Mining and mineral processing wastes should be managed in a way that is physically and chemically stable; (2) the waste that interacts with the environment should be inert; (3) waste that is not inert should be isolated in a form that limits interaction and subsequent mobilisation; (4) the waste should be contained, i.e., geographically bounded and managed with a minimal footprint that limits interaction with the surrounding environment; (5) the waste should be managed in a manner that considers local environmental and social conditions of each location; (6) the waste should be managed in a way that minimises water and energy use and the need to actively manage wastes after mine closure; and (7) preference should be given to technologies that improve the impacts of waste disposal on society and the environment and provide opportunities for re-use should they be pursued [19].

3.2. Geoenvironmental Aspects

Mine waste facilities mainly consist of waste rocks dumps and tailings dams, requiring appropriate management both in the short- and long-term.

3.2.1. Waste Rocks Dumps

Mining operations produce large quantities of waste rocks, which often has little or no valuable minerals. Waste rocks contain coarse, crushed, or blocky material covering a range of sizes, from very large rocks to fine particles. The current practices of dumping waste rocks by trucks, or draglines in the case of coal mines, create very large layered structures with preferential flow paths, which are conducive for the generation of acid and metalliferous drainage (AMD)—a significant challenge to mine owners, regulators, and other stakeholders. AMD, generated through the oxidation of sulphide minerals (mainly pyrite) in waste rocks, comprises poor-quality leachate typically characterised by acidic pH and high concentrations of sulphate, iron, and heavy metals.

Sometimes, due to the presence of carbonates or other gangue minerals, reactions do not lead to acidic conditions, however, they can still result in poor-quality drainage with high concentrations of sulphate, metals and metalloids (e.g., arsenic), and salts referred to as neutral or saline drainage. AMD can be detrimental to aquatic ecosystems and possibly human health, impact groundwater quality and limit the downstream water use, and results in long-term environmental liabilities. The measures to counter this, often only partially, include mined land rehabilitation programs and the installation of expensive water treatment systems, and require long-term monitoring of the impacted area and surrounding environment.

Remediation of acid producing mine waste and treatment of mine water can be costly. Management costs for AMD have been estimated globally at approximately US$ 1.5 billion per year [20], while the overall environmental liabilities are estimated to be in excess of US$ 100 billion [21]. Integration of an AMD management plan, from the early phases of exploration until final closure, can help to decrease the environmental impacts. This requires the commitment of the whole operation from mine planners to the sustainability managers, and it must be incorporated into the mine's financial models.

With detailed geological, mineralogical, and geochemical assessments at an early stage in the development of ore deposits, it is possible to develop a resource block model for waste rocks to allow selective handling of waste. This will substantially reduce the overall cost of the monitoring and disposal of AMD generating waste. The common practices used in integrating the AMD management plan into mine operations include, for example, selective mining and placement of sulphide waste, encapsulation of potentially acid-producing (PAF) materials, potential layering and blending during the construction of waste rocks dumps, dispose of PAF materials in the backfills, isolating the acid-producing pit floor using barriers, and preventing the interaction of groundwater with the pit floor [22].

Various low cost waste products have been used to manage AMD [23]. Fly ash has been used to help neutralise AMD, improve the quality of degraded soils, and as a part of a cover system designed to isolate potentially hazardous mining waste. However, a lack of information on the practical use of coal combustion by-products (including fly ash) and a lack of guidelines and regulations for their use has limited their application in mine site rehabilitation and backfilling [24]. Waste rocks can also be a resource of minerals and metals, or have other applications at the mine site or elsewhere, such as backfill for open voids and underground mines, landscaping, capping for waste facilities, soil components and soil additives (e.g., for neutralizing infertile alkaline agricultural soils), aggregate and construction materials, and alternative raw materials for cement and concrete [20].

3.2.2. Tailings Dams

Volumetrically, mine tailings impoundments are among the largest man-made structures in the world [25]. Furthermore, tailings dam failures account for the major mining-related environmental disasters [26]. A major recent failure happened in January 2019 when Vale's iron ore tailings dam in Brumadinho, Brazil collapsed and killed at least 206 people. Another catastrophic event happened in November 2015 when a tailings dam at the Samarco Mine, Brazil, collapsed, releasing more than 43.7 million cubic meters of water and mine waste, and the resulting mudflow reached the Atlantic coast through the Doce River, along more than 500 km of the river course. In September 2008, 277 people died in an accident caused by iron ore tailings release from a dam break in Shanxi Province, China [27].

Similar to waste rocks dumps, tailings may generate AMD and, apart from the catastrophic failure of dam walls, may pose chronic environmental and human health issues due to the dispersion of contaminants by dust and seepage. Tailings may also contain mineral processing reagents, including salts and cyanide.

Reactive strategies, such as remediation of tailings solids or tailings seepage water, can pose long-term legacies for companies, governments, and society after mine closure, and do not meet community expectations. Edraki et al. [28] compared different tailings disposal methods, including conventional disposal; paste and thickened tailings; tailings reuse, recycling, and reprocessing; and proactive management. The review of different tailings disposal methods indicated that an integrative, proactive approach to tailings management is needed for improved, environmental, social, and economic outcomes. Such an approach would involve using geo-metallurgical data from the orebody to predict the composition of tailings produced by different processing scenarios. The implications for environmental management and closure can then be predicted earlier and considered as part of the mine plan. The tailings revegetation and capping strategies for two Australian tailings storage facilities which had failed to prevent tailings seepage for several years after mine closure were discussed in [29] and [30].

There is also an increasing interest in new technologies to recycle and utilize mine tailings more effectively. For example, the alkali-activation of some mine tailings allows binders with sufficient compressive strength to be used as a mine backfill or raw material in the construction industry. The application of untreated tailings that contain trace elements in civil engineering projects is hindered by the potential leaching of some toxic elements. However, such materials can be used with cementitious binders, such as Portland cement, slag, lime, and gypsum [31–34]. Kim et al. [35] studied the feasibility of tailings with no pre-treatment in cement based low-strength materials. Liu et al. [36] fabricated a new type of porous ceramics by a conventional ceramic sintering process using lead-zinc mine tailings and fly ash as the raw material. Taha et al. [37] present a feasible approach of reusing the tailings from waste rock residual coal to produce ecofriendly fired bricks. An advantage associated with the re-use of tailings is that, as they have already undergone industrial processes, they usually exhibit good homogeneity and fine particle sizes. Other efforts to recycle mining waste have been described elsewhere [38].

Desulphurisation has been considered as an alternative method to avoid formation of AMD [39]. Depyritised tailings have low sulphide content so that they are not acid generating. The iron sulphide (pyrite) particles could be concentrated, and then back-filled in mines or deposited separately. Depyritised tailings could also potentially be used as cover on mine waste storage facilities. An approach where conventional tailings are separated by flotation into a sulphide-rich and a largely benign fraction of tailings has also been proposed [40].

3.3. Geometallurgy Specifications

The traditional approach for ore body evaluation focuses on the in-situ tonnes and grade as well as testwork to estimate recovery. However, with more complex and refractory ore bodies being exploited, there is an increasing emphasis on risk management and cost efficiency. The geometallurgical approach directly addresses this by using 3D block models that display the distribution of key metallurgical parameters through the orebody [41]. An effective geometallurgical model integrates the important geological information, including geological, geotechnical, geochemical, and mineralogical data, with metallurgical test work results.

Geometallurgy not only involves the integration of geological, mining, and metallurgical fields, but it can also be expanded to include environmental and economic information to produce a block model that can be used to obtain the best outcome in terms of plant response and plant sustainability [42]. There are several benefits in applying geometallurgical theory to reduce the technological and financial risk of a project. This can be through plant design, production forecasting, optimisation of the water balance, as well as modelling of tailings storage facilities and predicting the settlement behaviour and stability of different mine waste types under different conditions. Geometallurgy is also a useful method to better understand mining waste and the potential hazards from acid rock drainage [43].

Dunham and Vann [44] and Vann, et al. [45] outline a broad "whole-of-value-chain" view through integration of geology, mine planning, operational design, metallurgy, marketing, and environmental management to improve or maximise the economic value of mining projects and operations. Bye [46] discussed some industry case studies that demonstrate strategies for gaining value from geometallurgical initiatives. This highlighted that the geometallurgical approach is moving away from factored ore reserves to data-rich block models and in doing so provides reliable information for mining, metallurgical, and environmental considerations. Louwrens et al. [47] and Louwrens et al. [47] studied the usage of geometallurgical principles and methodologies for reprocessing of the Ernest Henry (Cu-Au) tailings storage facility in North-West Queensland, Australia. They established a new standard method of economic evaluation of the potential of reprocessing the tailings material. Edraki et al. [48] produced new tailings management models to predict the properties of tailings generated under various processing scenarios. These results were used to determine the implications for environmental management. Tungpalan et al. [49] developed a method to assess the variability of an ore deposit and its influence on metallurgical performance. Effective sample selection for mineralogical characterisation was used to link geometallurgy to circuit simulation. The drill core samples were classified according to the geological and mineralogical factors. Results of the classification were used to decide where more samples were needed, and which samples should undergo a more detailed analysis by a geologist. The improved characterisation of the ore body was used in a circuit simulation to predict the characteristics of the final concentrate and tailings. The knowledge of the characteristics of the concentrate and tailings can be used in mine planning so that parts of the ore body can be targeted for extraction that balance revenue from mineral recovery with tailings management costs.

A treatment methodology to improve the selectivity of enargite removal from a copper-gold ore and reduce arsenic content in the tailings using process mineralogy was proposed by Tayebi-Khorami et al. [50]. This methodology involved two steps:

- A de-sliming stage prior to the flotation process, where fine liberated enargite can be separated from the flotation feed; and
- A fine treatment stage, where the fine gangue minerals can be separated, the copper can be recovered, and the arsenic can be safely stored.

The separation of the fine enargite results in the exclusion of arsenic from the tailings.

3.4. Economic Drivers and Legal Implications

Economic drivers and legal implications play an important, if not critical, role in establishing new paradigms and allowing innovative approaches to be implemented. Without a proper understanding of the economics as well as the legislative and regulatory context, technically feasible solutions that deliver better outcomes can fail.

A pivotal report 30 years ago was "Our Common Future" by the World Commission on Environment and Development [51], which defined sustainable development as "development that meets the needs of the present without compromising the ability of future generations to meet their own needs". This highlights the concept of inter-generational equity relevant to mining legacies as well as economic development and legal requirements. While mineral and energy resources are obviously extracted without being replaced, it is possible to mitigate impacts and generate wealth for governments and communities from mining. Such examples include investing in community development and the regeneration of mined landscapes to create positive legacies.

The drive toward a more sustainable society will result in better worldwide standards and global regulatory and governance bodies. These will cover the spectrum of industrial and environmental impacts of all stages of a product's life cycle [52]. Better life-of-mine planning can reduce the amount of waste generated and help prevent acid mine drainage containing high concentrations of metals, and most likely reduce overall project costs [53].

With increasing reuse and recycling, resources will last longer and deliver on the aims of the circular economy. This transition requires fundamental changes, including appropriate infrastructure, regulation and legislation, and competitive economics. Recycling could be made mandatory for private and commercial companies, while product design should allow for easy dismantling and metals extraction, e.g., from computers and other electronic products [52].

Aspects related to licence to operate, social factors, and the regulatory environment depend heavily on the project jurisdiction. As a result, there is a strong contextual component in developing a new mining project that does not directly relate to technical aspects, but can affect the financial aspects [54].

3.5. Circular Economy Aspirations

The mining sector is represented mainly by linear activities, being the major supplier of resources to modern society, nevertheless the concept of circular economy can help to improve the sector's sustainability performance [55]. The aim would be to optimise the total material cycle from mining to manufacturing and to extend the product use phase, including the reuse and recycling of any waste streams arising in industrial and consumer activities to ensure overall resource efficiency and resilience [56]. Eco-efficiency and resilience have been identified as key characteristics of a sustainable mining operation, where optimising extraction and minimising the amount of valuable material in the waste would help to address problems, such as a declining ore grade, decreasing economic viability, and increasing mining legacies [4].

Under the conventional linear economy model, the current trends in mining, such as decreasing ore grades and higher tonnage rates, would continue escalating the problem with mining waste, and its associated inherent risks. Where feasible, replacing open cut mining with underground mining will make a significant reduction in waste generation. Alternative approaches, whether they relate to better waste disposal techniques, such as paste and thickened tailings, better mined land rehabilitation practices, or waste rocks and tailings reuse, recycling, and reprocessing, are urgently needed. In fact, the industry will need to move progressively to "closing the loop" strategies, which will dramatically reduce the quantities of wastes [1].

The circular economy articulates the importance of closed loop systems which reduce the need for the extraction and processing of new resources. This can be extended further to the overall impact from mining activities, with a particular focus on "getting more from less". As such, within the mining and metals sector, following the 3R waste reducing principle (reduce, reuse, recycle) can make a significant contribution [57]. This includes examples across two categories:

(a) Circular economy sensu stricto:

- Improving water and material reuse through cyclic systems and innovative technologies;
- Maximizing reuse of waste and by-products;
- Collaborating with the manufacturing sector to design adaptable and easy-to-repair products;
- Better marking of materials and alloys to aid identification at end-of-life and allow subsequent reuse and recycling.

(b) Efficiency measures as part of the circular economy in a wider sense:

- limiting the use of raw materials and balancing supply and demand;
- Improving recovery rates in mining and mineral processing;
- Minimizing waste generation such as tailings, gas emissions, and waste water;
- Developing feasible options for lower grade ores;
- Extending the life of a resource, material, product, or service through better planning for future applications and reuse [52].

Several different strategies for limiting mining waste and/or the associated environmental impacts have been classified by their ability to generate additional economic value and potentially decrease the environmental legacy of mining operations [58]. However, the best outcomes can be achieved with proactive waste management, which would combine ore body characterisation, mine planning, ore processing, waste disposal, re-processing, recycling and reuse, and finally land rehabilitation in one integrative approach [59]. This would be a crucial contribution that the mining industry can make towards the circular economy.

Another important and inherent part of the circular economy approach is the introduction of disruptive innovations [60]. In mining these can include, for example, the integration of tailings reprocessing with mined land rehabilitation, and using the post-mine landscape for new economic activities and development. Two recent examples of such innovative thinking originate from North Queensland in Australia. Kidston Renewable (solar and hydro) Energy Hub has been developed on the historic gold mine site, with the reuse of two open pits at different elevation levels as part of energy generation and storage [61]. The New Century mine project involves reviving zinc concentrate production after mine closure, based on the reprocessing of historic tailings and taking responsibility for the final rehabilitation of the mine site [62].

4. Re-Thinking Mining Waste

The mining industry is essential to global economic and social development, and it will continue to be the major resource supplier to our society for the foreseeable future [63]. A strong transition towards recycling and circularity is necessary, but requires some fundamental changes, including appropriate infrastructure, legislation, and favourable economics.

Despite the advancements in tailings management in the last few decades, there is a lack of optimal scenarios for mining waste that deliver overall sustainable societal benefits. Novel initiatives for dealing with mining waste and evidence-based best-practice guidelines for minimising environmental and human health risks are essential to reverse this trend. A transformational perspective to mining waste is necessary and should include a multidisciplinary and integrative approach for exploiting mining waste to create new economic value and move toward a zero environmental footprint. These elements speak to the aspirations of the circular economy and drive down impacts with residual waste management.

Applying circular economy thinking to mining waste presents a major opportunity to reduce the liability and increase the value of mining waste. However, many obstacles exist, such as current regulatory regimes and societal acceptance of products made from mining waste. Solutions to these obstacles are required across and between social, geoenvironmental, geometallurgical, engineering, economic, and legal/regulatory fields. By following the circular economy aspirations, the mining industry and regulators will engage in the full life cycle of projects, including final residual wastes, utilising natural cycles and transformations of metals in the environment, and, importantly, creating resilience to resources cycles (i.e., better responses to changes in the global supply and demand for different resources).

Knowledge of the industries that would use the products created with mining waste and what is required to connect the mining industry with downstream users is, however, far less developed. A greater understanding of the impact of companies that are likely to produce or use products from mining waste as well as the social perception and acceptability of those products is important and can only be properly determined by applying a multidisciplinary approach.

Human health and environmental impacts of metal toxicity will also be critical in determining whether products made from mining waste (e.g., construction materials) are safe. Effective regulation will be relevant to the design of new regulatory regimes that promote circular economy thinking and the management of mining waste in a way that increases its value and reduces risk posed by the waste.

A new approach driven by circular economy aspirations and sustainable development ideals, adhering to the highest available resource value principles, will help increase the value of mining waste. Specifically, it would help determine optimal scenarios for mining waste that would deliver overall societal benefits, notably including economic benefits for the industry and reducing the liabilities from mining legacies. These optimal scenarios are the catalyst for developing innovative processes, underpinned by rigorous, integrated analysis across the geometallurgical, geoenvironmental, engineering, social, legal, regulatory, and economic disciplines. The success would be measured by the delivery of transformational outcomes for the industry and government, including novel initiatives for dealing with mining waste and evidence-based best practice guidelines for minimising environmental and health risks.

5. Conclusions

In this paper, we identified and reviewed five key areas, which can form an integrative approach for exploiting or "re-thinking" mining wastes, framed around the circular economy. These include social dimensions, geoenvironmental aspects, geometallurgy specifications, economic drivers, and legal implications. While this dynamic and conceptual approach will be helpful for re-thinking mining waste, much work is still necessary in the research and development phases to identify efficient and effective solutions in each key area. Importantly, feasible solutions, like any credible sustainable development outcome, will require contextual knowledge for each mining operation and/or mining project under consideration.

Author Contributions: M.T.-K. led the writing—original draft as well as contributing to the conceptualization and methodology. M.E., G.C. and A.G. were involved in the conceptualization and methodology, and contributed to the writing, review, and editing.

Funding: This research received no external funding.

Conflicts of Interest: The authors declare no conflict of interest.

References

1. Rankin, W.J. Towards zero waste. *AusIMM Bull.* **2015**, *2015*, 32–37.
2. Adiansyah, J.S.; Rosano, M.; Vink, S.; Keir, G. A framework for a sustainable approach to mine tailings management: Disposal strategies. *J. Clean Prod.* **2015**, *108*, 1050–1062. [CrossRef]
3. Chryss, A.; Fourie, A.B.; Monch, A.; Nairn, D.; Seddon, K.D. Towards an integrated approach to tailings management. *J. South. Afr. Inst. Min. Metall.* **2012**, *112*, 965–969.
4. Lebre, É.; Corder, G. Integrating industrial ecology thinking into the management of mining waste. *Resources* **2015**, *4*, 765–786. [CrossRef]
5. Wellmer, F.W.; Buchholz, P.; Gutzmer, J.; Hagelüken, C.; Herzig, P.; Littke, R.; Thauer, R.K. *Raw Materials for Future Energy Supply*; Springer: Berlin, Germany, 2018.
6. Corder, G.D. Insights from case studies into sustainable design approaches in the minerals industry. *Miner. Eng.* **2015**, *76*, 47–57. [CrossRef]
7. ICMM. *Icmm Sustainable Development Framework*; International Council on Mining and Metals: London, UK, 2003; pp. 2–17.
8. UN. *Rio Declaration on Environment and Development*; Report of the United Nations conference on environment and development; United Nations: Rio de Janerio, Brazil, 3–14 June 1992.
9. GRI. The Global Reporting Initiative (gri) Standards. Available online: https://www.globalreporting.org/standards (accessed on 13 July 2018).
10. McLellan, B.C.; Corder, G.D.; Giurco, D.; Green, S. Incorporating sustainable development in the design of mineral processing operations - review and analysis of current approaches. *J. Clean Prod.* **2009**, *17*, 1414–1425. [CrossRef]
11. Kemp, D.; Owen, J.R.; van de Graaff, S. Corporate social responsibility, mining and "audit culture". *J. Clean Prod.* **2012**, *24*, 1–10. [CrossRef]
12. Davis, R.; Franks, D.M. Costs of company-community conflict in the extractive sector. *Corp. Soc. Responsib. Initiat. Rep.* **2014**, *66*, 1–56.
13. Franks, D.M.; Davis, R.; Bebbington, A.J.; Ali, S.H.; Kemp, D.; Scurrah, M. Conflict translates environmental and social risk into business costs. *Proc. Natl. Acad. Sci. USA* **2014**, *111*, 7576–7581. [CrossRef]
14. Esteves, A.M.; Franks, D.; Vanclay, F. Social impact assessment: The state of the art. *Impact Assess. Proj. A* **2012**, *30*, 34–42. [CrossRef]
15. Prenzel, P.V.; Vanclay, F. How social impact assessment can contribute to conflict management. *Environ. Impact Assess. Rev.* **2014**, *45*, 30–37. [CrossRef]
16. Lockie, S.; Franettovich, M.; Petkova-Timmer, V.; Rolfe, J.; Ivanova, G. Coal mining and the resource community cycle: A longitudinal assessment of the social impacts of the coppabella coal mine. *Environ. Impact Assess. Rev.* **2009**, *29*, 330–339. [CrossRef]
17. Kemp, D.; Vanclay, F. Human rights and impact assessment: Clarifying the connections in practice. *Impact Assess. Proj. Apprais.* **2013**, *31*, 86–96. [CrossRef]
18. Owen, J.R.; Kemp, D. Social licence and mining: A critical perspective. *Resour. Policy* **2013**, *38*, 29–35. [CrossRef]
19. Franks, D.M.; Boger, D.V.; Côte, C.M.; Mulligan, D.R. Sustainable development principles for the disposal of mining and mineral processing wastes. *Resour. Policy* **2011**, *36*, 114–122. [CrossRef]
20. Lottermoser, B.G. *Mine Wastes: Characterization, Treatment and Environmental Impacts*, 3rd ed.; Springer: Berlin, Germany, 2010.

21. MEND. *Mine Environment Neutral Drainage Manual*; Volume 3: Prediction. Mend 5.4.2c; Mine Environment Neutral Drainage (MEND) Program: Vancouver, Canada, 2000.
22. Skousen, J.; Rose, A.; Geidel, G.; Foreman, J.; Evans, R.; Hellier, W. *Handbook of Technologies for Avoidance and Remediation of Acid Mine Drainage*; West Virginia University: Morgantown, WV, USA, 1998.
23. Jones, S.N.; Cetin, B. Evaluation of waste materials for acid mine drainage remediation. *Fuel* **2017**, *188*, 294–309. [CrossRef]
24. Park, J.H.; Edraki, M.; Mulligan, D.; Jang, H.S. The application of coal combustion by-products in mine site rehabilitation. *J. Clean Prod.* **2014**, *84*, 761–772. [CrossRef]
25. Davies, M.P. Tailings impoundment failures are geotechnical engineers listening? *Geotech. News Vanc.* **2002**, *20*, 31–36.
26. MMSD. *Report of the Workshop on Finance, Mining and Sustainability: Exploring Sound Investment Decision Processes*; Mining, Minerals and Sustainable Development Project (MMSD); International Institute for Environment and Development (IIED): Paris, France, 14–15 January 2002.
27. Ke, X.; Zhou, X.; Wang, X.; Wang, T.; Hou, H.; Zhou, M. Effect of tailings fineness on the pore structure development of cemented paste backfill. *Constr. Build. Mater.* **2016**, *126*, 345–350. [CrossRef]
28. Edraki, M.; Huynh, T.; Wightman, E.; Tungpalan, K.; Palaniandy, S. An Integrated Approach to Proactive Tailings Management. In Proceedings of the Tailings and Mine Waste Management for the 21st Century, Sydney, Australia, 27–28 July 2015; The Australasian Institute of Mining and Metallurgy: Sydney, Australia; pp. 323–332.
29. Edraki, M.; Baumgartl, T.; Fletcher, A.; Mulligan, D.R.; Fegan, W.; Munawar, A. Hydrogeochemical Evolution of an Uncapped Gold Tailings Storage Facility. In Proceedings of the Enviromine 2009: First International Seminar on Environmental Issues in the Mining Industry, Santiago, Chile, 30 September–2 October 2009.
30. Edraki, M.; Forsyth, B.; Baumgartl, T.; Bradshaw, D. Geochemistry of tailings and seepage from three tailings storage facilities in australia – uncapped, capped and active tailings. In Proceedings of the Life-of-Mine 2012: Maximizing Rehabilitation Outcomes, Brisbane, Australia, 10–12 July 2012; The Australasian Institute of Mining and Metallurgy: Brisbane, Australia; pp. 269–278.
31. Fall, M.; Adrien, D.; Célestin, J.C.; Pokharel, M.; Touré, M. Saturated hydraulic conductivity of cemented paste backfill. *Miner. Eng.* **2009**, *22*, 1307–1317. [CrossRef]
32. Fang, Y.; Gu, Y.; Kang, Q.; Wen, Q.; Dai, P. Utilization of copper tailing for autoclaved sand–lime brick. *Constr. Build. Mater.* **2011**, *25*, 867–872. [CrossRef]
33. Argane, R.; Benzaazoua, M.; Hakkou, R.; Bouamrane, A. Reuse of base-metal tailings as aggregates for rendering mortars: Assessment of immobilization performances and environmental behavior. *Constr. Build. Mater.* **2015**, *96*, 296–306. [CrossRef]
34. Argane, R.; Benzaazoua, M.; Hakkou, R.; Bouamrane, A. A comparative study on the practical use of low sulfide base-metal tailings as aggregates for rendering and masonry mortars. *J. Clean Prod.* **2016**, *112*, 914–925. [CrossRef]
35. Kim, B.J.; Jang, J.G.; Park, C.Y.; Han, O.H.; Kim, H.K. Recycling of arsenic-rich mine tailings in controlled low-strength materials. *J. Clean Prod.* **2016**, *118*, 151–161. [CrossRef]
36. Liu, T.Y.; Tang, Y.; Han, L.; Song, J.; Luo, Z.W.; Lu, A.X. Recycling of harmful waste lead-zinc mine tailings and fly ash for preparation of inorganic porous ceramics. *Ceram. Int.* **2017**, *43*, 4910–4918. [CrossRef]
37. Taha, Y.; Benzaazoua, M.; Hakkou, R.; Mansori, M. Coal mine wastes recycling for coal recovery and eco-friendly bricks production. *Miner. Eng.* **2017**, *107*, 123–138. [CrossRef]
38. Driussi, C.; Jansz, J. Technological options for waste minimisation in the mining industry. *J. Clean Prod.* **2006**, *14*, 682–688. [CrossRef]
39. OKTediMining. 2009 Annual Environmental Report. OK Tedi Mining. 2014. Available online: https://www.oktedi.com/index.php/media-items/reports/environmental/annual-environmental-reports (accessed on 22 March 2019).
40. Bois, D.; Poirier, P.; Benzaazoua, M.; Bussière, B.; Kongolo, M. A feasibility study on the use of desulphurized tailings to control acid mine drainage. *C. Bull.* **2005**, *98*, 1.
41. Cropp, A.F.; Goodall, W.R.; Bradshaw, D.J. The influence of textural variation and gangue mineralogy on recovery of copper by flotation from porphyry ore: A review. In Proceedings of the GeoMet 2013: The Second AusIMM International Geometallurgy Conference, Brisbane, Australia, 30 September–2 October 2013.

42. Bridge, R.; Brosig, D.; Lozano, C.; Laurila, H. *Geometallurgy: An. Underutilised Technology*; Canadian Institute of Mining, Metallurgy and Petroleum: Quebec, Canada, 2013.
43. Dominy, S.C.; O'Connor, L. Geometallurgy—Beyond conception. In Proceedings of the Third AusIMM International Geometallurgy Conference (GeoMet) 2016, Perth, Australia, 15–17 June 2016; The Australasian Institute of Mining and Metallurgy: Perth, Australia; pp. 3–10.
44. Dunham, S.; Vann, J. Geometallurgy, geostatistics and project value—Does your block model tell you what you need to know. In Proceedings of the Project Evaluation Conference, Melbourne, Australia, 19–20 June 2007; pp. 189–196.
45. Vann, J.; Jackson, J.; Coward, S.; Dunham, S. The geomet curve—A model for implementation of geometallurgy. In Proceedings of the First AusIMM International Geometallurgy (GeoMet) Conference, Brisbane, Australia, 5–7 September 2011.
46. Bye, A.R. Case studies demonstrating value from geometallurgy initiatives. In Proceedings of the First AusIMM International Geometallurgy (GeoMet) Conference, Brisbane, Australia, 5–7 September 2011; Australasian Institute of Mining and Metallurgy: Brisbane, Australia; pp. 9–30.
47. Louwrens, E.; Napier-Munn, T.; Keeney, L. Geometallurgical characterisation of a tailings storage facility—A novel approach. In *Tailings and Mine Waste Management for the 21st Century, 27–28 July 2015*; Australasian Institute of Mining and Metallurgy: Sydney, Australia, 2015; pp. 125–132.
48. Edraki, M.; Huynh, T.; Baumgartl, T.; Huang, L.; Andrusiewicz, M.; Tungpalan, K.; Tayebi-Khorami, M.; Wightman, E.; Palaniandy, S.; Manlapig, E. Designer tailings: Improving the management of tailings through collaborative research. In Proceedings of the Triennial 8th Australian Workshop on Acid and Metalliferous Drainage, Adelaide, Australia, 29 April–2 May 2014; pp. 173–182.
49. Tungpalan, K.; Manlapig, E.; Andrusiewicz, M.; Keeney, L.; Wightman, E.; Edraki, M. An integrated approach of predicting metallurgical performance relating to variability in deposit characteristics. *Miner. Eng.* **2015**, *71*, 49–54. [CrossRef]
50. Tayebi-Khorami, M.; Manlapig, E.; Forbes, E. Relating the mineralogical characteristics of tampakan ore to enargite separation. *Miner. Basel* **2017**, *7*, 77. [CrossRef]
51. *Report of the World Commission on Environment and Development: Our Common Future*; General Assembly Resolution 42/187; United Nations: Rio de Janerio, Brazil, 11 December 1987.
52. WEF. *Mining & Metals in a Sustainable World 2050*; World Economic Forum: Geneva, Switzerland, 2015; Available online: http://www3.weforum.org/docs/WEF_MM_Sustainable_World_2050_report_2015.pdf (accessed on 22 March 2019).
53. Unger, C. Legacy issues and abandoned mines. In *Mining in the Asia-Pacific, the Political Economy of the Asia Pacific*; Callaghan, T., Graetz, G., Eds.; Springer: Berlin, Germany, 2017; pp. 333–369.
54. Corder, G.D.; McLellan, B.C.; Green, S.R. Delivering solutions for resource conservation and recycling into project management systems through susop (r). *Miner. Eng.* **2012**, *29*, 47–57. [CrossRef]
55. Lèbre, É.; Corder, G.; Golev, A. The role of the mining industry in a circular economy: A framework for resource management at the mine site level. *J. Ind. Ecol.* **2017**, *21*, 662–672. [CrossRef]
56. Corder, G.D.; Golev, A.; Fyfe, J.; King, S. The status of industrial ecology in australia: Barriers and enablers. *Resources* **2014**, *3*, 340–361. [CrossRef]
57. Wang, Y.M. China recycling economy development and its mineral resources' sustainable development. *Metal. Mine* **2005**, *2*, 1–3.
58. Golev, A.; Lebre, E.; Corder, G. The contribution of mining to the emerging circular economy. *AusIMM Bull.* **2016**, *2016*, 30–32.
59. Edraki, M.; Baumgartl, T.; Manlapig, E.; Bradshaw, D.; Franks, D.M.; Moran, C.J. Designing mine tailings for better environmental, social and economic outcomes: A review of alternative approaches. *J. Clean Prod.* **2014**, *84*, 411–420. [CrossRef]
60. Forum for the Future. *Circular Economy Business Model Toolkit*; Forum for the Future, Unilever; 2016; Available online: https://www.forumforthefuture.org/project/circular-economy-business-model-toolkit/overview (accessed on 22 March 2019).
61. Genex Power. Kidston Renewable Energy Hub. Available online: https://www.abc.net.au/news/2018-06-20/kidston-renewable-energy-hub/9890600 (accessed on 22 March 2019).

62. Barker, E. Century Mine Waste Dam to Become World's Fourth-Largest Zinc Operation Following Rehabilitation. Available online: https://www.abc.net.au/news/rural/2018-09-17/century-mine-from-waste-dam-to-zinc-mine/10253334 (accessed on 15 March 2019).
63. Elshkaki, A.; Graedel, T.E.; Ciacci, L.; Reck, B.K. Copper demand, supply, and associated energy use to 2050. *Glob. Environ. Chang.* **2016**, *39*, 305–315. [CrossRef]

© 2019 by the authors. Licensee MDPI, Basel, Switzerland. This article is an open access article distributed under the terms and conditions of the Creative Commons Attribution (CC BY) license (http://creativecommons.org/licenses/by/4.0/).

Article

Environmental Impact of Mine Exploitation: An Early Predictive Methodology Based on Ore Mineralogy and Contaminant Speciation

Aurélie Chopard [1,2], Philippe Marion [2], Raphaël Mermillod-Blondin [3], Benoît Plante [1] and Mostafa Benzaazoua [1,*]

1. Institut de Recherche en Mines et Environnement (IRME), Université du Québec en Abitibi-Témiscamingue (UQAT), Rouyn-Noranda, QC J9X 5E4, Canada
2. GeoRessources, ENSG, Université de Lorraine (UL), 54500 Vandoeuvre-lès-Nancy, France
3. Agnico Eagle Mines Limited, Rouyn-Noranda, QC J0Y 1C0, Canada
* Correspondence: mostafa.benzaazoua@uqat.ca; Tel.: 1-819-762-0971 #2404

Received: 27 April 2019; Accepted: 25 June 2019; Published: 28 June 2019

Abstract: Mining wastes containing sulfide minerals can generate contaminated waters as acid mine drainage (AMD) and contaminated neutral drainage (CND). This occurs when such minerals are exposed to oxygen and water. Nowadays, mineralogical work—when it is done—is independently and differentially done according to the needs of the exploration, geotechnics, metallurgy or environment department, at different stages in the mine development process. Moreover, environmental impact assessments (EIA) are realized late in the process and rarely contain pertinent mineralogical characterization on ores and wastes, depending on countries' regulations. Contaminant-bearing minerals are often not detected at an early stage of the mine life cycle and environmental problems could occur during production or once the mine has come to the end of its productive life. This work puts forward a more reliable methodology, based on mineralogical characterization of the ore at the exploration stages, which, in turn, will be useful for each stage of the mining project and limit the unforeseen environmental or metallurgical issues. Three polymetallic sulfide ores and seven gold deposits from various origins around the world were studied. Crushed ore samples representing feed ore of advanced projects and of production mines were used to validate the methodology with realistic cases. The mineralogical methodology consisted in chemical assays and XRD, optical microscopy, SEM and EPMA were done. Five of the ores were also submitted to geochemical tests to compare mineralogical prediction results with their experimental leaching behavior. Major, minor, and trace minerals were identified, quantified, and the bearing minerals were examined for the polluting elements (and valuables). The main conclusion is that detailed mineralogical work can avert redundant work, save time and money, and allow detection of the problems at the beginning of the mine development phase, improving waste management and closure planning.

Keywords: environmental mineralogy; exploration; contaminant; geochemical behavior

1. Introduction

With the growing demand for new technologies and in life quality of emerging countries, mineral and metal productions will become critical in years to come. The industry has to adapt its practices to be more efficient. Moreover, the mining industry, like other industries, produces waste that have negative impacts on the environment [1]. To maintain license-to-operate, companies must consider and address social issues and environmental concerns throughout the life cycle of the project [2]. Moreover, decisions in mining projects are mainly based on valuable metal grades without taking into account environmental issues. Contaminants are often not taken into account during the first stages (exploration or feasibility stages) of a mine project. In addition, if they are, geologists will not consider them at their potential contaminating value because their focus is on mineralization and deposit geology where trace elements are less informative. This can lead to major environmental challenges underestimation. Frequently, problems regarding the release of toxic elements are discovered after the mine starts up. Consequently, mining companies should modify their treatment techniques, their effluent management (liquid and solid) and their closure plan. This can generate unexpected and supplementary costs. In the case of a belated discovery of a contaminant, the issue is generally explained by the mineralogy of the ore or of the waste [3,4].

The contaminants are mostly brought about through weathering of sulfides containing metals and metalloids such as As, Cu, Ni, Sb, Zn, etc. Indeed, when solid mine wastes contain sulfides or sulfosalts, and are exposed to air and oxygen, chemical reactions will occur and the mine wastes will generate poor water quality. When the pH is acidic, the phenomenon is called acid mine drainage (AMD). The sulfides react at different oxidation rates according to their nature and crystallography and differently contribute to water pollution [5–12]. Sulfosalt minerals are considered as "trash minerals", as they contain numerous toxic elements (As, Bi, Co, Hg, Ni, Tl, Sb, Se, etc.), and dissolve easily when exposed to weathering conditions and are more likely to generate contaminants. Another phenomenon, called contaminated neutral drainage (CND), can occur when neutralizing minerals are present or when not enough sulfides are found in the solids. The pH is circumneutral, but metals and metalloids concentrations are present in the leachates above regulation levels.

Microscale mineral dissolution in mine wastes will determine the weathering processes and the minerals are the sources of metal concentrations in the leachates [13]. Mineral dissolution, oxidation, or precipitation depend on several mineralogical parameters: amount, composition, particle size distribution, liberation degree, other mineral associations, ore texture, and surface properties [14]. For example, the associations of minerals can play a significant role in the acidity and contaminants generation, as their association can change the reactivity rates of the minerals. In the case of sulfides, which are semiconducting minerals, galvanic interactions occur and will accelerate or inhibit the oxidation of the affected sulfides [15–17]. Over the last few years, some authors have demonstrated that the mineralogy is decisive in AMD and CND generation [18–21]. Applied mineralogy in environmental studies would help to predict the water quality and to characterize the problematic minerals in a sample [22,23]. The various elements dispersed into the environment from a given sample are strongly influenced by the major and trace element composition of the minerals present in this sample.

Mineralogical information is important for all the mining departments: geology, metallurgy, and environment [24]. Mineralogical investigations are necessary to identify and quantify major, minor and trace minerals and can detect valuable elements as by-products that could improve the recovery efficiency of the ore and increase the value of the deposit. Valuable elements can be found as the main element of trace minerals like platinum in sperrylite ($PtAs_2$), or as a trace element in major, minor, or trace minerals like indium in sphalerite (ZnS). Moreover, valuable minerals in small amounts can be detected and beneficiated as well. If their detection is early in the development process, the appropriate treatment can be implemented.

Likewise, AMD and CND prediction studies can be conducted on the wastes to assess their environmental behavior. They mainly consist in geochemical testing. A few mineralogical characterizations can be done in specific cases (e.g., contaminant detection in the leachates of the geochemical tests) [25]. But tailing samples are only available after metallurgical pilot tests are preformed, which means that the project is in an advanced stage. In order to predict the environmental impact of the mining deposit at an earlier stage (like exploration), the studies could be performed directly on the ore. Unfortunately, the quality of the mine drainage is not always predictable with ore products. As Dold [21] recommends, refined methodologies must be applied to characterize ore deposits before mine start up and to assess the geochemical behavior of mine products. This might give us the opportunity to design optimized processes for efficient metal recovery of the different mineral assemblages in an ore deposit and, at the same time, to minimize the future environmental impact and costs for mine waste management.

This study proposes to implement a reliable and simple methodology for ore characterization, combining accurate chemical analysis as well as rigorous mineralogical analysis and observations, as early as the exploration stage. The procurement of an ore profile will allow detection of the contaminating elements as well as the valuable elements, related to metallurgical and environmental challenges. This approach is based on the mineralogy of the ores and will take into account bearing minerals and the element speciation. Geochemical tests have been also completed on five ores to link the mineralogical results with the geochemical results. The final aim is to develop tools to acquire information providing companies with a better assessment of the environmental impacts of a future exploitation and a better preparation for metallurgical issues that they will have to face, as early as possible in the mining development.

2. Materials and Methods

2.1. Materials

Three polymetallic and seven gold deposits were studied, mainly from Canada, but also from Argentina, Finland and Mexico. Table 1 briefly describes the deposits. Operators collected the samples of these ores on the mine sites, except for Westwood, where the collection was done by the authors. Some samples collected were already prepared in the mines' lab. All samples were then submitted to a meticulous preparation to remain representative of the initial sample. The initial sample was considered a part of the feed. The feed ore is the first solid material to meticulously characterize environmental and ore processing behaviors. Inline Appendix A Table A1 describes in detail the different sample collection and the mine's preparation. The preparation described in Figure 1 aimed to obtain two 500 g samples of different particle size. The first one is a 2–5 mm sample, prepared in polished section, to allow observation of the ore texture. The second one is a milled sample with a D80 approximately the same as the plant's final product. The latter was used for all chemical and mineralogical analyses and to be quantitatively representative when observed on polished sections by microscopy. The samples already with a particle size of ≤2 mm were directly screened at 1 mm and kept despite their fine particle size distribution, and then rod milled to obtain the chosen D80. The representativeness of the initial material was kept thanks to successive homogenization and dividing with appropriate techniques (Figure 1). The ore powders containing less than 5 wt. % sulfur were also concentrated with a heavy liquid (bromoform) by centrifugation to maximize sulfide grain observation. The polished sections were prepared with epoxy resin, hardener and black carbon powder. The black carbon powder aims to avoid particles sedimentation and improve the representativeness of the analyzed surface, to limit differential removal of particles by polishing, and to reduce the particles in contact in the polished section [26]. This preparation has the advantage of making the surface conductive without carbon coating, which allows successive observations by optical and electron microscopy.

Table 1. Deposits description: country, geology, mineralization, and products.

Mine's Name	Country	Ore Type	Host Rock	Mineralization	Main Product	By-product(s)
LaRonde	Canada	Volcanogenic massive sulfide (VMS)	Metamorphic volcanic rocks	Massive and disseminated sulfide lenses	Au	Ag, Cu, Pb, Zn
Pirquitas	Argentina	Epithermal Ag-Sn	Low-grade metamorphosed marine sandstone, siltstone and minor shale beds	Sulfide and quartz sulfide vein systems	Ag	Sn, Zn
Raglan	Canada	Komatiite Ni-Cu-Platinum-group element	Mafic and ultramafic volcanic rocks (peridotite)	Magmatic immiscibility sulfides lenses associated with ultramafic flows	Ni	Co, Cu, Pd, Pt, Rh
C. Malartic	Canada	Archean porphyry gold system	Potassic-altered, silicified greywackes, altered porphyry and gabbro dykes and ultramafic rocks	Widespread shell of disseminated gold-bearing pyrite	Au	
Goldex	Canada	Shear zone gold	Granodiorite	Stockwork veins; microscopic gold within pyrite and coarse native gold	Au	
Kittilä	Finland	Shear zone gold	Mafic volcanic and sedimentary rocks	Structural gold in arsenopyrite and pyrite	Au	
Lapa	Canada	Shear zone gold	Volcanic rocks	Quartz veins, tabular zones, biotite-altered zones	Au	Ag
Meliadine	Canada	Shear zone gold	Sedimentary and volcanic sequences	Sulfide quartz veins, quartz lodes and sulfide replacement	Au	
Pinos Altos	Mexico	Epithermal Au-Ag	Volcanic and intrusive rocks (Andesite and ignimbrite)	Low sulfidation epithermal type hydrothermal quartz veins, stockworks and breccias	Au	Ag
Westwood	Canada	Volcanogenic massive sulfide (VMS)	Metamorphic volcanic rocks	Sulfide quartz veins, disseminated and semi-massive to massive sulfide lenses	Au	

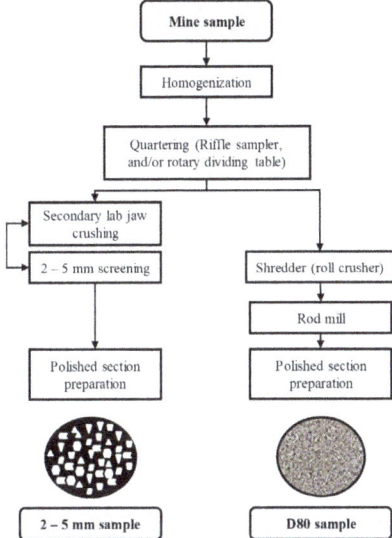

Figure 1. Lab preparation procedure to obtain the two different polished sections.

2.2. Methods

2.2.1. Particle Size Distribution

The particle size distributions of the successive milling step samples and of the final powders were measured with a S3500 laser grain size analyzer (Microtrac) [27].

2.2.2. Chemical Characterization

Several chemical analysis techniques were tested to obtain a reliable chemical composition of the samples. The selected methods are described here. The sulfur and carbon content were analyzed by a CS-2000 induction furnace ELTRA (Haan, Germany) coupled with an infrared analyzer. The major elementary composition (Na, Mg, Al, Si, P, Cl, K, Ca, Ti, V, Cr, Mn, Fe) was determined with a whole rock analysis by fused discs (lithium metaborate fusion) [28–31] and by multi-acid digestion in the microwave which was followed by analysis using an Optima 3100 ICP-AES instrument (Perkin Elmer, Waltham, MA, USA) to obtain more accurate results for the smaller concentrations of these elements. The elementary chemical composition (As, Ba, Bi, Cd, Co, Cu, Mo, Ni, Pb, Sb, Se, Sn, and Zn) was determined by an Optima 3100 ICP-AES analysis following a multi-acid digestion (HNO_3-Br_2-HF-HCl) of 500 mg of a pulverized aliquot in the microwave and by a combined ICP-AES and ICP-MS following a sodium peroxide fusion. The trace elements (rare earths and others) were analyzed by ICP-MS following the sodium peroxide fusion. The analysis of gold was performed using a 50 g subsample by a fire assay procedure combined with both gravimetric and atomic absorption spectroscopy (AAS). The PGEs (Pd and Pt) were analyzed by fire assay combined with AAS. The mercury content was determined by thermal decomposition, amalgamations and AAS.

2.2.3. Mineralogical Characterization

The powder samples were micronized to be analyzed by X-ray diffraction with an AXS Advance D8 system (Bruker) equipped with a copper cathode, acquired at a rate of $0.02°$ s^{-1} between 2θ values of $5°$ and $70°$. The DiffracPlus EVA software (Version 9.0) was used to identify the mineral species, and the quantitative mineralogical compositions were evaluated using the TOPAS software (Version 2.1) with a Rietveld refinement [32]. The absolute precision of this quantification method is ± 0.5 to 1% [33,34], but in practice, 1% to 5% should be considered. The identification of the minerals and the texture observations were performed in reflected light mode using an AxioImager M2m optical microscope (Zeiss, Oberkochen, Germany) equipped with the AxioVision (Version 4.8) software. Scanning electron microscopy (SEM) observations using backscattered electrons (BSE) for texture observations were made on a S-3500N microscope (Hitachi, Chiyoda, Tokyo, Japan) with the INCA software (450 Energy). The major element chemical composition of individual minerals was determined using an energy dispersive spectrometer (EDS; Oxford SDD, X-Max 20mm^2). The SEM observations and microanalyses were performed with 15 mm working distance and 100-µA current at 20 kV. Then, the more precise composition of some minerals and trace elements quantification in arsenopyrite, chalcopyrite, galena, pentlandite, pyrite, pyrrhotite and sphalerite were achieved by analyzing a minimum of ten particles from each mineral using a SX100 Electron Probe Micro-Analyzer (EPMA, Cameca) coupled with four WDS (Wavelength-dispersive X-ray Spectrometry) spectrometers. All quantitative analyzes were realized using an accelerating voltage of 20 kV and a constant beam current of 20 nA. The counting time on each peak/background was 10 s on the major elements and 40 s on the trace elements: Ag, As, Au, Cd, Co, In, Mn, Ni, Pb, Se, Sn, Te, and Tl. Depending on the sample studied, pyrite was analyzed for S, Fe, As, Au, Ag, Tl, and Se; pyrrhotite for S, Fe, Co, and Ni; arsenopyrite for S, Fe, As, and Au; chalcopyrite for S, Cu, Fe, Ag, As, Sn, Tl, Te, and Cd; sphalerite for S, Zn, Fe, Ag, In, Cd, Mn, Pb, and Tl; galena for S, Pb, Fe, and Ag; and pentlandite for S, Fe, Ni, and Co.

2.2.4. Geochemical Testing

The geochemical behavior of five ores was evaluated using modified weathering cell tests. These weathering cells are small-scale humidity cells that render similar results for the rates of reactions [35–37]. A 100 mm Büchner funnel with a 0.45 µm glass-fiber filter is held with 67 g of material. The sample is leached twice a week with 50 mL deionized water. The cells were placed in a controlled-weather box to maintain the samples under optimal saturation conditions ranges and avoid extreme drying [38]. Leachates from the weathering cells were analyzed for pH, conductivity, Eh, sulfur and elemental concentrations. The element concentrations in the leachates were analyzed by ICP-AES on an aliquot acidified to 2% HNO_3 for preservation. The weathering cells were run for thirty cycles.

2.3. General Methodology

The approach used in this study is mainly based on the chemical assay, obtained with a thorough methodology taking into account the mineralogy of the sample. In this paper, the reliability of the chemical assay is considered. Each element, valuable or contaminant, present in a higher concentration than one or several defined thresholds, are examined for their bearing minerals. The thresholds chosen in this study correspond to the elemental grade either superior at ten times the Clarke value [39,40], or superior at the threshold values determined in Quebec Directive 019, for the regulation of mine solid materials. As most of the elements sought are contained in sulfides or sulfosalts, the identification of these opaque minerals is made by optical microscopy in reflected light mode. Then, the minerals of interest are circled to be further analyzed by electron microscopy, to get a higher magnification or EDS-microanalyses with a detection limit around 5000 ppm. For the detection of suspected trace elements in certain minerals, EPMA analyses were performed with different detection limits according to the counting time of the detectors. Figure 2 presents the logical diagram of the methodology. The greyed-out rectangles represent the results obtained and necessary for the following steps of exploitation and management.

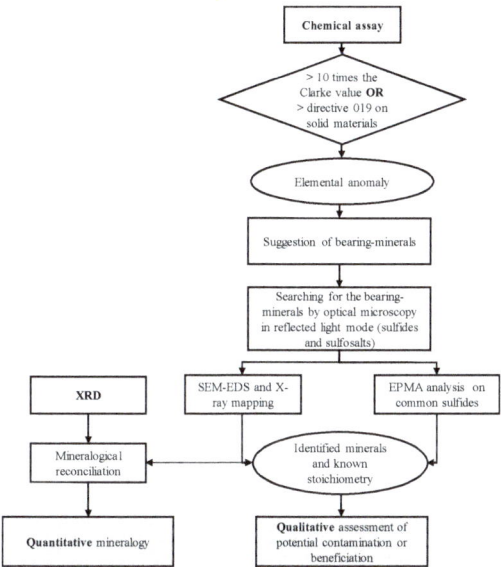

Figure 2. Logical diagram of the methodology of environmental and processing prediction applied at the stage of exploration. Then, a reconciliation of all the results is made by spreadsheet calculations to obtain a quantitative mineralogical characterization. The quantitative mineralogy was calculated with X-ray powder diffraction (XRD) mineralogy, chemical assay, and stoichiometry of the minerals. This stoichiometry was either theoretical or determined by electron probe micro-analyzer (EPMA).

3. Results and Discussion

3.1. Chemical Characterization

Major, minor, and trace elements contained in the ore samples are presented in Appendix A Tables A2 and A3. Elements were classified per grade value in the ore. The interest may be either for contamination or for recovery. In this paper, the focus is given on the contaminating elements. All the studied ores contain sulfur in significant amounts (>0.3% S), except for Pinos Altos. Arsenic (As), mercury (Hg) and antimony (Sb) are largely above the defined thresholds in all the ores. The most abundant polluting elements in the ores studied are: As, Hg, Sb, bismuth (Bi), cadmium (Cd), copper (Cu), lead (Pb), selenium (Se) and zinc (Zn). To a lesser extent, barium (Ba), cobalt (Co), chromium (Cr), manganese (Mn), molybdenum (Mo), nickel (Ni), tin (Sn), strontium (Sr), tellurium (Te), and thallium (Tl) are present in the ores. The rare-earth elements were also analyzed, but no significant amount was identified. We chose to exclude these elements from the study. The valuable elements Ag, Au, In, Li, Pd, and Pt were identified. Gold was found in all the ores, even in small amount in Raglan and Pirquitas. Silver was found in most of the ores, except in Canadian Malartic, Goldex, and Kittilä. A significant amount of indium (30 ppm) was found in the Pirquitas sample and a non-negligible amount in LaRonde and Westwood (1 and 0.5 ppm, respectively). It is difficult to state the amount of lithium in the ores. Palladium and platinum were logically found in Raglan, but platinum was also found in small amounts in Meliadine and Pinos Altos.

3.2. Mineralogical Characterization and Identification of the Bearing Minerals

After selecting the potential contaminating elements, the polished sections were observed under the optical microscope. The optical microscope is very useful to identify accessory minerals. As most of the contaminants identified are contained in accessory minerals, the optical microscope was used to detect the bearing minerals of these contaminants. Then, a SEM-EDS analyze allowed us to confirm the approximate stoichiometry of the mineral and/or to recognize the minerals not identifiable only by the physical and optical properties under optical microscopy. As the mineralogical state of a given element determines its geochemical behavior during dissolution and further transport, it is of importance to know the elemental speciation to assess the contaminating potential (or recovery potential). Table 2 presents the bearing minerals of the elements of interest for all the ores, classified in a decreasing order of elemental concentrations. The LaRonde sample presents a complex mix of sulfide minerals, incorporating varied metals (Cu, Zn, Pb, Bi, Sn, Ag, Cd, Au, Tl, In) and metalloids (As, Sb, Te). The Pirquitas sample presents a very complex mix of sulfides and sulfosalts. The sulfosalts of Pirquitas (Aramayoite, Freibergite, Matildite, Owyheeite, Pyrargyrite (see Table 2 for formulae) contain Ag, Cu, Sb, As, Bi, Pb, and Co. The Raglan sample shows lower concentrations (Zn < 88 ppm; As, Se < 20 ppm; Sb, Bi, Te, Ag, Cd < 10 ppm; and Pt, Hg, Au < 1 ppm). Therefore, it was more difficult to identify the bearing minerals and to detect the elements in the minerals in a relatively short time, since the mineral phase detection limit depends on the number of observed grains and the elemental detection limit depends on photon counting time. Only pentlandite, pyrrhotite, chalcopyrite, chromite, and sperrylite were observed and analyzed. The principal assemblage in the sample consists of pyrrhotite and pentlandite, with pentlandite in "flames" (Figure 3).

The gold ore samples mainly display pyrite as sulfide, except for Lapa, where pyrrhotite is dominant. In general, all the gold ores contain Zn in too high concentrations, except for Goldex and Canadian Malartic. The main bearing mineral for Zn is sphalerite. Cd and Tl were also found in sphalerite. Mo is present in several gold ores (from 1.7 to 8 ppm). Sb is present in all samples, except for Pinos Altos, where the geological context is different. The bearing minerals for Sb are often difficult to identify as a unique grain of this mineral is enough to give concentrations above 10 times the Clarke value. Ullmannite (NiSbS), berthierite ($FeSb_2S_4$) and tetrahedrite (($Cu,Fe)12Sb_4S_{13}$) sulfosalts were identified. As and Cu are present most of the time as arsenopyrite and chalcopyrite, but sulfosalts are also present, as gersdorffite (NiAsS) and tetrahedrite. Cubanite ($CuFe_2S_3$) was also found in Westwood, As-pyrite in Pinos Altos, Kittilä, and Meliadine, and As-ullmannite in Lapa. Ni was present only in two samples, as argentopentlandite ($Ag(Fe,Ni)_8S_8$) in Pinos Altos and pentlandite (($Fe,Ni)_9S_8$) and pyrrhotite ($Fe_{0.93}S$) in Lapa and Raglan.

Lead tellurides were found in Westwood, however, galena is the main bearing mineral for Pb. Se and Bi are present in almost all the samples, in amount varying from 0.33 ppm to 139 ppm, and from 2.8 ppm to 169 ppm. No bearing minerals were identified for Se. Bi was found as kochkarite ($PbBi_4Te_7$) in Westwood; as native bismuth and kawazulite ($Bi_2(Te,Se,S)_3$) in LaRonde as matildite and in galena; aramayoite and owyheeite in Pirquitas. Co was found in Lapa as gersdorffite. Finally, Tl and Hg were found in very low amount (from 0.02 to 1.4 ppm) in Pinos Altos, Goldex, Kittilä, and Westwood. These two elements are very toxic for ecosystems. Gold is present as free grains, electrum and inclusions in arsenopyrite. Structural gold was not analyzed in the present study, as the focus is on contaminating elements. The Pinos Altos sample is different from the others as its sulfur content is below 0.3 wt. % and the sample contains oxides and sulfates.

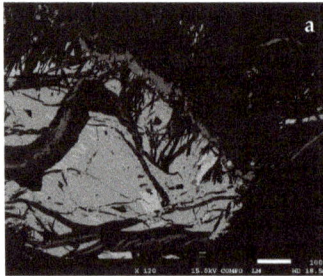 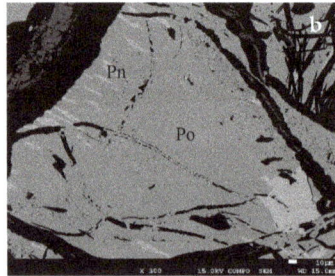

Figure 3. (**a**) Flames of pentlandite in a pyrrhotite grain, Raglan sample; (**b**) Zoom on the flames of pentlandite (Pn) in pyrrhotite (Po).

Table 2. Observed bearing minerals of the elements of interest for the ten ores. S, 16.2%; Fe, 15.5%. * Products; ** By-products.

		LaRonde	
Elements	Grade (ppm)	Observed Bearing Minerals	Formula
Cu **	3039	Chalcopyrite	$CuFeS_2$
		Stannite	Cu_2FeSnS_4
		Ferrokesterite	$Cu_2(Fe,Zn)SnS_4$
		Freibergite	$(Ag,Cu,Fe)_{12}(Sb,As)_4S_{13}$
		Tetrahedrite	$(Cu,Fe)_{12}Sb_4S_{13}$
Zn **	946	Sphalerite	$(Zn,Fe)S$
		Ferrokesterite	$Cu_2(Fe,Zn)SnS_4$
Pb	219	Galena	PbS
Bi	169	Bismuth	Bi
		Kawazulite	$Bi_2(Te,Se,S)_3$
As	143	Arsenopyrite	$FeAsS$
		Freibergite	$(Ag,Cu,Fe)_{12}(Sb,As)_4S_{13}$
Sn	22	Stannite	Cu_2FeSnS_4
		Ferrokesterite	$Cu_2(Fe,Zn)SnS_4$
Ag *	14	Electrum	(Ag,Au)
		Freibergite	$(Ag,Cu,Fe)_{12}(Sb,As)_4S_{13}$
		Chalcopyrite	$CuFeS_2$
Te	14	Benleonardite	$Ag_8(Sb,As)Te_2S_3$
		Kawazulite	$Bi_2(Te,Se,S)_3$
Sb	5	Freibergite	$(Ag,Cu,Fe)_{12}(Sb,As)_4S_{13}$
		Tetrahedrite	$(Cu,Fe)_{12}Sb_4S_{13}$
		Benleonardite	$Ag_8(Sb,As)Te_2S_3$
Cd	3.5	Sphalerite	$(Zn,Fe)S$
Au *	2.5	Electrum	(Ag,Au)
Tl	2.4		
In	1	Sphalerite	$(Zn,Fe)S$
		Pirquitas	
Elements	Grade (ppm)	Observed Bearing Minerals	Formula
Zn	8380	Sphalerite	$(Zn,Fe)S$
		Freibergite	$(Ag,Cu,Fe)_{12}(Sb,As)_4S_{13}$
Sn	2280	Cassiterite	SnO_2
		Kesterite	$Cu_2(Zn,Fe)SnS_4$
		Solid solution hocartite-pirquitasite	$Ag_2(Fe,Zn)SnS_4$
As	1160	Arsenopyrite	$FeAsS$
		Pyrite	$(Fe,As)S$
		Freibergite	$(Ag,Cu,Fe)_{12}(Sb,As)_4S_{13}$
Sr	460		
Ba	360		
Cu	307	Freibergite	$(Ag,Cu,Fe)_{12}(Sb,As)_4S_{13}$

Table 2. Cont.

	Pirquitas		
Elements	Grade (ppm)	Observed Bearing Minerals	Formula
Ag *	141	Galena Aramayoite Acanthite Matildite Pyrargyrite Freibergite Owyheeite	$(Pb,Ag)S$ $Ag_3Sb_2(Sb,Bi)S_6$ Ag_2S $AgBiS_2$ Ag_3SbS_3 $(Ag,Cu,Fe)_{12}(Sb,As)_4S_{13}$ $Pb_7Ag_2(Sb,Bi)_8S_{20}$
Pb	134	Owyheeite	$Pb_7Ag_2(Sb,Bi)_8S_{20}$
		Galena	$(Pb,Ag)S$
Bi	123	Galena Aramayoite Owyheeite Matildite	$(Pb,Ag)S$ $Ag_3Sb_2(Sb,Bi)S_6$ $Pb_7Ag_2(Sb,Bi)_8S_{20}$ $AgBiS_2$
Sb	54.8	Freibergite Aramayoite Pyrargyrite Owyheeite	$(Ag,Cu,Fe)_{12}(Sb,As)_4S_{13}$ $Ag_3Sb_2(Sb,Bi)S_6$ Ag_3SbS_3 $Pb_7Ag_2(Sb,Bi)_8S_{20}$
Cd	40	Sphalerite	$(Zn,Fe)S$
		Kesterite	$Cu_2(Zn,Fe)SnS_4$
In **	30.2	Sphalerite	$(Zn,Fe)S$
		Cassiterite	SnO_2
Se	25.4		
Tl	11		
Te	1.88		
Hg	0.07		
Au	0.04		
	Raglan		
Elements	Grade (ppm)	Observed Bearing Minerals	Formula
Ni *	24,460	Pentlandite Pyrrhotite	$(Fe,Ni)_9S_8$ $Fe_{0.86}S$
Cu **	6033	Chalcopyrite	$CuFeS_2$
Cr	2300	Chromite	$Fe^{2+}Cr_2O_4$
Co	548	Pentlandite	$(Fe,Ni)_9S_8$
Zn	88		
As	16		
Se	14.2		
Sb	7.8		
Bi	4.9		
Te	4		
Ag	2		
Cd	1.55		
Pt **	0.69	Sperrylite	$PtAs_2$
Hg	0.09		
Au	0.06		

Table 2. Cont.

Canadian Malartic			
Elements	Grade (ppm)	Observed Bearing Minerals	Formula
Ba	700	Barite	$BaSO_4$
Sr	570	Barite	$BaSO_4$
Cu	122	Chalcopyrite	$CuFeS_2$
Pb	25	Galena	PbS
Sb	23		
As	17		
Mo	5.3		
Bi	5.1		
Se	1.6		
Au	0.56	Free grains	Au
Goldex			
Elements	Grade (ppm)	Observed Bearing Minerals	Formula
Sr	460		
Sb	446		
Ni	126	Crusher pieces	
Cu	79	Chalcopyrite	$CuFeS_2$
As	36	Arsenopyrite	FeAsS
Bi	10.4		
Mo	2.6		
Au	2.1		
Hg	0.51		
Lapa			
Elements	Grade (ppm)	Observed Bearing Minerals	Formula
As	1865	Arsenopyrite Ullmannite Gersdorffite-Fe	FeAsS Ni(Sb,As)S (Ni,Fe,Co)AsS
Ni	623.1	Pentlandite Pyrrhotite	$(Fe,Ni)_9S_8$ $Fe_{0.93}S$
Sb	140.7	Ullmannite Berthierite	Ni(Sb,As)S $FeSb_2S_4$
Se	139		
Zn	115	Sphalerite	(Zn,Fe)S
Co	52	Gersdorffite	(Ni,Fe,Co)AsS

Table 2. *Cont.*

		Pinos Altos	
Elements	Grade (ppm)	Observed Bearing Minerals	Formula
Ba	730	Barite Coronadite	$BaSO_4$ $Pb(Mn^{4+},Mn^{2+})_8O_{16}$
Zn	197	Sphalerite	$(Zn,Fe)S$
Li	90		
Ni	89	Argentopentlandite	$Ag(Fe,Ni)_8S_8$
Ag	88	Lenaite Argentopyrite Argentojarosite Free grains	$AgFeS_2$ $AgFe_2S_3$ $AgFe^{3+}(SO_4)_2(OH)_6$ Ag
Pb	66	Coronadite	$Pb(Mn^{4+},Mn^{2+})_8O_{16}$
As	44	Pyrite	$(Fe,As)S_2$
Au	3.2	Free grains	Au
Bi	2.8		
Mo	1.7		
Tl	1.4		
Cd	1		
Hg	0.82		
Se	0.43	Lenaite	$AgFeS_2$
Pt	0.03		
		Kittilä	
Elements	Grade (ppm)	Observed Bearing Minerals	Formula
As	12,200	Arsenopyrite Pyrite Gersdorffite	$FeAsS$ $(Fe,As)S_2$ $(Ni,Co)AsS$
Mn	2400	Clinochlore	
Zn	197	Sphalerite	$(Zn,Fe)S$
Cu	104	Chalcopyrite Tetrahedrite	$CuFeS_2$ $(Cu,Fe)_{12}Sb_4S_{13}$
Sb	77	Ullmannite Tetrahedrite	$NiSbS$ $(Cu,Fe)_{12}Sb_4S_{13}$
Au	6.63	Free grains Arsenopyrite	Au $FeAsS$
Se	1.3		
Tl	1		
Hg	0.4		
Cd	0.75		

Table 2. Cont.

Meliadine			
Elements	Grade (ppm)	Observed Bearing Minerals	Formula
As	13,600	Arsenopyrite Pyrite	FeAsS $(Fe,As)S_2$
Ba	450	Barite	$BaSO_4$
Zn	213	Sphalerite	$(Zn,Fe)S$
Pb	101	Galena	PbS
Cu	92	Chalcopyrite	$CuFeS_2$
Sb	17		
Mo	8		
Bi	5		
Au	4.0	Arsenopyrite Free grains Electrum	FeAsS Au (Ag,Au)
Ag	1	Galena Electrum	PbS (Ag,Au)
Cd	0.77	Chalcopyrite	$CuFeS_2$
Se	0.33		
Pt	0.02		
Westwood			
Elements	Grade (ppm)	Observed Bearing Minerals	Formula
Zn	3244	Sphalerite	$(Zn,Fe)S$
Ba	810	Barite	$BaSO_4$
Cu	245	Chalcopyrite Cubanite	$CuFeS_2$ $CuFe_2S_3$
Pb	186	Altaite Plumbotellurite Kochkarite	PbTe $PbTe^{4+}O_3$ $PbBi_4Te_7$
Sb	57		
As	42	Arsenopyrite	FeAsS
Bi	10	Kochkarite	$PbBi_4Te_7$
Cd	7.6	Sphalerite	$(Zn,Fe)S$
Se	3.6		
Au	3.6	Free grains Lenaite	Au $AgFeS_2$
Ag	3	Argentopyrite Lenaite	$AgFe_2S_3$ $AgFeS_2$
Mo	1.7		
In	0.5		
Hg	0.02		

As a general remark, it was difficult to detect the bearing minerals of elements present in too small concentrations, except when the element is coupled with another major element present in higher concentrations (e.g., Cd in sphalerite as Zn is sought and Cd is suspected to be included in sphalerite).

For some major elements not identified and suspected in minor minerals, longer SEM work would be necessary with the applications of the X-ray mapping on polished sections.

EPMA Results

EPMA results are presented in Appendix A Table A4. The major sulfides known for bearing the elements of interest only were analyzed by EPMA, as this mineralogical technique is less available and more time-consuming than others are. The methodology has to stay simple and affordable. However, EPMA is very useful to find the exact concentration of trace elements in a known mineral and the exact stoichiometry of a rarer mineral (like a sulfosalt). In this study, the major sulfides (cited later) are the main bearing minerals of the main contaminants (As, Cd, Co, Cu, Mn, Zn) except for sulfosalts and for certain valuable elements (Ag, In). So, grains of pyrite, pyrrhotite, arsenopyrite, chalcopyrite, sphalerite, galena, and pentlandite were analyzed. The pyrite grains were arseniferous in the Pirquitas and Kittilä samples. The pyrite of Pirquitas also contains Ag and Tl. The pyrrhotite grains were nickeliferous in the Raglan and Meliadine samples. No cobalt was found in the pyrrhotite grains. For arsenopyrite, according to the percentage of arsenic, it is possible to estimate the amount of gold possibly present in the grains [41]. According to the atomic percentage of arsenic in Kittilä (30.9) and Meliadine (31.1), no more than 4000 ppm in weight of gold could be found in these arsenopyrite grains. The detection limit was of 1300 ppm and no gold was detected. The chalcopyrite grains of Meliadine contain Cd, but no other trace element in chalcopyrite were found in any sample. All the grains of sphalerite analyzed were iron sphalerite and contain Cd: 1.16 ± 0.09 wt. % in the LaRonde sample, 0.37 ± 0.21 wt. % in the Pirquitas sample, and 1.00 ± 0.12 wt. % in the Westwood sample. Tl was also found in sphalerite in no negligible amount: 0.1 ± 0.09 wt. % in LaRonde and 0.33 ± 0.22 wt. % in Pirquitas. The LaRonde and Pirquitas grains of sphalerite contain In, 0.45 ± 0.10 wt. % and 0.25 ± 0.18 wt. % respectively. Galena was only detected in the Meliadine sample. Galena exhibits a composition close to the theoretical stoichiometry, but contains Fe (1.37 ± 0.20 wt. %) and Ag (0.09 ± 0.04 wt. %). These elements have already been found in traces in galena [42,43]. Finally, the pentlandite of Raglan contains Co, 0.75 ± 0.08 wt. %.

3.3. Chemical and Mineralogical Data Reconciliation

The quantification of the major gangue minerals and of the bearing minerals of major and minor elements was determined thanks to the combination of XRD analysis and data reconciliation with the chemical analyses [44,45]. The trace elements are not considered in the reconciliation, as their amount in not significant enough to be quantified. Thus, the bearing minerals of precious metals are not quantified, as well as the bearing minerals of sparse elements. Twenty-six elements were used for the data reconciliation: Al, As, Ba, Bi, C, Ca, Cd, Co, Cr, Cu, Fe, K, Mg, Mn, Mo, Na, Ni, Pb, S, Sb, Si, Ti, Zn, and Sn. Se and Sr were not used as their bearing minerals and were not identified, likely because of their presence in very few grains. The EPMA results were used for a better quantification of the sulfides. The results are presented in Table 3. The main concern with the reconciliation between the chemical and mineralogical data was about the Al content. For example, for the Raglan sample, the chemistry calculated by the mineralogical composition and the whole rock result are not corresponding. Moreover, senaite and chromite were not taken into consideration for the Raglan sample. Another issue is with the pyrite and pyrrhotite quantification, as well as the hematite and magnetite quantification, where linear systems of N equations with N unknowns should be solved, and several solutions can be found. The content of pyrite, pyrrhotite, hematite, or even magnetite can then be misestimated. These issues can be resolved by analyzing the polished sections under optical microscopy [46].

Table 3. Quantitative mineralogical composition determined by chemical/mineralogical reconciliation.

Dana Class		Polymetallic Ores		
		LaRonde	Pirquitas	Raglan
Group	Minerals	(wt. %)	(wt. %)	(wt. %)
Sulfides	Arsenopyrite	0.03	0.08	
	Chalcopyrite	0.89		1.72
	Galena	0.03		
	Pentlandite			6.86
	Pyrite	26.82	9.66	
	Pyrrhotite	3.99		18.34
	Sphalerite	0.16	1.08	
	Stannite		0.10	
Plagioclases	Albite			
	Labradorite	2.19		
Chlorites	Chamosite	0.93		
	Clinochlore		0.90	10.40
Sulfates	Anhydrite		0.32	
	Barite		0.06	
Micas	Biotite	0.48		
	Muscovite	7.55	18.22	
	Paragonite	7.5		
Oxides	Cassiterite		0.25	
	Hematite			
	Magnetite			5.29
	Quartz	46.27	49.75	
	Rutile	0.52	0.58	
	Ilmenite			
Serpentine	Antigorite			16.77
	Lizardite			40.62
Kaolinite	Dickite		15.51	
	Kaolinite	2.4		
Phosphates	Apatite	0.25		
Zeolite	Gobbinsite		3.51	
	TOTAL	100.01	100.02	100.00

Dana class		Gold Ores						
		Canadian Malartic	Goldex	Kittilä	Lapa	Meliadine	Pinos Altos	Westwood
Group	Minerals	(wt. %)	(wt. %)	(wt. %)	(wt. %)	(wt. %)	(wt. %)	(wt. %)
Sulfides	Arsenopyrite			2.45	0.25	3.12		
	Chalcopyrite	0.04		0.03		0.03		0.07
	Galena					0.01		0.02
	Pentlandite				0.06			
	Pyrite	1.71	1.41	4.46	0.10	0.84	0.02	10.03
	Pyrrhotite			2.83	0.86	1.00		
	Sphalerite			0.03	0.02	0.04	0.03	0.57
	Ullmannite			0.01	0.02			
Sulfosalt	Gersdorffite				0.10			
Plagioclases	Albite	42.25	39.34	14.65		16.38	1.66	12.82
	Labradorite			0.00				
Chlorites	Chamosite	7.85	6.73	6.19	10.47			3.97
	Clinochlore			0.00		7.64	2.34	
Carbonates	Ankerite		1.08	9.14	13.83	1.20		
	Calcite	3.24	9.22	0.00		5.14	0.23	2.30
	Dolomite	1.34		14.35				
	Siderite			0.10		7.31		

Table 3. Cont.

Dana class			Gold Ores						
			Canadian Malartic	Goldex	Kittilä	Lapa	Meliadine	Pinos Altos	Westwood
Group	Minerals		(wt. %)	(wt. %)	(wt. %)	(wt. %)	(wt. %)	(wt. %)	(wt. %)
Sulfates	Barite		0.12					0.12	0.14
Epidote	Epidote								4.93
Pyrophyllite	Talc					26.46			
Micas	Biotite Muscovite Siderophyllite		7.08	0.23 4.25	5.84 11.91	10.64	15.02	7.15	20.56
Oxides	Hematite Magnetite Quartz Rutile		22.24 0.48	20.74 0.32	26.05 1.63	16.93 0.52	2.85 1.65 36.94	2.57 66.36 0.25	43.47 0.67
Tourmaline	Dravite			16.68					
Smectite	Montmorillonite							5.51	
Feldspaths	Microcline Orthoclase		13.19					12.53	
Phosphates	Apatite				0.48				0.45
Cordierite	Cordierite					20.02			
Sodalite	Lazurite							1.22	
	TOTAL		99.54	100.00	100.15	100.00	99.17	99.99	100.00

3.4. Geochemical Results

The pH and electrical conductivity results from the modified weathering cells are shown in Figure 4a,b, respectively. The pH is neutral and relatively stable for Kittilä, Meliadine, and Westwood, varying from 7.2 to 8.5, 7.2 to 8.4, and 7.1 to 8.1, respectively. The electrical conductivity of Kittilä and Meliadine is quite low at an average of 230 µS/cm and 240 µS/cm, respectively, whereas Westwood presents higher electrical conductivity values with an average of 750 µS/cm and median of 470 µS/cm. The pH of the LaRonde leachates has been found to decrease from 6.9 in the beginning, to 3.4 after a hundred days of testing, with an average of 4.2. The electrical conductivity (median of 490 µS/cm) decreases from 2770 µS/cm to 400 µS/cm after sixty days of testing and increases up to 600 µS/cm after one hundred days. The pH of the Pirquitas sample is acidic and relatively stable, varying from 3.7 to 4.6, with a slight drop all along the test. Its electrical conductivity is very high for the first three flushes (2250, 970, and 540 µS/cm) and then stabilizes (median of 440 µS/cm). This behavior is due to the dissolution of initial oxidation products and probably to the sulfosalts at the beginning of leaching. Sulfur concentrations are in the same order of magnitude for Kittilä and Meliadine (average of 30 mg/L) and for Pirquitas, LaRonde, and Westwood it averages 110 mg/L, and 190 mg/L, respectively. Moreover, Westwood presents the highest [Ca + Mg + Mn] concentrations, averaging at 230 mg/L, which is representative of the neutralization potential [44], and mostly due to Ca concentrations. Ca is present in calcite in the Westwood sample. Although Kittilä contains the highest Ca concentration in solid, the leachates of the sample do not contain the highest concentrations in Ca (average of 25 mg/L) as Ca is present as dolomite and ankerite, which are dissolved more slowly than calcite. The neutralizing element concentrations [Ca + Mg + Mn] decreases quickly until 10 mg/L for LaRonde. Meliadine presents concentrations decreasing from 170 mg/L to 30 mg/L with Ca and Mg as the major contributing elements. Calcite, siderite and ankerite are present in the Meliadine sample. Pirquitas has the highest Mn concentration and [Ca + Mg + Mn] concentration decreases from 190 to 2 mg/L.

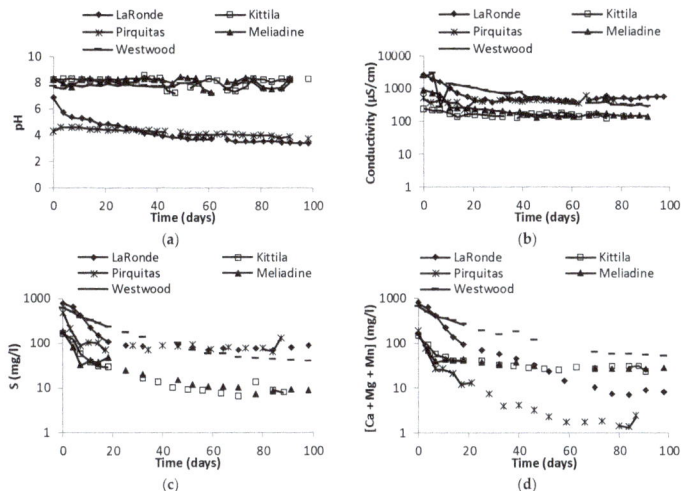

Figure 4. Geochemical results from the weathering cells: (**a**) pH and (**b**) electrical conductivity; (**c**,**d**) concentrations (mg/L) of metals associated with sulfide oxidation in the ore samples: (**c**) S, (**d**) Ca + Mg + Mn.

The concentrations of elements associated with metal sulfide oxidation (sulfates and metals) are shown in Figure 5. Despite their content in the solid samples, no Bi concentrations were found in the leachates. Pb was found in the first flush for Pirquitas, and in the leachates of LaRonde (average of 0.6 mg/L). Mo was found in the leachates of Meliadine. Hg, Sn and Tl were not analyzed. Table 4 presents the average and median leachate concentrations in mg/l and the initial content in the solid samples. Although all the ores contain Sb in significant amount, it was only found in the leachates with an acidic pH (LaRonde and Pirquitas).

For the other elements, the concentrations are generally higher in the first leachates, except for elements such as Cd, Cu, Fe, or Te. Cd is relatively stable all along the test for LaRonde, Pirquitas (average of 0.05 mg/L, 0.7 mg/L, respectively). These concentrations are above the resurgence norm of the directive 019 of Quebec (0.0021 mg/L). In the LaRonde leachates, Cu and Fe follow the same behavior, which could signalize galvanic interactions between the sulfides [47] and a preferential oxidation of chalcopyrite. Although the Sb and As content are lower in Pirquitas and Kittilä than in Meliadine, their concentrations in the leachates are higher (average of 0.26 mg/L and 0.20 mg/L for Pirquitas and Kittilä, respectively, and average of 0.07 mg/L for Meliadine) than in the leachates of Meliadine, because As and Sb are present as sulfosalts in Pirquitas and Kittilä.

The mineralogy of the various ores in this study has influenced the water quality results. Metals have been leached from both acid generating and buffering reactions. However, these metals would stay under soluble form depending on the solubility product on the compound and of the sorption mechanisms at different pH. Table 5 presents the yield of the release of elements that are possible to follow. Therefore, the release of Zn, Cd, Cu, Pb, Co, As, and Sb was assessed by calculating the yield according to the following equation:

$$\eta = \frac{E_{l100}}{E_i} \qquad (1)$$

where η = yield in percentage, E_{l100}: cumulative values of the element E after a hundred days of testing, E_i: initial amount of the element E in the weathering cell.

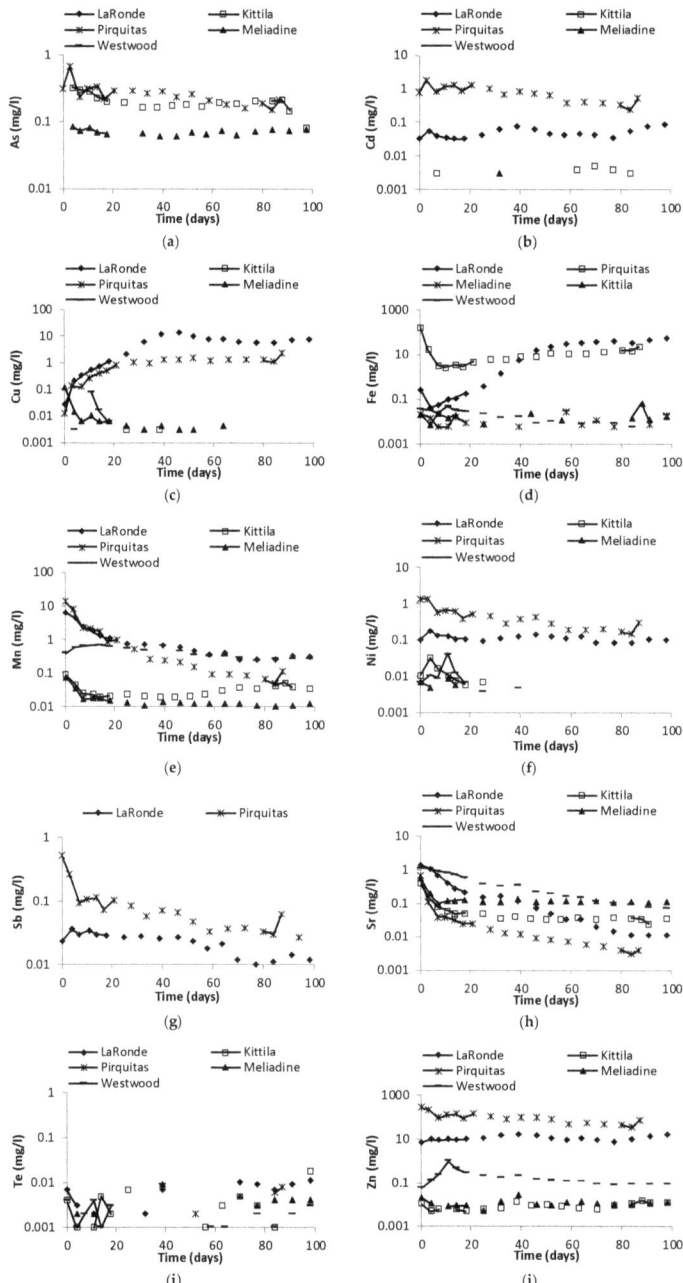

Figure 5. Geochemical results from the weathering cells: concentrations (mg/L) of metals associated with sulfide oxidation in the ore samples; (**a**) As, (**b**) Cd, (**c**) Cu, (**d**) Fe, (**e**) Mn, (**f**) Ni, (**g**) Sb, (**h**) Sr, (**i**) Te, and (**j**) Zn.

Table 4. Initial content and metal concentrations in the leachates of geochemical testing in the Kittilä, Meliadine, Westwood, LaRonde and Pirquitas ore samples.

	Kittilä	Average pH = 8.0			Meliadine	Average pH = 8.1		Westwood	Average pH = 7.7	
Element	Solid content (in ppm)	Leachates Average Content (in mg/L)	Leachates Median Content (in mg/L)	Resurgence Criterion (mg/L)	Solid Content (in ppm)	Leachates Average Content (in mg/L)	Leachates Median Content (in mg/L)	Solid Content (in ppm)	Leachates Average Content (in mg/L)	Leachates Median Content (in mg/L)
S	39,600	31.2	12.1		16,600	32.8	15.3	56,600	186	104
As	12,200	0.2	0.2	0.34	13,600	0.07	0.07	42	0	0
Cd	0.75	0	0	0.0021	0.77	0	0	7.6	0	0
Cu	104	0.003	0.003	0.0073	92	0.01	0.01	245	0.03	0.01
Fe	96,700	0.02	0.02		97,800	0.01	0.01	70,600	0.02	0.02
Mn	2430	0.03	0.03		320	0.02	0.01	660	0.45	0.42
Ni	32	0.01	0.01	0.26	0	0	0	0	0	0
Sb	77	0	0	0.088	17	0	0	57	0	0
Sr	120	0.06	0.04		260	0.14	0.12	110	0.42	0.28
Te	NA	0.003	0.001		NA	0.002	0	NA	0.002	0.001
Zn	153	0.01	0.01	0.067	213	0.012	0.011	3244	0.21	0.13

	LaRonde	Average pH = 4.2			Pirquitas	Average pH = 4.2	
Element	Solid Content (in ppm)	Leachates Average Content (in mg/L)	Leachates Median Content (in mg/L)	Resurgence Criterion (in mg/L)	Solid Content (in ppm)	Leachates Average Content (in mg/L)	Leachates Median Content (in mg/L)
S	162,000	189	88		55,000	114	85.5
As	143	0	0	0.34	1160	0.26	0.24
Cd	3.5	0.05	0.04	0.0021	40	0.73	0.73
Cu	3039	5.09	5.67	0.0073	307	0.92	1.07
Fe	155,000	18.35	11.07		47,400	18.04	10.6
Mn	401	1.25	0.62		32	1.62	0.24
Ni	40	0.12	0.11	0.26	0	0.47	0.4
Sb	5	0.02	0.02	0.088	55	0.1	0.07
Sr	110	0.26	0.1		460	0.05	0.01
Te	14	0.004	0.002		1.9	0.002	0
Zn	946	10.85	9.88	0.067	8380	97.3	86.7

NA = Non-Applicable.

Table 5. Geochemical parameters of the LaRonde, Kittilä, Pirquitas, Meliadine, and Westwood samples, and pure samples of sphalerite, chalcopyrite, galena, and arsenopyrite.

		Yields after 100 Days of Testing					
Elements	Pure mineral	Ore	LaRonde	Kittilä	Pirquitas	Meliadine	Westwood
SO_4^{2-}	-	pH_{100}	3.4	8.2	3.7	8.3	8
		η pure mineral	1.5%	1.0%	2.4%	2.8%	5.8%
Zn	Sphalerite	1.0%	20.6%	0.08%	14.1%	0.09%	0.12%
Cd	Sphalerite	0.17%	25.4%	2.4%	22%	-	-
Cu	Chalcopyrite	3.5%	3.4%	-	3.7%	0.1%	0.03%
Pb	Galena	0.27%	5%	-	-	-	-
Co	-	-	3.3%	-	-	-	-
As	Arsenopyrite	7.4%	-	0.02%	0.28%	0.01%	-
Sb	-	-	7.5%	-	2.1%	-	-

		Reactivity Rates, from 40 to 100 Days					
Elements	Pure mineral	Ore	LaRonde	Kittilä	Pirquitas	Meliadine	Westwood
SO_4^{2-}	-	r pure mineral	8.30×10^{-5}	3.54×10^{-5}	2.09×10^{-4}	9.49×10^{-5}	1.85×10^{-4}
Zn	Sphalerite	8.73×10^{-5}	1.87×10^{-3}	8.71×10^{-6}	9.20×10^{-4}	7.57×10^{-6}	5.80×10^{-6}
Cd	Sphalerite	2.16×10^{-5}	2.48×10^{-3}	5.22×10^{-4}	1.46×10^{-3}	-	-
Cu	Chalcopyrite	4.21×10^{-5}	4.00×10^{-4}	-	6.01×10^{-4}	2.23×10^{-6}	-
Pb	Galena	3.28×10^{-5}	4.82×10^{-4}	-	-	-	-
Co	-	-	2.23×10^{-4}	-	5.65×10^{-4}	-	1.78×10^{-5}
As	Arsenopyrite	5.77×10^{-4}	-	2.37×10^{-4}	2.35×10^{-5}	8.44×10^{-7}	-
Sb	-	-	5.32×10^{-4}	-	1.02×10^{-4}	-	-

The yield was calculated for the elements in the five ores and in the weathering cells of pure materials [12], to compare the reactivity of these minerals according to their associations and their occurrence in a material. Similarly, the reactivity rates of the same elements were calculated on the stabilized portion of the geochemical tests, which means from forty to one hundred days, normalized by the initial amount in the sample (Table 5). As expected, the release of elements is strongly correlated with the pH of the solution. For instance, the yield of sphalerite for Zn and Cd is twenty times higher in LaRonde and Pirquitas than in the pure sphalerite sample. That can be explained in part with the pH as it is of 6.3 for the leachates of pure sphalerite. However, arsenopyrite has a different behavior in ores than it does in pure samples. Despite the content in the initial solid, the LaRonde sample does not release arsenic. This may be due to the liberation parameters of the arsenopyrite grains. In contrast, Kittilä (As-pyrite and gersdorffite) and Pirquitas (As-pyrite and freibergite) release arsenic since the bearing minerals differ, and the grains are accessible for the reaction. The yields for chalcopyrite dissolution are the same for the pure mineral geochemical test than in the LaRonde and Pirquitas tests.

The reactivity rate is ten times higher in LaRonde and Pirquitas than in the pure chalcopyrite sample. That may be due to galvanic interactions occurring in the ore samples [47]. For sphalerite, it is less evident as the pH plays a great role in the reaction mechanisms.

3.5. Acid Mine Drainage Estimation

After determining the precise mineralogy of the ores, the quality of the drainage generated by the wastes can be assessed. The differences in mineral textures and rock competency bring complications, as do microbial activity and the different oxidation rates of the sulfide minerals according to their trace-element content. Approximations are unavoidable [48]. Moreover, static tests are used in this study, as they are a first screening to determine which sample and which part of the deposit will need careful geochemical testing and further mineralogical characterization (liberation degree and mineralogical associations). Appendix A Table A5 exposes the different estimations of the acid-generation potential (AP) and the neutralization potential (NP) according to different mineralogical methods: Sobek, Schuller [49], Paktunc [25], Bouzahzah, Benzaazoua [44] and unpublished work from the authors for the AP calculation and Lawrence and Scheske [50], Paktunc [51], Plante, Bussière [52] and the standard carbonate NP (CNP) and corrected CNP (CCNP) methods [53] for the NP calculations. These static tests, their characteristics, their advantages and disadvantages are fully explained in Plante, Bussière [52] and in Bouzahzah, Benzaazoua [44]. In the case of the polymetallic ores, the minerals of economic interest are, in a first approximation, considered to be 95% recovered in the plant. Therefore, the contribution of chalcopyrite, sphalerite, and galena has been removed from the calculations in the AP determination for the LaRonde sample, the contribution of sphalerite and stannite for the Pirquitas sample, and of pentlandite and chalcopyrite for the Raglan sample. This estimation can change according to the ore processing method and its particular extraction yields. The net neutralization potential (NNP) and the ratio AP/NP were calculated with all possibilities of AP and NP results according to the various methods, by the matrix. Two combinations were used to calculate NNP and AP/NP: the AP and NP calculations by the Paktunc [51] methods, and the AP by Chopard, Benzaazoua [54] with the NP by Plante, Bussière [52]. The classification used here is from Ferguson and Morin [55] for the NNP value and from Price, Morin [56] for the NP/AP ratio value. If the NNP is below −20 kg CaCO$_3$/t, the material is considered acid generating; if the NNP is comprised between −20 and +20 kg CaCO$_3$/t, the material is classified uncertain, and if the NNP is above 20 kg CaCO$_3$/t, therefore, the material is not considered acid generating. If NP/AP < 1, the wastes will be acid generating; if 1 < NP/AP > 2, the wastes are likely to be acid generating. If 2 < NP/AP < 4, the wastes are not likely to be acid generating. Finally, if NP/AP > 4, the wastes are considered as not acid generating. Bouzahzah, Benzaazoua [44] recommends the use of the NNP as a classification's criterion, whereas Sherlock, Lawence [57] recommend the use of the NP/AP ratio. Figure 6 shows the results for the NNP calculation with the methods of Plante, Bussière [52] and Chopard, Benzaazoua [54].

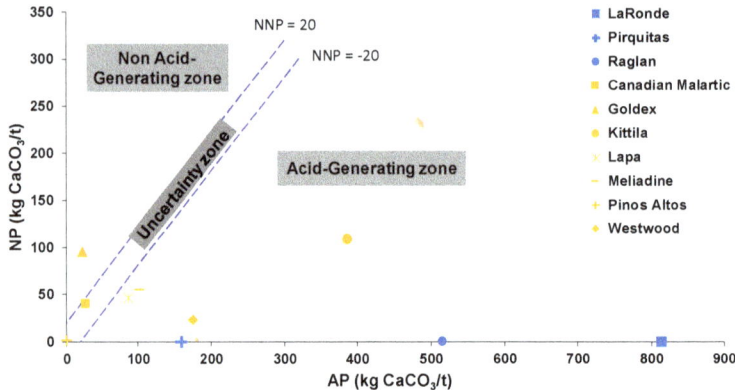

Figure 6. Net neutralization potential values for the ten ore samples calculated from the neutralization potential (NP) by Plante [52] and from the acid-generation potential (AP) by Chopard et al. [54].

Despite the possible removal of certain minerals by flotation, all polymetallic samples studied will generate acid-generating wastes. Therefore, contaminants will also be released. This is confirmed by the geochemical tests on the LaRonde and Pirquitas samples, as their pH is acidic (average of 4.2) and by the elemental concentrations found in the leachates. For gold ores, it was evaluated that all sulfides would go into the waste materials. Westwood, Kittilä and Meliadine would be acid generating, however the kinetic test results are in contradiction with this estimation. On the first hundred days of testing, as the pH of the leachates is comprised between 7.0 and 8.2 for these samples. This contradiction may be explained by the texture of the ores and by the short testing time. The lag time for acid generation is indeed an important consideration in acid rock drainage (ARD) prevention and the early results of geochemical testing may not be representative of long-term behavior [58]. Conversely, the SEM observations have shown partially or totally liberated sulfides. Thus, the oxidation-neutralization curve [59,60] was plotted to compare the cumulative extracted amounts of the main sulfide oxidation products (sulfates) and the main acid neutralization products (calcium, magnesium, and manganese) (Figure 7) to attempt to explain predictable results. The acid-generating potential can be assessed by extrapolating the oxidation-neutralization curve on a longer period and by projecting the initial sulfur and Ca + Mg + Mn initial contents of the sample.

For Kittilä, the initial composed projection is over the oxidation-neutralization curve, which means that the sample contains neutralizing minerals in sufficient amount to neutralize the acidity generated by sulfide oxidation. In this case, the material should not be acid generating in the long-term. This statement should be established by taking into account the ore's texture and other mineralogical parameters than modal mineralogy. Contrariwise, for Meliadine and Westwood, the initial composition projections are under the oxidation-neutralization curve. The samples would be acid generating in the long term, and thus the results of the static test would be relevant and the initial geochemical behavior of the two samples would not reflect the whole behavior in the long run. Lapa would be acid generating too, but according to the methods of Paktunc [25,51], it would be classified as uncertain. Canadian Malartic and Pinos Altos will be uncertain. Only the Goldex wastes will be classified as not acid generating.

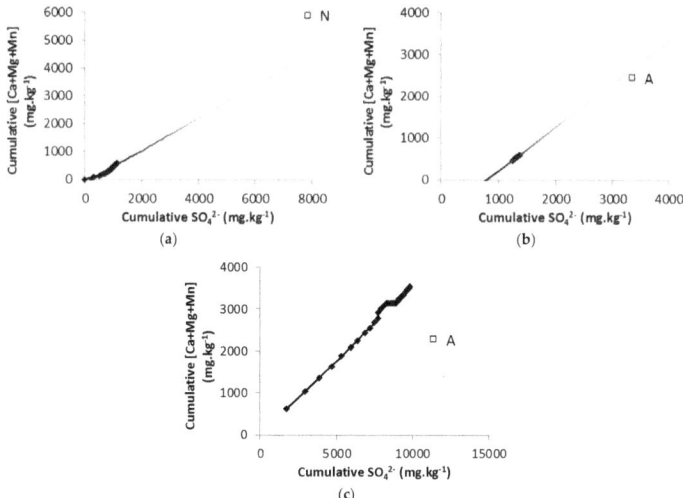

Figure 7. Extrapolation of the oxidation-neutralization curves and projection of the samples initial compositions for (**a**) Kittilä, (**b**) Meliadine, and (**c**) Westwood. A: acid-generating samples; N: neutralizing samples.

4. Conclusions and Perspectives

Ten samples were submitted to a detailed mineralogical characterization to identify the possible issues of their future exploitation, for environmental and ore processing considerations. This work allowed to determine the specification of valuable elements (Ag, Au, Li, In, Pd, Pt) and potential contaminants (As, Ba, Bi, Cd, Co, Cr, Cu, Hg, Mn, Mo, Ni, Se, Sn, Sr, Te, Tl, and Zn). Regarding the methodology itself, firstly, it is important to be confident in the assay, using the appropriate chemical methods. Then, the reconciliation of mineralogical data (XRD, optical microscopy, SEM-EDS, and EPMA) and the assay is necessary to obtain a reliable quantitative mineralogy (as XRD is only a semi-quantification technique) and improve the estimation of the acid-generation potential. The mineralogical difficulties occurred when the concentrations of an element were too low to detect its bearing mineral. The geochemical interpretation difficulties were causes by the changing pH of the leachates. The possible contaminants will be found in the leachates only if they are soluble under the pH conditions of the water. It was difficult to consider this parameter here, as three of the five ore tests have presented a neutral pH at the beginning of the geochemical test but have not presented high levels of concentrations of the possible contaminants identified by the mineralogical characterization. The pH effects on the geochemical and environmental behavior of samples (like the release of contaminants) will be investigated in a later study.

The strength of this study is due to the availability and relatively cost-effectiveness of the main mineralogical techniques used to identify the elements and the minerals (optical and electron microscopy), and the possibility to choose if EPMA will be necessary to bring benefit for upcoming decisions. The cost of the methodology is very low compared with the time and costs saved by the mineralogical information obtained on the ores. This knowledge will enable a complete planning and integrated waste management strategy at the stage of exploration. Regarding the identification of secondary valuable elements, a detailed mineralogical characterization would also allow junior companies to minimize the risks associated with a deposit and then to increase the economic value of a potential site after its sale.

The goal of this study was to propose a simple and low-cost methodology based on a mineralogical characterization. However, other work could be done at the stage of exploration to still better estimate the environmental and mineral processing challenges, particularly, automated mineralogy could be done in addition to this methodology to bring supplementary information. Samples must be carefully chosen, as automated mineralogy is largely more expensive. The information on particle size, distribution, liberation and minerals association should be integrated in the methodology to assess for instance the availability of the minerals for reactions or the elements to be leached (or recovered). More work could be done on thermochemical equilibrium calculations with PhreeqC to perform mineral speciation and saturation index calculations to determine the major sources of ionic elements in water.

Author Contributions: Conceptualization, R.M.-B. and M.B.; Data curation, A.C.; Formal analysis, A.C.; Funding acquisition, M.B.; Investigation, A.C.; Methodology, A.C.; Project administration, B.P. and M.B.; Resources, R.M.-B. and M.B.; Supervision, P.M., B.P. and M.B.; Validation, P.M., B.P. and M.B.; Visualization, M.B.; Writing—original draft, A.C.; Writing—review & editing, P.M., R.M.-B., B.P. and M.B.

Funding: This research received no external funding.

Acknowledgments: The authors greatly thank to the technical staff of the URSTM for helping in the preparation samples and making the chemical digestion and analysis. Thank you to the IRME partners: Agnico Eagle Mines Limited, Canadian Malartic Mine, IamGold Corporation, Raglan Mine – Glencore, and Rio Tinto Fer and Titane Inc. The analysis services of the GeoRessources laboratory at the University of Lorraine in France are also gratefully acknowledged.

Conflicts of Interest: The authors declare no conflict of interest.

Appendix A

Table A1. Initial sample description and final D80 desired.

Mine's Name	Sample	Mine's Lab Preparation	Initial Mass (g)	Initial Particle Size	Final Target D80 (µm)
LaRonde	Blasted ore	No	2150	D80 ≈ 80 mm	68
Pirquitas	JIG feed ore	Unknown	900	<10 mm	48
Raglan	Blasted ore	No	39,000	D80 ≈ 80 mm	79
C. Malartic	Stockpile	No	12,000	D80 ≈ 100 mm	64
Goldex	Jaw crusher sample	Crushed, split	1630	D80 ≈ 7.1 mm	66
Kittilä	GTK composite	Crushed, split	2650	<1 mm	85
Lapa	Composite	Crushed, split	1000	<2 mm	80
Meliadine	Tiriganiaq open pit Composite	Crushed, split	1000	<2 mm	63
Pinos Altos	Santo Niño Composite	Crushed, split	4300	<2 mm	74
Westwood	Stockpile	No	35,000	D80 ≈ 100 mm	64

Table A2. Precious elements in the ores' samples. The values above ten times the Clarke value are in gray.

	Precious Metals (ppb)					
	Ag	Au	In	Li	Pd	Pt
Clarke value	70	1.1	50	60,000	15	1
LaRonde	14,000	2534	1000	20,000		
Pirquitas	141,000	42	30,200	60,000		
Raglan	2000	60			1627	694
C. Malartic		560		20,000		
Goldex		2097				
Kittilä		6628		20,000		
Lapa						
Meliadine	1000	4018		20,000		22
Pinos Altos	88,000	3207		90,000		30
Westwood	3000	3582	500	20,000		

Table A3. Major, minor and trace elements of the ores' sample. For minor and trace elements, and for Cr, Mn, and S, the values above the Directive 019 values are in light gray, the values above ten times the Clarke values or above both the defined thresholds are in dark gray; ND: No determined.

	Minor and Trace Elements (ppm)																
	As	Ba	Bi	Cd	Co	Cu	Hg	Mo	Ni	Pb	Sb	Se	Sn	Sr	Te	Tl	Zn
Clarke value	5	250	0.2	0.15	25	70	0.007	1.5	80	16	0.2	0.05	2.2	375	0.005	0.45	75
Directive 019	5	200	ND	0.9	20	50	0.1	6	50	40	ND	3	5	ND	ND	ND	120
LaRonde	143	150	168.8	3.47	11	3039	0.04	0.7	39	219	5.0	<0.0004	ND	110	14	2.4	946
Pirquitas	1160	360	122.5	40.0	10	307	0.07	1.1	<0.0007	134	54.8	1.88	2280	460	2.0	11	8380
Raglan	16	20	4.9	1.55	548	6033	0.09	0.6	24460	13	7.8	14.2	ND	20	4	<0.5	88
C. Malartic	17	700	5.1	0.09	16	122	<0.01	5.3	20	25	22.7	1.56	ND	570	<2	<0.5	65
Goldex	36	250	10.4	0.08	14	79	0.51	2.6	126	7	445.9	0.05	ND	460	<2	<0.5	54
Kittilä	12,200	220	1.9	0.75	25	104	0.40	1.5	32	15	76.9	1.25	ND	120	<2	1	153
Lapa	1865	250	<0.1	0.48	52	65	<0.01	1.2	623	13	141	<0.0004	1.3	<10	<2	<0.5	114
Meliadine	13,600	450	5.0	0.77	15	92	<0.01	8.0	<0.0007	101	16.8	0.33	ND	260	<2	<0.5	213
Pinos Altos	44	730	2.8	1.00	6	67	0.82	1.7	<0.0007	89	65.9	0.43	ND	100	<2	1.4	197
Westwood	42	810	10.1	7.62	18	245	0.02	1.7	<0.0007	186	57.4	3.55	ND	110	<2	<0.5	3244

	Major Elements (wt. %)										
	Al	Ca	Cr	Fe	K	Mg	Mn	Na	Si	Ti	S
Clarke value	0.08	3.6–4.1	0.01–0.02	4.1–5	2.1–2.5	2.1–2.3	0.04–0.16	2.4–2.8		0.45	0.03–0.05
Directive 019	ND	ND	0.0085	ND	ND	ND	0.1	ND	ND	ND	ND
LaRonde	5.79	0.30	0.0069	15.5	0.78	0.12	0.04	0.53	26.17	0.31	16.2
Pirquitas	7.46	0.16	0.0079	4.74	2.00	0.14	0.00	0.26	32.81	0.35	5.50
Raglan	1.20	0.25	0.2294	21.7	0.39	17.49	0.07	0.01	13.15	0.05	10.1
C. Malartic	7.79	1.91	0.0169	3.67	2.49	1.62	0.04	3.51	26.31	0.29	0.94
Goldex	8.22	3.55	0.0099	2.91	0.44	1.72	0.03	3.67	26.22	0.19	0.75
Kittilä	5.88	5.19	0.0155	9.67	1.69	3.39	0.24	1.22	21.24	0.98	3.91
Lapa	4.80	3.97	0.1008	6.35	0.96	9.33	0.10	0.90	24.01	0.31	0.45
Meliadine	5.66	2.42	0.0079	9.78	1.47	1.21	0.03	1.36	26.69	0.19	1.66
Pinos Altos	3.80	0.25	0.0074	2.07	2.46	0.36	0.07	0.36	38.51	0.15	0.11
Westwood	7.40	1.96	0.0066	7.06	2.02	1.39	0.07	1.06	28.16	0.40	5.66

Table A4. EPMA analysis of the elements of interest in certain minerals observed.

Mineral	Elt (%)	DDL (ppm)	Polymetallic Ores			Gold Ores				
			LaRonde	Pirquitas	Raglan	Canadian Malartic	Goldex	Kittilä	Meliadine	Westwood
Pyrite	S		53.53 ± 0.31	51.8 ± 3.71		53.2 ± 3.07	53.3 ± 3.07	50.4 ± 3.92	53.2 ± 3.08	54.1 ± 0.31
	Fe		46.47 ± 0.44	46.9 ± 1.79		46.8 ± 2.15	46.7 ± 2.14	46.3 ± 2.37	46.8 ± 2.15	45.9 ± 0.44
	As	650	<2000 ppm	0.83 ± 0.26		<650 ppm	<650 ppm	3.26 ± 0.49	<650 ppm	<2000 ppm
	Au	1100	NA	NA		NA	NA	<1100 ppm	NA	NA
	Ag	1400	NA	0.40 ± 0.25		NA	NA	NA	NA	NA
	Tl	1500	NA	0.45 ± 0.27		NA	NA	NA	NA	NA
	Se	650	<650 ppm	NA		NA	<650 ppm	<650 ppm	NA	<650 ppm
Pyrrhotite	S				40.0 ± 1.82			39.8 ± 1.81	39.3 ± 1.76	
	Fe				59.7 ± 1.94			60.2 ± 1.95	60.6 ± 1.92	
	Co	330			<330 ppm			<330 ppm	<330 ppm	
	Ni	550			0.37 ± 0.22			<550 ppm	0.10 ± 0.10	
Arsenopyrite	S							21.1 ± 1.88	21.1 ± 1.89	
	As							43.9 ± 3.08	43.6 ± 3.08	
	Fe							35.0 ± 1.90	35.2 ± 1.92	
	Au	1300						<1300 ppm	<1300 ppm	
Chalcopyrite	S		35.5 ± 1.74						34.7 ± 1.64	NA
	Cu		34.2 ± 1.85						34.5 ± 1.22	NA
	Fe		30.2 ± 1.39						30.5 ± 1.21	NA
	Ag	430	NA						<430 ppm	<430 ppm
	As	1700	<1700 ppm						NA	NA
	Sn	700	<700 ppm						NA	NA
	Tl	2000	<2000 ppm						NA	NA
	Te	1600	<1600 ppm						NA	NA
	Cd	500	<500 ppm						0.37 ± 0.10	NA
Sphalerite	S		33.1 ± 1.57	33.6 ± 2.53						33.4 ± 1.62
	Zn		60.7 ± 2.33	63.8 ± 2.13						61.2 ± 2.64
	Fe		4.39 ± 0.43	2.06 ± 0.30						4.31 ± 0.42
	Ag	950	NA	0.84 ± 0.27						NA
	In	680	0.45 ± 0.10	0.25 ± 0.18						NA
	Cd	475	1.16 ± 0.09	0.37 ± 0.21						1.00 ± 0.12
	Mn	550	0.25 ± 0.09	NA						0.16 ± 0.09
	Pb	400	NA	0.71 ± 0.65						NA
	Tl	2600	0.1 ± 0.09	0.33 ± 0.22						NA
		1500								

Table A4. Cont.

Mineral	Elt (%)	DDL (ppm)	Polymetallic Ores				Gold Ores			
			LaRonde	Pirquitas	Raglan	Canadian Malartic	Goldex	Kittilä	Meliadine	Westwood
Galena	S								13.2 ± 0.32	
	Pb								85.4 ± 1.06	
	Fe								1.37 ± 0.20	
	Ag	430							0.09 ± 0.04	
Pentlandite	S				33.8 ± 1.67					
	Fe				30.0 ± 1.22					
	Ni				35.5 ± 1.34					
	Co	330			0.75 ± 0.08					

Table A5. Different estimations of the acid-generation potential (AP) and the neutralization potential (NP) according to different mineralogical methods; and neutralization net potential (NNP) and NP/AP ratios.

		LaRonde	Pirquitas	Raglan
NP (kg CaCO3/t)	Lawrence & Scheske [50]	2	2	88
	CNP	0	6	8
	CCNP	0	6	8
	Paktunc [51]	0	0	0
	Plante [52]	0	0	0
AP (kg CaCO3/t)	Sobek et al. [49]	496	160	225
	Paktunc [51]	483	158	205
	Bouzahzah et al. [44]	483	159	205
	Chopard et al. [54]	814	159	516
NNP	Paktunc [51]	−483	−158	−205
	Chopard et al. [54] & Plante [52]	−814	−159	−516
NP/AP	Paktunc [51]	0	0	0
	Chopard et al. [54] & Plante [52]	0	0	0

Table A5. Cont.

		Canadian Malartic	Goldex	Kittilä	Lapa	Meliadine	Pinos Altos	Westwood
NP (kg CaCO$_3$/t)	Lawrence & Scheske [50]	55	105	134	101	127	5	31
	CNP	47	103	302	101	126	0	23
	CCNP	47	103	302	101	63	0	23
	Paktunc [51]	40	95	103	38	55	2	23
	Plante [52]	40	96	109	47	55	2	23
AP (kg CaCO$_3$/t)	Sobek et al. [49]	29	23	122	14	52	3	177
	Paktunc [51]	28	23	120	13	44	1	171
	Bouzahzah et al. [44]	28	23	142	15	73	0	166
	Chopard et al. [54]	28	23	387	88	103	1	176
NNP	Paktunc [51]	11	72	−17	25	10	2	−148
	Chopard et al. [54] & Plante [52]	12	73	−278	−41	−48	1	−153
NP/AP	Paktunc [51]	1.40	4.12	0.86	2.89	1.23	3.66	0.13
	Chopard et al. [54] & Plante [52]	1.41	4.16	0.28	0.53	0.54	2.48	0.13

References

1. EPA. *Acid Mine Drainage Prediction*; U.S. Environmental Protection Agency Office of Solid Waste Special Waste Branch: Washington, DC, USA, 1994.
2. Natural Resources Canada NR. *Evaluation Report: Green Mining Initiative*; Canada NR: Ottawa, ON, Canada, 2015.
3. Plante, B.; Benzaazoua, M.; Bussière, B. Predicting Geochemical Behaviour of Waste Rock with Low Acid Generating Potential Using Laboratory Kinetic Tests. *Mine Water Environ.* **2010**, *30*, 2–21. [CrossRef]
4. El Adnani, M.; Plante, B.; Benzaazoua, M.; Hakkou, R.; Bouzahzah, H. Tailings Weathering and Arsenic Mobility at the Abandoned Zgounder Silver Mine, Morocco. *Mine Water Environ.* **2015**, *35*, 508–524. [CrossRef]
5. Steger, H.; Desjardins, L. Oxidation of sulfide minerals, 4. Pyrite, chalcopyrite and pyrrhotite. *Chem. Geol.* **1978**, *23*, 225–237. [CrossRef]
6. Steger, H.; Desjardins, L. Oxidation of sulfide minerals; V, Galena, sphalerite and chalcocite. *Can. Mineral.* **1980**, *18*, 365–372.
7. Rimstidt, J.D.; Chermak, J.A.; Gagen, P.M. *Rates of Reaction of Galena, Sphalerite, Chalcopyrite, and Arsenopyrite with Fe (III). in Acidic Solutions*; ACS symposium series; American Chemical Society: Washington, DC, USA, 1994; pp. 2–13.
8. Nicholson, R. Iron-sulfide oxidation mechanisms: Laboratory studies. *Environ. Geochem. Sulphide Mine Wastes* **1994**, *22*, 163–183.
9. Morin, K.; Hutt, N.; Ferguson, K. Measured rates of sulfide oxidation and acid neutralization in humidity cells: Statistical lessons from the database. In Proceedings of the Sudbury '95: Mining and the Environment, Sudbury, ON, Canada, 28 May–1 June 1995.
10. Thomas, J.E.; Smart, R.S.C.; Skinner, W.M. Kinetic factors for oxidative and non-oxidative dissolution of iron sulfides. *Miner. Eng.* **2000**, *13*, 1149–1159. [CrossRef]
11. Frostad, S.; Klein, B.; Lawrence, R.W. Evaluation of Laboratory Kinetic Test Methods for Measuring Rates of Weathering. *Mine Water Environ.* **2002**, *21*, 183–192. [CrossRef]
12. Chopard, A.; Benzaazoua, M.; Plante, B.; Bouzahzah, H.; Marion, P. Kinetic tests to evaluate the relative oxidation rates of various sulfides and sulfosalts. In Proceedings of the 10th ICARD Conference on Acid Rock Drainage, and IMWA, Santiago, Chile, 21–24 April 2015.
13. Diehl, S.; Hageman, P.L.; Smith, K.S. What's weathering? Mineralogy and field leach studies in mine waste, Leadville and Montezuma mining districts, Colorado. In Proceedings of the 7th International Conference on Acid Rock Drainage (ICARD 7), St. Louis, MO, USA, 26–30 March 2006; pp. 507–527.
14. Petruk, W. *Applied Mineralogy in the Mining Industry*; Elsevier: Amsterdam, The Netherlands, 2000.
15. Kwong, Y.J.; Swerhone, G.W.; Lawrence, J.R. Galvanic sulphide oxidation as a metal-leaching mechanism and its environmental implications. *Geochem. Explor. Environ. Anal.* **2003**, *3*, 337–343. [CrossRef]
16. Cruz, R.; Luna-Sánchez, R.M.; Lapidus, G.T.; González, I.; Monroy, M. An experimental strategy to determine galvanic interactions affecting the reactivity of sulfide mineral concentrates. *Hydrometallurgy* **2005**, *78*, 198–208. [CrossRef]
17. Liu, Q.; Li, H.; Zhou, L. Galvanic interactions between metal sulfide minerals in a flowing system: Implications for mines environmental restoration. *Appl. Geochem.* **2008**, *23*, 2316–2323. [CrossRef]
18. Lottermoser, B. *Mine Wastes: Characterization, Treatment and Environmental Impacts*; Springer: Berlin, Germany, 2010.
19. Parbhakar-Fox, A.K.; Edraki, M.; Walters, S.; Bradshaw, D. Development of a textural index for the prediction of acid rock drainage. *Miner. Eng.* **2011**, *24*, 1277–1287. [CrossRef]
20. Brough, C.P.; Warrender, R.; Bowell, R.J.; Barnes, A.; Parbhakar-Fox, A. The process mineralogy of mine wastes. *Miner. Eng.* **2013**, *52*, 125–135. [CrossRef]
21. Dold, B. Pre-mining Characterization of Ore Deposits: What Information Do We Need to Increase Sustainability of the Mining Process? In Proceedings of the 10th International Conference on Acid Rock Drainage (ICARD) & IMWA Annual Conference, Santiago, Chile, 21–24 April 2015.
22. Parbhakar-Fox, A.; Lottermoser, B.; Bradshaw, D. Evaluating waste rock mineralogy and microtexture during kinetic testing for improved acid rock drainage prediction. *Miner. Eng.* **2013**, *52*, 111–124. [CrossRef]

23. Bouzahzah, H.; Benzaazoua, M.; Bussière, B.; Plante, B. Revue de Littérature Détaillée sur les Tests Statiques et les Essais Cinétiques Comme Outils de Prédiction du Drainage Minier Acide. Déchets Sciences et Techniques Techniques. 2014, pp. 14–31. Available online: http://lodel.irevues.inist.fr/dechets-sciences-technique/docannexe/file/340/2_bouzahzah.pdf (accessed on 27 June 2019).
24. Goodall, W.R.; Cropp, A. *Integrating Mineralogy into Everyday Solutions*; MinAssist: Carlton, Australia, 2013.
25. Paktunc, A.D. *Characterization of Mine Wastes for Prediction of Acid Mine Drainage Environmental Impacts of Mining Activities*; Springer: Berlin, Germany, 1999; pp. 19–40.
26. Bouzahzah, H.; Benzaazoua, M.; Mermillod-Blondin, R.; Pirard, E. A novel procedure for polished section preparation for automated mineralogy avoiding internal particle settlement. In Proceedings of the 12th International Congress for Applied Mineralogy (ICAM), Istanbul, Turkey, 10–12 August 2015.
27. Merkus, H.G. *Particle Size Measurements: Fundamentals, Practice, Quality*; Springer Science & Business Media: Berlin, Germany, 2009.
28. Couture, R.A. An improved fusion technique for major-element rock analysis by XRF. *Adv. X-ray Anal.* **1989**, *32*, 233–238.
29. Alvarez, M. Glass disk fusion method for the X-ray fluorescence analysis of rocks and silicates. *X-ray Spectrom.* **1990**, *19*, 203–206. [CrossRef]
30. Spangenberg, J.; Fontbote, L.; Pernicka, E. X-Ray fluorescence analysis of base metal sulphide and iron-manganese oxide ore samples in fused glass disc. *X-ray Spectrom.* **1994**, *23*, 83–90. [CrossRef]
31. Claisse, F. *Glass Disks and Solutions by Fusion in Borates for Users of Claisse Fluxers*; Corporation Scientific Claisse Inc.: Sainte-Foy, QC, Canada, 1995.
32. Young, D.S.; Sachais, B.S.; Jefferies, L.C. *The Rietveld Method*; International union of crystallography: Chester, UK, 1993.
33. Raudsepp, M.; Pani, E. Application of Rietveld analysis to environmental mineralogy. *Environ. Asp. Mine Wastes* **2003**, *31*, 165–180.
34. Bouzahzah, H.; Califice, A.; Benzaazoua, M.; Mermillod-Blondin, R.; Pirard, E. Modal analysis of mineral blends using optical image analysis versus X ray diffraction. In Proceedings of the International Congress for Applied Mineralogy ICAM08, Brisbane, Australia, 8–10 September 2008; AusIMM: Carlton, Australia, 2008.
35. Cruz, R.; Bertrand, V.; Monroy, M.; González, I. Effect of sulfide impurities on the reactivity of pyrite and pyritic concentrates: A multi-tool approach. *Appl. Geochem.* **2001**, *16*, 803–819. [CrossRef]
36. Villeneuve, M.; Bussière, B.; Benzaazoua, M.; Aubertin, M. Assessment of interpretation methods for kinetic tests performed on tailings having a low acid generating potential. In Proceedings of the 8th International Conference on Acid Rock Drainage was held in Conjunction with Securing the Future, Skellefteå, Sweden, 23–26 June 2009.
37. Plante, B.; Benzaazoua, M.; Bussière, B. Kinetic Testing and Sorption Studies by Modified Weathering Cells to Characterize the Potential to Generate Contaminated Neutral Drainage. *Mine Water Environ.* **2011**, *30*, 22–37. [CrossRef]
38. Bouzahzah, H.; Benzaazoua, M.; Bussiere, B.; Plante, B. Prediction of acid mine drainage: Importance of mineralogy and the test protocols for static and kinetic tests. *Mine Water Environ.* **2014**, *33*, 54–65. [CrossRef]
39. Clarke, F.W. *The Data of Geochemistry*; US Government Printing Office: Washington, DC, USA, 1920.
40. Taylor, S. Abundance of chemical elements in the continental crust: A new table. *Geochimica et Cosmochimica Acta* **1964**, *28*, 1273–1285. [CrossRef]
41. Benzaazoua, M.; Marion, P.; Robaut, F.; Pinto, A. Gold-bearing arsenopyrite and pyrite in refractory ores: Analytical refinements and new understanding of gold mineralogy. *Mineral. Mag.* **2007**, *71*, 123–142. [CrossRef]
42. Blackburn, W.H.; Schwendeman, J.F. Trace-element substitution in galena. *Can. Mineral.* **1977**, *15*, 365.
43. George, L.; Cook, N.J.; Cristiana, L.; Wade, B.P. Trace and minor elements in galena: A reconnaissance LA-ICP-MS study. *Am. Mineral.* **2015**, *100*, 548–569. [CrossRef]
44. Bouzahzah, H.; Benzaazoua, M.; Bussière, B. Acid-generating potential calculation using mineralogical static test: Modification of the Paktunc equation. In Proceedings of the 23rd World Mining Congress (WMC 2013), Montréal, QC, Canada, 11–15 August 2013.

45. Mermillod-Blondin, R.; Benzaazoua, M.H.B.; Leroux, D. Development and calibration of a reconciliated mineralogy method based on multitechnique analyses: Application to acid mine drainage prediction. In Proceedings of the 28th International Mineral Processing Congress (IMPC), Québec, QC, Canada, 11–15 September 2016.
46. Chopard, A.; Marion, P.; Royer, J.J.; Taza, R.; Bouzahzah, H.; Benzaazoua, M. Automated sulfides quantification by multispectral optical microscopy. *Miner. Eng.* **2019**, *131*, 38–50. [CrossRef]
47. Chopard, A.; Plante, B.; Benzaazoua, M.; Bouzahzah, H.; Marion, P. Geochemical investigation of the galvanic effects during oxidation of pyrite and base-metals sulfides. *Chemosphere* **2017**, *166*, 281–291. [CrossRef] [PubMed]
48. Kwong, Y.-T.J. Prediction and Prevention of Acid Rock Drainage from a Geological and Mineralogical Perspective. MEND, 1993. Available online: http://mend-nedem.org/wp-content/uploads/1.32.1.pdf (accessed on 27 June 2019).
49. Sobek, A.A.; Schuller, W.; Freeman, J.; Smith, R. *Field and Laboratory Methods Applicable to Overburdens and Minesoils*; US Environmental Protection Agency: Cincinnati, OH, USA, 1978; Volume 45268, pp. 47–50.
50. Lawrence, R.W.; Scheske, M.A. Method to calculate the neutralization potential of mining wastes. *Environ. Geol.* **1997**, *32*, 100–106. [CrossRef]
51. Paktunc, A.D. Mineralogical constraints on the determination of neutralization potential and prediction of acid mine drainage. *Environ. Geol.* **1999**, *39*, 103–112. [CrossRef]
52. Plante, B.; Bussière, B.; Benzaazoua, M. Static tests response on 5 Canadian hard rock mine tailings with low net acid-generating potentials. *J. Geochem. Explor.* **2012**, *114*, 57–69. [CrossRef]
53. Frostad, S.R.; Price, W.A.; Bent, H. Operational NP determination—Accounting for iron manganese carbonates and developing a site-specific fizz rating. In Proceedings of the Sudbury '95: Mining and the Environment III, Sudbury, ON, Canada, 28 May–1 June 1995.
54. Chopard, A.; Benzaazoua, M.; Bouzahzah, H.; Plante, B.; Marion, P. A contribution to improve the calculation of the acid generating potential of mining wastes. *Chemosphere* **2017**, *175*, 97–107. [CrossRef]
55. Ferguson, K.D.; Morin, K.A. The prediction of acid rock drainage—Lessons from the database. In Proceedings of the Second International Conference on the Abatement of Acidic Drainage, Montreal, QC, Canada, 16–18 September 1991; Quebec Mining Association: Quebec City, QC, Canada.
56. Price, W.A.; Morin, K.; Hutt, N. Guidelines for the prediction of acid rock drainage and metal leaching for mines in British Columbia: Part II. Recommended procedures for static and kinetic testing. In Proceedings of the 4th International Conference on Acid Rock Drainage, Vancouver, BC, Canada, 31 May–6 June 1997; pp. 15–30.
57. Sherlock, E.J.; Lawence, R.W.; Poulin, R. On the neutralization of acid rock drainage by carbonate and silicate minerals. *Int. J. Rock Mech. Min. Sci. Geomech. Abstr.* **1995**, *32*, 43–54. [CrossRef]
58. GARD G 2.4. *The Acid Generation Process*; INAP: Atlanta, GA, USA, 2016.
59. Benzaazoua, M.; Bussière, B.; Dagenais, A. Comparison of kinetic tests for sulfide mine tailings. In Proceedings of the English International Conference on Tailings and Mine waste 01, Fort Collins, CO, USA, 16–19 January 2001; Balkema: Danvers, MA, USA, 2001; pp. 263–272.
60. Benzaazoua, M.; Bussière, B.; Dagenais, A.-M.; Archambault, M. Kinetic tests comparison and interpretation for prediction of the Joutel tailings acid generation potential. *Environ. Geol.* **2004**, *46*, 1086–1101. [CrossRef]

© 2019 by the authors. Licensee MDPI, Basel, Switzerland. This article is an open access article distributed under the terms and conditions of the Creative Commons Attribution (CC BY) license (http://creativecommons.org/licenses/by/4.0/).

Article

Spatial Mapping of Acidity and Geochemical Properties of Oxidized Tailings within the Former Eagle/Telbel Mine Site

Abdellatif Elghali *, Mostafa Benzaazoua *, Bruno Bussière and Thomas Genty

Research Institute on Mines and Environment (RIME), Université du Québec en Abitibi Témiscamingue, 445 Boul. Université, Rouyn-Noranda, QC J9X 5E4, Canada; bruno.bussiere@uqat.ca (B.B.); thomas.genty@agnicoeagle.com (T.G.)
* Correspondence: abdellatif.elghali@uqat.ca (A.E.); mostafa.benzaazoua@uqat.ca (M.B.)

Received: 14 January 2019; Accepted: 12 March 2019; Published: 14 March 2019

Abstract: At some orphaned and abandoned mine sites, acid mine drainage can represent a complex challenge due to the advanced tailings' oxidation state as well as the combination of other factors. At the field scale, several parameters control sulfides' oxidation rates and, therefore, the acidity generation. The objective of this paper is to map the acidity and geochemical properties of oxidized tailings within a closed tailings storage facility. Based on systematic sampling, various geochemical parameters were measured within the oxidized Joutel tailings, including the: Neutralization potential, acid-generating potential, net neutralization potential, neutralization potential ratio, paste pH, thickness of oxidized, hardpan, and transition zones. The different parameters were integrated in geographical information system (GISs) databases to quantify the spatial variability of the acidity and geochemical properties of oxidized tailings. The oxidized tailings were characterized by low sulfide (mainly as pyrite) and carbonate (mainly as siderite/ankerite) contents compared to unweathered tailings. Acidic zones, identified based on paste pH, were located in the eastern portion of the southern zone and at the northern tip of the northern zone.

Keywords: acid mine drainage; geographical information systems; paste pH; siderite; multivariate analysis; spatial mapping

---

1. Introduction

Mining operations generate large volumes of finely ground non-economic rock that is referred to as tailings. These materials are characterized by fine particle size distributions (PSD), especially compared to waste rocks [1]. Tailings from base and precious metal mines often contain iron sulfide minerals [2], most commonly as pyrite (FeS_2) and pyrrhotite ($Fe_{1-x}S$, x = 0 to 0.2). While some of the tailings at underground mine operations are used as backfill to support underground excavations (25% of total tailings produced in Canadian operations) [3–5], these solid wastes are mostly stored at the surface in tailings storage facilities (TSFs).

Under humid climatic conditions, such as those present in the southern and central latitudes of Canada, TSF designs usually rely on maintaining high degrees of saturation to control the diffusion of oxygen into the tailings and thus reduce sulfide oxidation. However, if the water table drops and the tailings desaturate, the sulfides present in the unsaturated zones will be exposed to atmospheric oxygen. Under these conditions, the oxidative dissolution of sulfides could potentially result in the production of acid and the leaching of metals [6]. When the acid-generating potential (AP) in tailings is higher than their neutralization potential (NP) [2,7,8], acidic drainage waters can result. Furthermore, high concentrations of sulfates, iron, and other potentially toxic metal(oid)s (e.g., As, Co, Ni) can be released [9–11]. This phenomenon is known as acid mine drainage (AMD) or acid rock drainage (ARD).

In some cases, drainage waters can be near-neutral in pH and still contain concentrations of elements that surpass environmental standards and regulations. This phenomenon is known as contaminated neutral drainage (CND) [12] or metal leaching (ML). The cost of mine site reclamation is influenced by the type and quality of drainages produced in TSFs [13].

The process of sulfide oxidation can result in the generation of protons. This increases the acidity of pore waters and leads to the dissolution of carbonates, which buffer the acid. Depending on the geochemical conditions in the pore water, various reactions may occur, such as: Hydrolysis, precipitation, co-precipitation, and sorption [14–17]. During these reactions, the tailings' mineralogical composition can change significantly depending on the dissolution rates of sulfides, carbonates, and, to a lesser extent, silicates. As a result of oxidation, neutralization, and hydrolysis reactions, novel phases may precipitate. The most common secondary phases are ferric oxyhydroxides (e.g., goethite, ferrihydrite, lepidocrocite), gypsum, and iron sulfates. These secondary phases affect the geochemical behavior of tailings considerably [10].

Ferric oxyhydroxides are known for their high affinity to limiting contaminant mobility through mechanisms, including co-precipitation, adsorption, and substitution. Under some specific conditions, the occurrence of secondary phases may also affect the hydrogeochemical behavior in TSFs by modifying the tailings' hydrogeological properties, such as porosity. For example, the formation of hardpan within TSFs affects the water balance and flow paths [18,19]. Hardpan formation is more likely to occur at orphaned or abandoned mine sites.

Predicting and assessing the geochemical behavior of oxidized tailings is challenging relative to unweathered, fresh tailings. The geochemical behavior of oxidized tailings is determined by the reactivity and nature of novel secondary species and the residual reactive phases. Moreover, in the case of an already closed mine site, the uppermost tailings layer is generally oxidized, and the acidity is mainly already produced in some localized areas. Consequently, acidic leachates can result from two sources: (1) Ongoing sulfide oxidation, and (2) latent acidity present due to the oxidation of the ferrous iron. Due to the low permeability of tailings, the hydraulic residence time of porewaters is usually high. Thus, it may take up to several decades for the residual acidity to be flushed out completely. This phenomenon can be more complex due to the presence of hardpan close to the surface that has different hydrogeological properties [19–23]. The main objective of the study is to map the acidity and geochemical properties of the oxidized tailings. Elghali et al. (2019) [19] showed that Joutel's tailings are an acid-generating base on punctual sampling.

2. Mine Site Description

Joutel is a closed gold mine site located at the north of Val d'Or (Quebec, Canada). The mine was operated by Agnico Eagle Mines Ltd. (Toronto, ON, Canada) between 1974 and 1994 [11,19,24,25]. The gold was associated to sulfide deposit, mainly as pyrite. The gold was extracted using bulk flotation followed by concentrate regrinding-cyanidization [24]. The produced tailings were deposited in an approximately 120 ha tailings storage facility (TSF). The TSF is divided into two zones. The northern zone, which was used from 1974 to 1986, is the first zone of the tailings deposition and is characterized by a highly elevated topography compared the southern zone. The southern zone was used from 1986 to 1993. The input locations for the tailings were located at the eastern side of the south zone and that for the northern zone is unknown.

The oxidized tailings from the north and south zone showed a saturated hydraulic conductivity of about 2.42×10^{-5} cm/s and 1×10^{-4} cm/s, respectively [19]. Moreover, the air entry values analyzed by Elghali et al. (2019) [19] were between 3.5 and 75 KPa and between 25 and 80 KPa for the tailings from the south and north zone, respectively. In addition, Joutel tailings are characterized by a heterogeneous particle size; their D_{90} (which corresponds to 90% passing on the cumulative particle size distribution curve) varied between 37 and 164 µm [19].

Joutel's tailings were recently studied by Elghali et al. (2019) [19] to evaluate their acid generating potential. The main conclusions of this study showed that Joutel's oxidized tailings could be acid generating if the leachates are not mixed with the unweathered tailings pore water, which means that the geochemistry of Joutel's tailings is mainly controlled by the reactivity of the oxidized tailings. This study showed that the hardpan formed at the interface between oxidized and unweathered tailings may limit vertical water infiltration.

3. Materials and Methods

The methodologies used in this study combined geochemical, spatial, and statistical techniques to better understand the geochemical processes occurring within Joutel's TSF. The methodologies used in this study are illustrated in Figure 1.

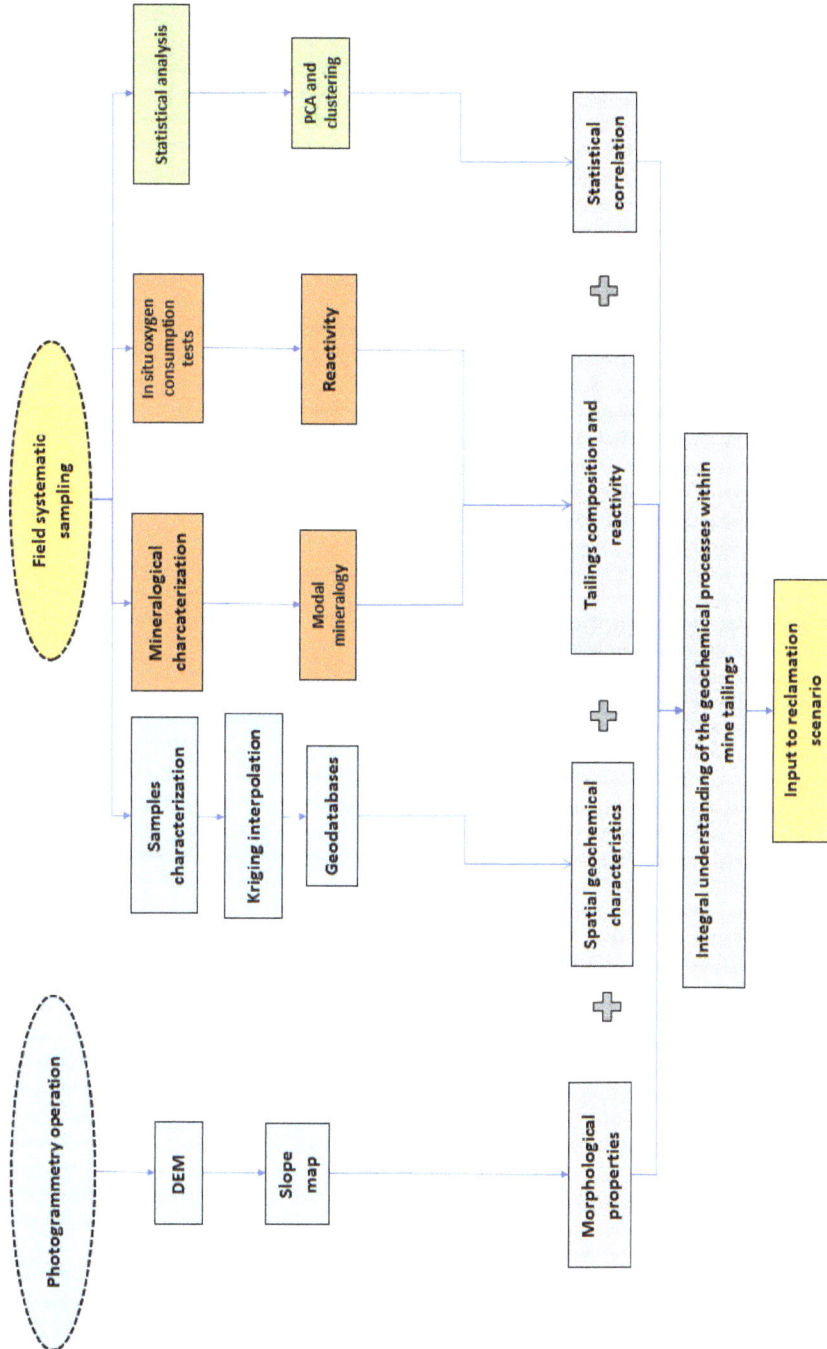

Figure 1. Schematic illustration of the methodological approaches used in this study.

3.1. Materials Sampling and Methodology

Sampling is a critical first step in the process of predicting the acid-generation potential (AGP) of mine wastes. The choice of a sampling strategy depends principally upon the objectives of the study. This may account for differences in the sampling strategies used in different studies on evaluating the AGP of mine wastes [11,12,24,26,27]. Various strategies have also been developed for evaluating soil and water pollution [28–32] and could be used for assessing the AGP of oxidized tailings considering that the particle sizes of mine tailings are comparable to fine soils. The sampling strategy used for this study was systematic [30,31,33]. A square sampling grid was constructed with lines oriented east-west and a line spacing of about 100 m; this configuration balanced restrictions due to cost with requirements for sampling resolution (Figure 2). Depending on the field constraints (e.g., presence of vegetation, streams), a predefined sampling point could be slightly moved. Oxidized tailings were sampled using a manual auger at each sampling point. Trenches were dug into the unweathered tailings to measure the thickness of the oxidation zone, hardpan, and transition zone. Each sample point was homogenized separately and submitted to several analyses. Approximately 1 kg of oxidized tailings was collected over the entire oxidized horizon from each point and each sample was spatially referenced using a GPS system. The sampling depth was variable depending on the thickness of the oxidized horizon; the depth of each sampling point is indicated in Table S1. Finally, a total of 122 samples were collected from the two tailings' storage facilities. The main reason of sampling only the oxidized tailings for the spatial mapping is that the geochemistry of Joutel's tailings is mainly controlled by the upper oxidized tailings due to the presence of hardpan layers [19]. Some areas at the south zone were not sampled due to the presence of more than 50 cm of water.

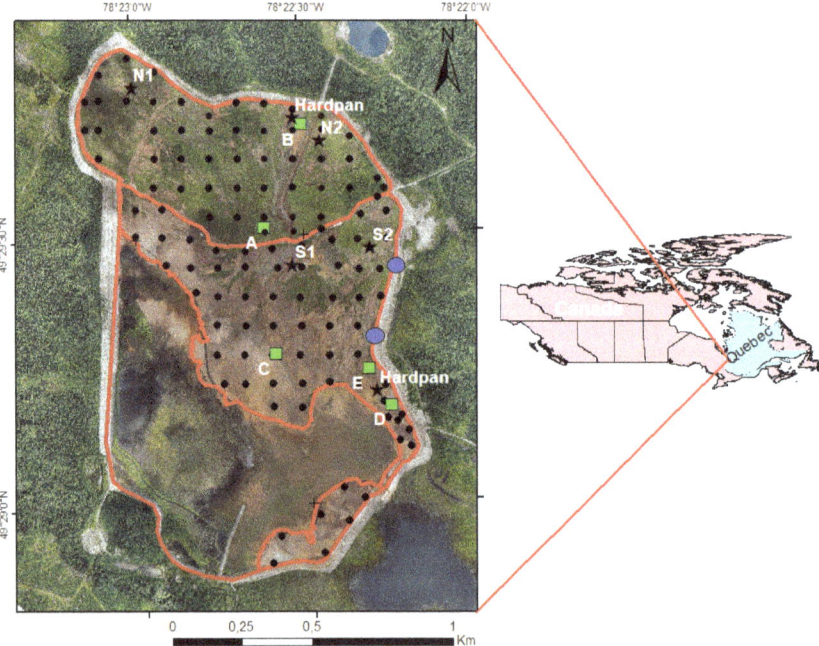

Figure 2. Map showing the sampling grid at the Joutel mine site. N1, N2, S1, and S2 refer to oxygen consumption test locations done on tailings, hardpan (north) and hardpan (south) correspond to oxygen consumption tests done on hardpan, and the green tiles correspond to samples analyzed using QEMSCAN. Red polygon corresponds to the TSF dam and blue circles correspond to the input locations of tailings.

3.2. Methods

3.2.1. Geochemical and Mineralogical Characterization

Total sulfur and total carbon were measured using an induction furnace (ELTRA CS-2000) with a detection limit of 0.09%. Sulfates within solid samples were analyzed after solid digestion using 40% HCl. Mineralogy of samples was investigated using the quantitative evaluation of materials by scanning electron microscopy (QEMSCAN). QEMSCAN is an automated system that produces particle maps (color coded by mineral) through the collection of rapidly acquired X-rays. The maps and corresponding data files quantify the modal mineralogy, texture, grain size, elemental deportment, and liberation of samples analyzed. The polished sections were analyzed by PMA (particle mineralogy analysis) mode. More than 10,000 grains were analyzed for each polished section to ensure enough data for good statistical representation. The measurement resolution varied from 2.5 µm to 6 µm depending on the particle size. A species identification protocol (SIP) specifically designed for the mineralogy was used for data processing [34,35]. The data enabled quantification of the mineralogical composition and mineral liberation of carbonates and sulfides. Five locations (A, B, C, D, and E) were sampled using a trench to identify the different tailings' horizons (oxidized tailings, hardpan, and unweathered tailings). Depending on the thickness of the oxidized horizon, one or two samples were collected. A total of 18 samples were analyzed using QEMSCAN. The different locations and samples are described in Table 1. The modal mineralogy was analyzed for the 18 samples and the mineral liberation of sulfide and carbonate was analyzed for oxidized tailings, hardpan, and unweathered tailings from station B and E.

Table 1. Description of the samples collected for the mineralogical characterization using QEMSCAN.

Tailings Storage Facility	Station	X	Y	Horizon	Depth (cm)	Description
North zone	A	689,948	5,485,450	A-Oxy	0–12	Oxidized tailings
				A-hard	12–27	Hardpan
				A-Unw1	27–47	Unweathered tailings
				A-Unw2	47–100	Unweathered tailings
	B	690,078	5,485,806	B-Oxy	0–18	Oxidized tailings
				B-hard	18–33	Hardpan
				B-Unw1	33–65	Unweathered tailings
				B-Unw2	65–110	Unweathered tailings
South zone	C	689,993	5,485,014	C-Oxy1	0–10	Oxidized tailings
				C-Oxy2	10–20	Oxidized tailings
				C-hard	20–40	Hardpan
				C-Unw	40–55	Unweathered tailings
	D	690,402	5,484,841	D-Oxy	0–15	Oxidized tailings
				D-hard	15–30	Hardpan
				D-Unw	30–45	Unweathered tailings
	E	690,320	5,484,967	E-Oxy	0–15	Oxidized tailings
				E-hard	15–30	Hardpan
				E-Unw	30–45	Unweathered tailings

3.2.2. Acid Generation Potential Assessment

The paste pH of solid samples was analyzed using a pH meter after adding 5 mL of deionized water to 2.5 g of tailings; this method has a precision of ±0.02 units. The acid generation potential (AP) of each sample was calculated using the sulfide content (AP = 31.25 × %S-sulfide) [36,37]. The neutralization potential (NP) was calculated using the carbon content and using the Sobek method (NP = 83.3 × %C-carbonates) [7,36,38–40] with a correction factor of 50% applied due to presence of Fe-Mn carbonates. These phases overestimate the NP as calculated based on the carbon content (Figure S1); further details are provided in Section 3. The relative error associated with NP values determined by titration is approximately ±12 kg CaCO$_3$/t [41] and that related to the AP calculation is approximately ±3 kg CaCO$_3$/t. The net neutralization potential (NNP) is defined as the difference between the NP and the AP. The neutralization potential ratio (NPR) is defined as the ratio between the NP and the AP. Net acid generation (NAG) tests were performed on 2.5 g of pulverized tailings in 250 mL of 15% H_2O_2. In these tests, the samples are allowed to react until effervescence ceases [42–44]. The pH of the liquor is then analyzed and the sample is considered acid-generating if the final pH is <4.5 and non acid-generating if the final pH is >4.5 [45].

3.2.3. Field Oxygen Consumption Tests

Oxygen consumption tests were used to determine the sulfide oxidation rates in the tailings [26,46–48]. High oxygen consumption means a high sulfide oxidation rate. In this study, six locations were chosen to evaluate in situ oxygen consumption rates: Four locations within the oxidized tailings (Figure 2) and two locations within hardpan layers (Figure 2: Hardpan). These tests involved installing aluminum cylinders with known dimensions in the tailings to form a closed system (~10 cm deep) (Figure 3A). For the hardpan locations, the cylinder was embedded using a drill (Figure 3C). The cylinders were covered with a plastic cap equipped with an oxygen sensor [46] (Figure 3B) and oxygen concentrations were logged for five days. The data was interpreted only for a short duration (3 h). Oxygen fluxes were calculated using the graphical method described in Mbonimpa et al. (2011) [48].

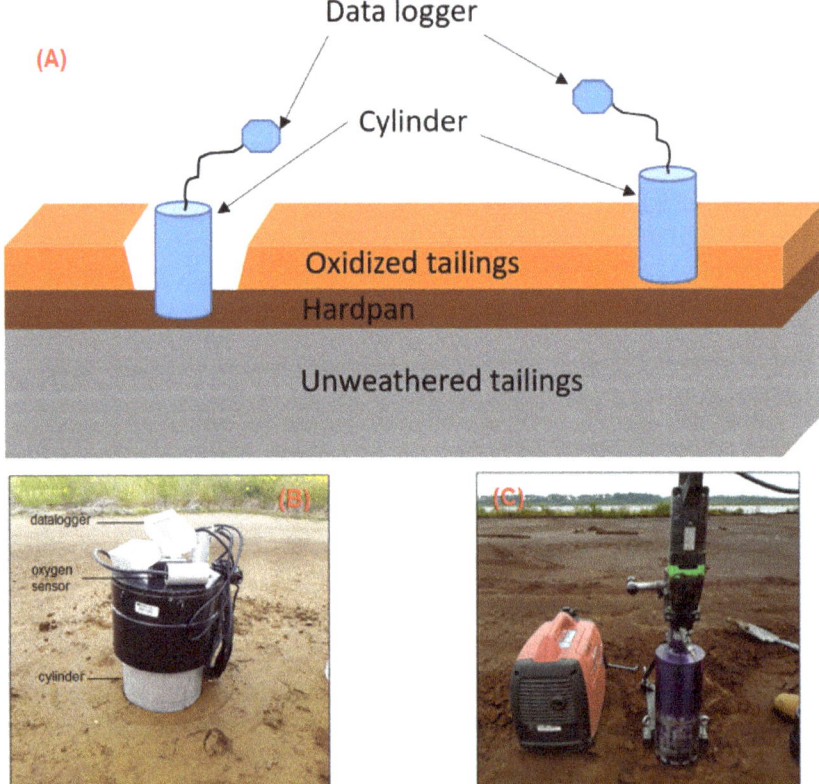

Figure 3. Images showing design of oxygen consumption tests. (**A**) schematic illustration of oxygen consumption tests of tailings and hardpan, (**B**) datalogger and oxygen sensor, (**C**) drill used to the cylinder in the hardpan.

3.2.4. Spatial Mapping and Statistical Analysis

Due to differences in the deposition ages of the tailings in the northern and southern portions of the Joutel TSF, each zone was interpolated separately based on the systematic sampling approach (Section 3.1). During this study, various analyses were performed within the oxidized tailings zones. A total of 11 geodatabases were established using ArcGIS 10.3.1. The geodatabases produced for each zone were: C wt.%, total S wt.%, $S_{sulfides}$ wt.%, paste pH, NP, AP, NNP, NPR, oxidized horizon thickness, transition zone thickness, and hardpan thickness. Other geodatabases, such as the slope map and the flow accumulation map, were produced for both zones using a digital elevation model with a 21-cm per pixel resolution produced using photogrammetry.

The included parameters were interpolated through kriging [29,49–51] with the geostatistical wizard of ArcGIS 10.3.1. Kriging, or interpolation method accuracy, was performed using the validation technique. Random sampling was done after data interpolation. Data were plotted on the interpolated map, and then the values estimated by kriging were compared to the measured values. Validation was performed only for the paste pH, as it was one of the most important parameters and the best reflection of the acidity of the oxidized tailings. A total of 25 points were used to validate the interpolation method. Descriptive statistics and principal component analysis (PCA) were done using the XL-stat extension for Microsoft Excel [52]. The results of the kriging validation are presented in Figure 4 and confirmed a reliable data interpolation. Among the 25 measured points, only two samples were considered as outliers ($\alpha = 0.05$).

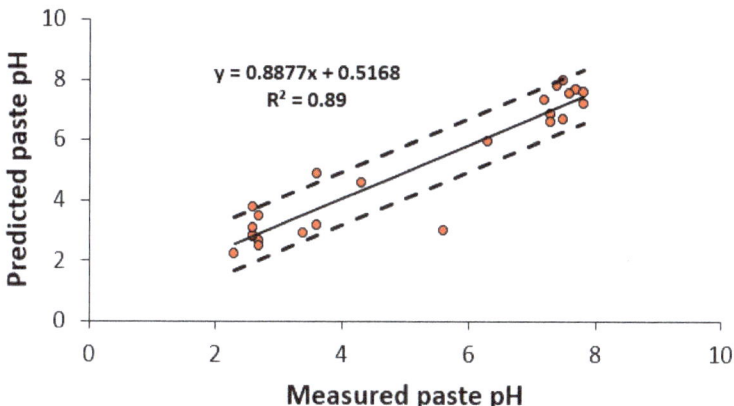

Figure 4. Results of kriging validation: measured paste pH vs predicted paste pH by kriging.

4. Results and Discussion

4.1. Results

4.1.1. Mineralogical Characteristics and the Effect of Siderite/Ankerite on the Tailings' Neutralization Potential

Mineralogical Characteristics

The results of the modal mineralogy analysis of the studied samples are shown in Figure 5. The studied samples show high spatial variability in their mineralogical composition. Sulfide species detected within the different samples included pyrite as the main sulfide mineral, while pyrrhotite, chalcopyrite, and arsenopyrite were in trace concentrations. Carbonate species analyzed within the different samples were mainly Fe-Mn carbonates (siderite and ankerite) and, to a lesser extent, Ca-Mg carbonates (calcite and dolomite).

Unweathered samples showed sulfide contents between 7 and 27 wt.% for the north zone (stations A and B) and between 11 and 21 wt.% for the south zone (stations C, D, and E). Fe-Mn carbonate content was between 21 and 39 wt.% for the north zone and between 31 and 45 wt.% for the south zone. However, Ca-Mg carbonate contents, which are the main minerals responsible for acidity buffering during AMD formation, were ≤7 wt.% for the analyzed samples. The hardpan samples showed a sulfide content around 6 wt.% for the north zone and between 6 and 20 wt.% for the south zone. Gypsum content, which is the result of calcium and sulfates precipitation, was between 2 and 15 wt.% and Fe-oxy-hydroxides content was between 2 and 26 wt.%. The oxidized tailings showed a different mineralogical composition compared to that of the hardpan and unweathered tailings. The oxidized tailings contained high concentrations of Fe-oxy-hydroxides (6–35 wt.%). Their sulfide content was between 0.5 and 27 wt.% and Ca-Mg-carbonate was ≤12 wt.%.

Other minerals detected within the oxidized, hardpan, and unweathered tailings were mainly plagioclase, sericite/muscovite, quartz, and chlorite/clays. The detailed mineralogical composition of these samples is presented in Figure 5.

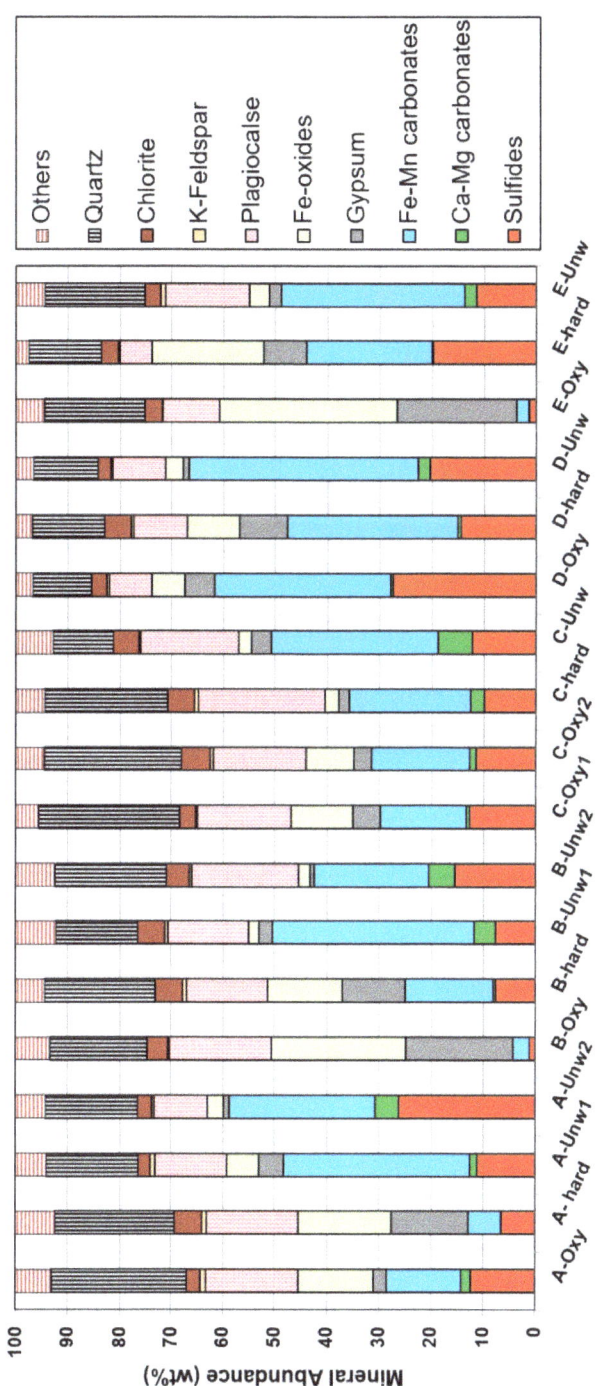

Figure 5. Mineralogical composition of the different stations' horizons.

Mineral Liberation

Mineral liberation is a textural parameter, recognized as a key factor that influences the reactivity of mine waste [34,35,53]. Figure 6 presents results of the pyrite, pyrrhotite, calcite-dolomite, and ankerite-siderite liberation degree. Pyrite was almost liberated (exposed) for all analyzed samples except for oxidized tailings from stations B and E (Figure 6A), where the exposed pyrite was about 10 wt.%. Pyrrhotite was less liberated compared to pyrite (Figure 6B); exposed pyrrhotite was less than 25 wt.% for all the analyzed samples. The exposed part of the calcite-dolimite minerals varied between 2 and 62 wt.% for the analyzed samples and their lowest liberation degree was analyzed within the oxidized tailings from stations B and E. Ankerite-sedirite liberation dergree is illustrated in Figure 6D; their exposed samples varied between 7 and 72 wt.%.

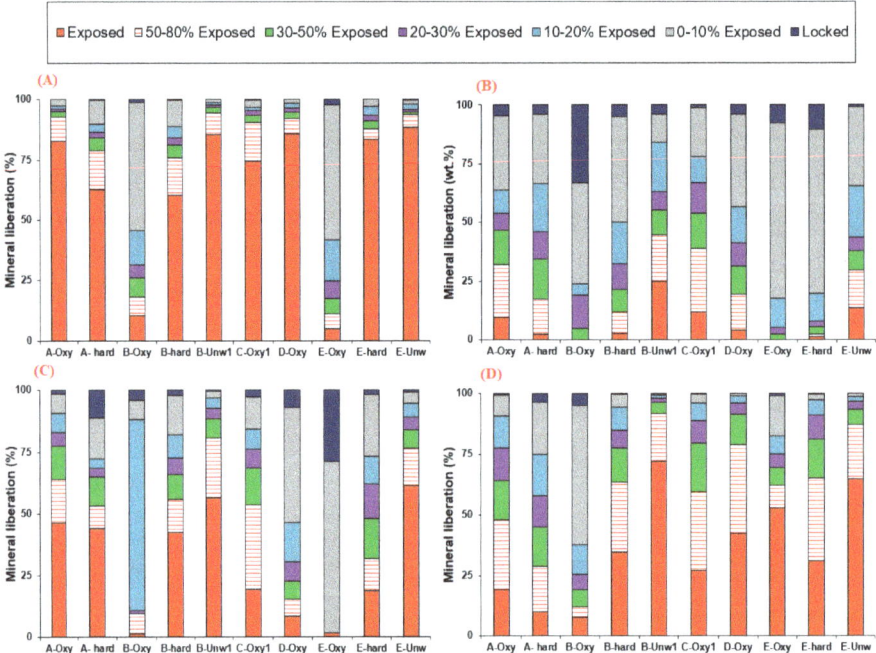

Figure 6. Mineral liberation of pyrite (**A**), pyrrhotite (**B**), calcite-dolomite (**C**), and ankerite-siderite (**D**).

Effect of Fe-Mn Carbonates on the Tailings' Neutralization Potential

The presence of siderite and ankerite leads to the overestimation of the tailings' NP when it is calculated based on the carbonate content. This was confirmed by comparing the NP calculated based on carbonate contents with that determined using the Sobek method (1978) [36] as modified by [54] (Figure S1). The overestimation of NP based on carbonate contents ranged between 10% and 54%. This difference was due to the presence of siderite and ankerite as the major carbonates. The neutralization potential of Fe-rich carbonates is balanced by the acidity due to the oxidation of Fe^{2+} to Fe^{3+} and subsequent precipitation of Fe^{3+} as ferric oxyhydroxide phases [37,55–57]. Therefore, Fe-rich carbonates do not provide an additional buffering capacity [37,55]. For this reason, the NP of the Joutel tailings was corrected conservatively using a factor of 50% for all samples in this study; i.e., $NP = 0.5 NP_{carbonates}$.

4.1.2. Site Topography and Slopes

The results of the DEM and slope map are presented in Figure 7. The northern zone is characterized by high elevations compared to the southern zone (Figure 7A); the maximum elevation in the northern and southern zones are ~306 m and 280 m, respectively. The elevation decreases from north to south and from east to west (Figure 7). The slope within the Joutel TSF (Figure 7B) is weak and mostly lower than 8°, except in a few areas where the slope values are higher than 80° (streams and channels). Therefore, surface runoff is greatly influenced by the slope. These higher slope values correspond to streams. Combining the DEM and slope maps, the surface runoff flow directions are oriented NE-SW. Moreover, surface runoff is enhanced by the presence of several streams.

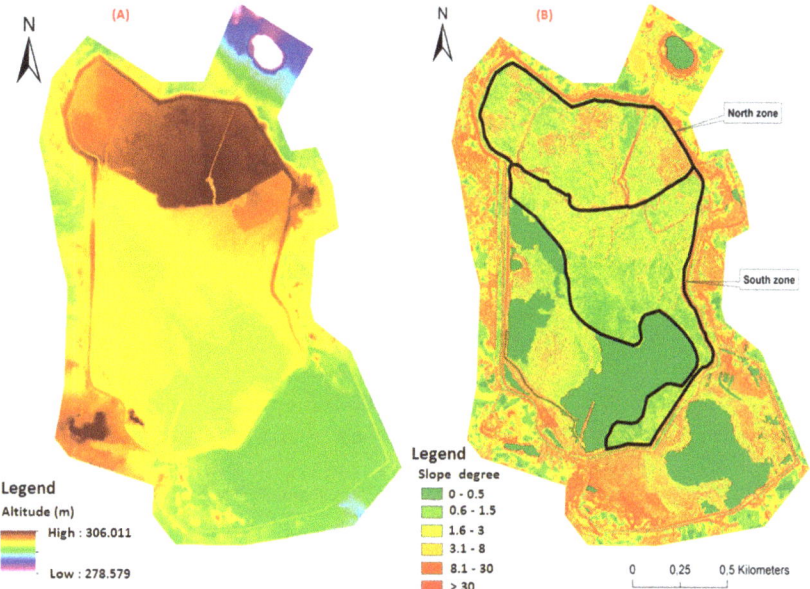

Figure 7. (**A**) Digital elevation model and (**B**) slope map within the Joutel TSF.

4.1.3. Paste pH, Oxidized Horizon, Hardpan, and Transition Horizon Thickness

The paste pH mapping, illustrated in Figure 8A, showed high spatial variability within the oxidized tailings. In the northern zone, only some localized areas are characterized by acidic paste pH values (<3.2), while most of the northern zone is characterized by neutral paste pH values (>6). The southern zone is characterized by acidic paste pH values located at the eastern part. In general, the extent of the acidity is much higher in the southern zone than in the northern zone.

Hardpan is formed at the interface between oxidized tailings and unweathered tailings. At the Joutel site, hardpan is observable throughout the northern zone and in parts of the southern zone (Figure 8B). In the southern zone, hardpan appears in the eastern portion only and varies in thickness between 1 cm and 15 cm. In the north zone, the hardpan occurrence was observed at all the sampling stations and its thickness was between 1 cm and 10 cm. The occurrence of hardpan appears to be associated with elevation and, more specifically, the water table level. Higher topographic levels lead to deeper water table levels and more unsaturated conditions at the surface of the tailings. This results in increased sulfide oxidation and secondary phase precipitation [58]. Consequently, the western portion of the southern zone, which is characterized by lower elevations and nearly saturated conditions, shows less hardpan formation.

The oxidized layer thickness ranged from a few centimeters to more than 25 cm (Figure 8C). The maximum observed oxidized layer thickness occurred in western portions of the northern zone. In general, the oxidation layer thickness was higher in the northern zone than it was in the southern zone. This could potentially be explained by two factors: (i) The age of the tailings deposition, and (ii) the thickness of the unsaturated zone, which is influenced by the site's topography. The northern zone is older than the southern zone and is characterized by higher elevations (Figure 7A), which favor unsaturated conditions and sulfide oxidation. The transition zone, which corresponds to the unsaturated zone, is the layer between oxidized and unweathered tailings. The transition zone thickness ranged between 2 and more than 26 cm (Figure 8D). Generally, the transition layer was thicker in the northern zone than in the southern zone.

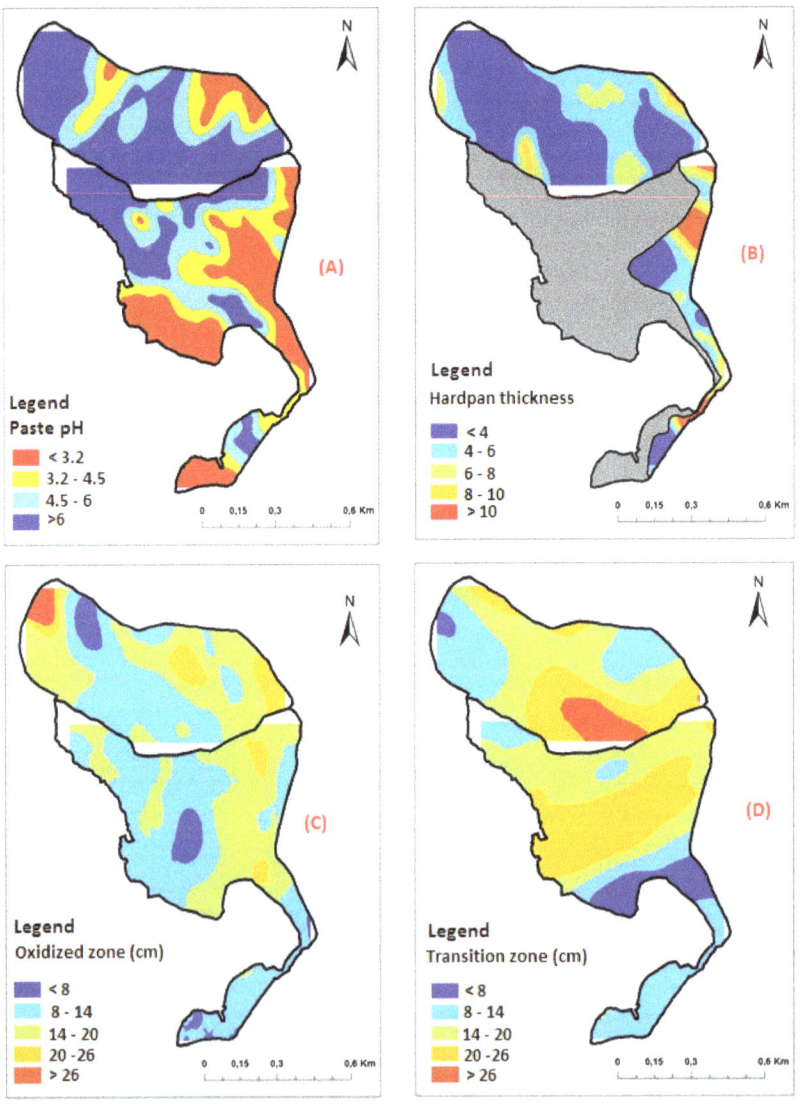

Figure 8. Mapping of: (**A**) paste pH, (**B**) hardpan thickness, (**C**) oxidized zone, and (**D**) transition zone.

4.1.4. S-Sulfide, C-Carbonates, NNP, and NPR

Maps of the total $C_{inorganic}$ and $S_{sulfide}$ analyses are presented in Figure 9A,B, respectively. The distribution patterns of $C_{inorganic}$ and $S_{sulfide}$ contents agree with the observations of the oxidized layer thickness. Carbon occurs primarily in carbonates, which were nearly depleted in the eastern portion of the southern zone and the western portion of the northern zone, where the carbon content was less than 1 wt.%. Carbon content in the northern zone decreased from north to south. This spatial distribution was not the same in the southern zone; carbon content decreased from south-east to north-west. Carbonate depletion appears to also be associated with the topography of the TSF (Figure 7A). Carbon content ranged between ≤0.09 wt.% and more than 4 wt.%.

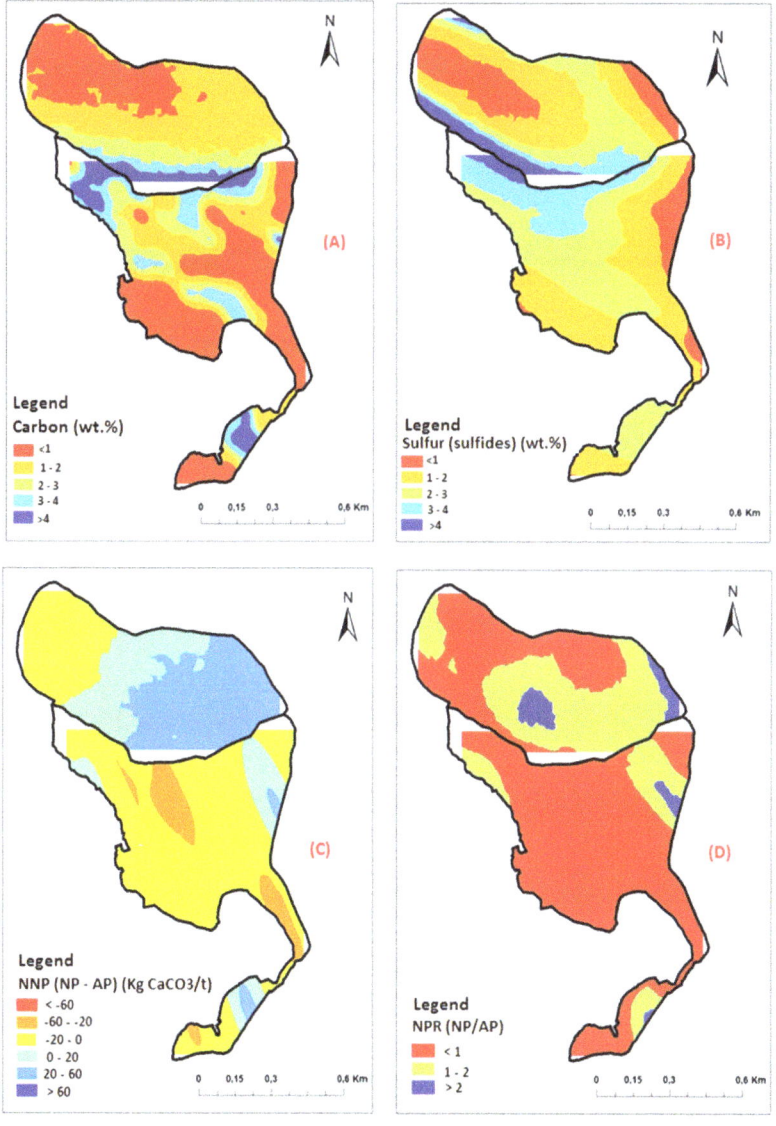

Figure 9. Results of the mapping of: (**A**) $C_{inorganic}$, (**B**) $S_{sulfide}$, (**C**) NNP, and (**D**) NPR.

The $S_{sulfide}$ distribution, which corresponds to the residual acid potential of oxidized tailings, is illustrated in Figure 9B. This distribution can be compared to that of the total $C_{inorganic}$. The eastern portion of the southern zone and the western portion of the northern zone are characterized by high sulfide depletion (<1 wt.%). In other locations, the $S_{sulfide}$ content (mainly pyrite) is higher than 2 wt.%. In the southern zone, the $S_{sulfide}$ content increases from south-east to north-west, while in the northern zone, it increases from north to south. In general, sulfide oxidation seems to be faster and more complete within the eastern portion of the southern zone and the extreme western portion of the northern zone. This could possibly be attributed to the tailings' degree of saturation, which is affected by the irregular topography of the TSF, as well as the heterogeneous particle size distributions of tailings across the site (Figure 7). It is recognized that the particle size influences the reactivity of mine wastes [34,35,59].

The NNP and NPR were used to classify the acid-generation potential of the oxidized Joutel tailings. The mapping of the NNP is illustrated in Figure 9C. In the southern zone, almost all samples displayed NNP values ranging between −20 and 0 kg $CaCO_3$/t, with the exception of a few locations that had positive values. Using this criterion, samples from the southern zone were uncertain samples; only a few locations could be classified as acid generating (NNP lower than −25 kg $CaCO_3$/t). The northern zone showed higher NNP values; this zone could be divided into three parts from east to the west. The first part, at the far east, is characterized by positive NNP values ranging between 20 and 60 kg $CaCO_3$/t; the second part, in the center, is characterized by NNP values ranging between 0 and 20 kg $CaCO_3$/t; and the third part, at the far west, is characterized by NNP values ranging between −20 and 0 kg $CaCO_3$/t.

A map of the NPR values is shown in Figure 9D. The southern zone showed NPR values <1 except in a few locations. In general, tailings in the southern zone could be classified as acid-generating. The northern zone is characterized by a higher extent of zones with NPR values ranging between 1 and 2. Acid-generating tailings in the northern zone are mostly located in the east.

4.1.5. Oxygen Consumption Tests

Oxygen consumption tests allow for the in situ evaluation of sulfide reactivity [26]. Calculation of oxygen fluxes consisted of oxygen consumption tests [60]. Decreases in oxygen concentrations in a sealed chamber were converted to oxygen fluxes using the fundamental gas diffusion law, or Fick's law [26,46,47,60]. The relationship between the decrease in oxygen concentration and oxygen flux is expressed as:

$$F_l = C_0 \times (K_r \times D_e)^{0.5} \quad (1)$$

where K_r is the first order reaction rate coefficient, D_e is the effective diffusion coefficient, and C_0 is the initial oxygen concentration at $t = 0$ [60]. Solving this equation is expressed as followed:

$$Ln\left(\frac{C}{C_0}\right) = -t(K_r \times D_e)^{0.5} \times \frac{A}{V} \quad (2)$$

where A and V are the area and volume of the cylinder headspace, respectively.

The slope of the plot of $Ln (C/C_0)$ versus time is $(K_r \times D_e)^{0.5}$. The interpretation procedure, as described in Mbonimpa et al. (2003) [47], requires relatively short-duration tests (~180 min). The results of the oxygen flux calculations are shown in Table 2. Oxygen fluxes in oxidized tailings from the southern zone varied from 83 to 162 mol/m²/year, while fluxes in oxidized tailings from the northern zone varied between 13 and 71 mol/m²/year. Oxygen fluxes in hardpan from the southern and northern zones were 63 and 42 mol/m²/year, respectively. Oxygen fluxes at the six locations were low compared to observations of other similar sulfidic tailings [61–63]. This was expected due to the low sulfide content within the oxidized tailings. In contrast, the hardpan locations also presented low oxygen fluxes despite their high sulfide contents. This is likely due to their microstructure (low porosity) and texture [19]. Sulfides within the hardpan are almost entirely coated by ferric oxyhydroxides, which protects them by limiting oxygen diffusion to their unreacted cores. Hardpans are also characterized by low porosities and cementitious textures [58].

Table 2. Results of in situ oxygen consumption tests on oxidized and hardpan samples.

	Slope	R^2	A (m^2)	V (m^3)	$(K_r \times D_e)^{0.5}$	F_l (mol/m^2/year)
S1	-8.00×10^{-5}	0.99	2.19×10^{-2}	4.89×10^{-2}	3.58×10^{-5}	162
S2	-7.00×10^{-5}	0.95	2.19×10^{-2}	8.28×10^{-2}	1.85×10^{-5}	83
N1	-1.00×10^{-5}	0.96	2.19×10^{-2}	7.64×10^{-2}	2.87×10^{-6}	13
N2	-5.00×10^{-5}	1.00	2.19×10^{-2}	6.99×10^{-2}	1.57×10^{-5}	71
Hardpan (South)	-6.00×10^{-5}	0.99	2.19×10^{-2}	9.37×10^{-2}	1.40×10^{-5}	63
Hardpan (North)	-4.00×10^{-5}	0.95	2.19×10^{-2}	9.37×10^{-2}	9.35×10^{-6}	42

4.1.6. Multivariate Analysis

In this study, paste pH in this study was used as a criterion to indicate the acidity of the upper oxidized tailings. To determine the key parameters that affect paste pH and the relationships among these different geochemical parameters, PCA was performed using four other parameters: AP, NP, NNP, and NPR. Correlation matrices for the northern and southern zones are shown in Table 3. Paste pH positively correlated with both AP and NP; the highest correlation was with NP. Linear correlation values between NP and paste pH were 0.86 and 0.62 for the southern and the northern zones, respectively. Despite the calculations for NNP and NPR being based on NP and AP values, their correlations with paste pH were lower (maximum of 0.45). Using only the two first factors (F1 and F2), the cumulative variability was approximately 94.50% and 83.70% for the southern and northern zones, respectively. Factor F1 explained 64.60% and 47.40% of the total variance in the paste pH in the southern and northern zones, respectively. Similarly, factor F2 explained 29.90% and 36.30% of the total variance in the paste pH in the southern and northern zones. The results of the correlation values were more significant within PCA (Figure 10A,B). Paste pH, NP, and AP are in the same quadrant of the circle of PCA. There are high values of paste pH and NP in F1 and significant loading of AP, NNP, and NPR in F1 within the data of the two zones, which is confirmed by the correlation matrix between the factors and variables (Table 4). Concerning F2, there is a high value of AP in this component within the two zones and negative values of NNP and NPR in this component (Table 4). The high value of the paste pH and NP in the F1 indicates as excepted that they are interdependent. High values of the paste pH involve high values of NP and vice versa; acidic paste pH values are due to sulfide oxidation and subsequent carbonate dissolution. Consequently, the principal component, F1, could be interpreted as the NP. The high loading of AP in the F2 and the negative loading of the NNP and NPR indicates that these parameters are negatively correlated. This is explained by the initial calculation of NNP (NP-AP) and NPR (NP/AP). Consequently, the second principal component, F2, could be interpreted as the AP.

The biplots in Figure 10C,D show all tailings samples on a plot of F1 versus F2. Samples were split into three groups for each zone. The differences between samples are due to variations in their NP (F1) and AP (F2). Some samples in the northern and southern zones do not fit into the identified groups and could be considered outliers. Group 1 (Gr1) is comprised of samples with AP values greater than 80 kg CaCO$_3$/t, NPR values lower than 0.5, NNP values lower than -65 kg CaCO$_3$/t, and acidic paste pH values. Group 2 (Gr2) is comprised of samples with AP values lower than 30 kg CaCO$_3$/t, NPR values lower than 0.45, NNP values lower than -30 kg CaCO$_3$/t, and paste pH values lower than 3.30. Group 3 is comprised of samples with paste pH values greater than 6, AP values greater than 90 kg CaCO$_3$/t, NNP values ranging between -40 and 3 kg CaCO$_3$/t, and an average NPR of 1. The group 1 is located at the eastern and northern part of south and north zone of the TSF, respectively. The group 2 is located at the central part of the south zone and the north zone. The group 3 is located at western zone and the south part of the south and the north zone, respectively.

Table 3. Correlation matrices relating paste pH with AP, NPR, NNP, and NP for the southern and northern zones ($p < 0.05$).

	Paste pH (-)	AP (kg CaCO$_3$/t)	NPR (-)	NNP (kg CaCO$_3$/t)	NP (kg CaCO$_3$/t)
Southern Zone (n = 66)					
Paste pH (-)	1				
AP (kg CaCO$_3$/t)	0.777	1			
NPR (-)	0.457	0.025	1		
NNP (kg CaCO$_3$/t)	0.408	−0.056	0.868	1	
NP (kg CaCO$_3$/t)	0.861	0.681	0.655	0.693	1
Northern Zone (n = 56)					
Paste pH (-)	1				
AP (kg CaCO$_3$/t)	0.517	1			
NPR (-)	0.200	−0.495	1		
NNP (kg CaCO$_3$/t)	0.297	−0.200	0.531	1	
NP (kg CaCO$_3$/t)	0.623	0.539	0.105	0.717	1

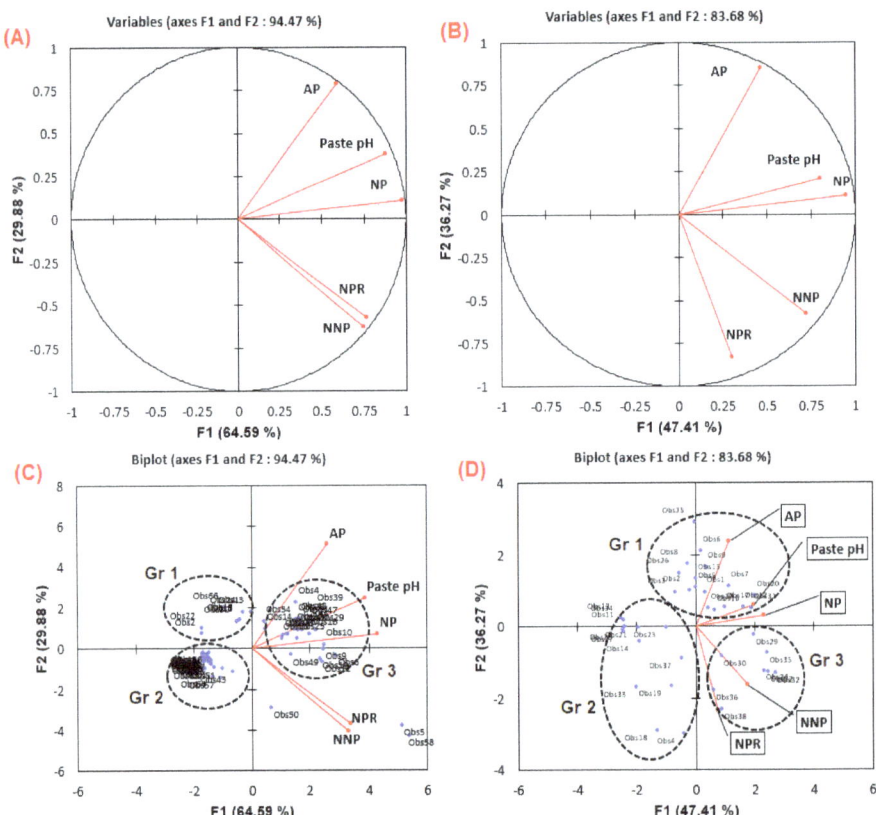

Figure 10. PCA plots for: (**A**) the southern zone and (**B**) the northern zone. Clustering of samples for: (**C**) the southern zone and (**D**) the northern zone.

Table 4. Correlations between variables and factors for the southern and northern zones ($p < 0.05$).

	Southern Zone (n = 66)			
	F1	F2	F3	F4
Paste pH (-)	0.879	0.379	0.128	0.259
AP (kg CaCO$_3$/t)	0.595	0.787	−0.011	−0.163
NPR (-)	0.765	−0.571	0.268	−0.131
NNP (kg CaCO$_3$/t)	0.748	−0.628	−0.207	0.058
NP (kg CaCO$_3$/t)	0.978	0.107	−0.160	−0.075
	Northern Zone (n = 56)			
	F1	F2	F3	F4
Paste pH (-)	0.805	0.211	0.499	−0.242
AP (kg CaCO$_3$/t)	0.463	0.854	0.042	0.233
NPR (-)	0.302	−0.832	0.389	0.256
NNP (kg CaCO$_3$/t)	0.720	−0.580	−0.369	−0.096
NP (kg CaCO$_3$/t)	0.948	0.109	−0.288	0.083

4.2. Discussion

The sampling strategy used in this study consisted of systematic sampling with a sample spacing of about 100 m. This allowed the mapping of the spatial variability of the geochemical properties of the oxidized tailings at a resolution of 100 × 100 m. However, depending on the objective of the study and sampling costs, adjustments could be made to the spatial resolution of the sampling without compromising the mapping resolution. This methodology allowed delimiting the problematic zones, which are characterized by an acidic paste pH. Furthermore, corresponding to the conceptual model discussed in Elghali et al. (2019b) [19], which stated that the geochemistry of Joutel's tailings is mainly controlled by the reactivity of the oxidized tailings (Figure S2), mapping the oxidized tailings' depth could be helpful to calculate the volume of the acidic tailings to be treated.

Joutel's oxidized tailings showed high spatial variability regarding their chemical and mineralogical properties. The TSF displayed zones with completely depleted sulfur-sulfide and carbon-carbonates and other zones with moderate carbon-carbonates and sulfur-sulfide (≥2 wt.%). The analyzed samples using QEMSCAN showed a variable mineralogical composition regarding their sulfide and carbonate contents, which explains the chemical variability of these samples. Furthermore, sulfides within Joutel tailings are not completely liberated (Figure 6). The sulfide and carbonate associations with non-sulfide gangue minerals reduce their reactive amount [34,35]. An average of 11% and 16% of Ca-Mg carbonates and pyrite are less than 50% liberated, respectively. During sulfide oxidation and carbonate dissolution, iron oxy-hydroxides may precipitate at the surface of sulfides and carbonates, which considerably reduces their liberation degree. This phenomenon of sulfide and carbonate coating is widely observed within reactive and oxidized tailings [10,18].

The chemical and mineralogical variability could be explained based on two parameters, which are: (i) The heterogeneity of tailings' physical characteristics (particle size distribution and specific surface area) caused by tailings segregation during their deposition [64], and (ii) the initial heterogeneity of the mineralogical characteristics of the ore body. Moreover, oxidized tailings and hardpans samples showed a relatively similar mineralogical composition regarding secondary minerals contents, but a different texture. As stated in Elghali et al. (2019) [19], the hardpan is a cementitious layer, with low saturated hydraulic conductivity values, formed at the interface of oxidized and unweathered tailings.

Paste pH is generally considered to be a static test and used to assess the instantaneous acid-base balance. These tests do not provide any information about reaction kinetics for unweathered tailings [65]. To better understand the long-term behavior of oxidized samples, NAG tests were performed on several randomly selected samples. The combination of paste pH and NAG pH could indicate the long-term behavior of these samples [65,66]. Samples with extremely acidic paste pH values (lower than 4) were classified as immediately acid-generating. In contrast, several samples with neutral paste pH values, but acidic NAG pH values were classified as potentially acid-generating, meaning that they will generate acidity after a lag time (Figure 11). This lag time corresponds to the period during which the Fe and Mn released by carbonate dissolution becomes oxidized. Based on these tests, Joutel's TSF can be considered potentially acid-generating in the long-term.

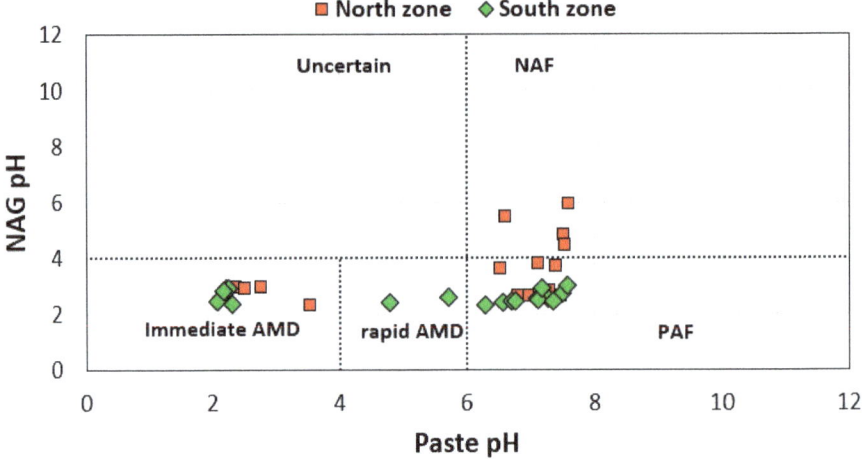

Figure 11. Paste pH vs NAG pH values for selected samples from the southern and northern zones (NAF = Non-acid forming, PAF = Potentially acid forming).

In general, the eastern side of the south zone is characterized with acidic paste pH and almost depleted sulfide and carbonate content. This could be explained by the tailing's deposition history. The input locations for the tailings were located at the eastern part of this zone. Consequently, heavy minerals, such as sulfides and coarse fractions, will preferentially settle next to the deposition points. Furthermore, the eastern zone will be a reactive one compared to the other locations in the south zone.

The implementation of georeferenced databases can: (i) Improve the understanding of the geochemical processes within TSFs, and (ii) help in the design of site-wide reclamation strategies. Based on the mapping of paste pH values, the Joutel TSF was divided into two zones: An acidic zone with paste pH values lower than 6.5, and a neutral zone with paste pH values with greater than 6.5. The acidic zones correspond to areas where carbonates are almost completely depleted; this was confirmed by the high coefficient of correlation between paste pH values and $C_{inorganic}$ contents. The neutral zones are characterized by non-negligible carbonate and sulfide contents. This means that not all the tailings on the Joutel site are acid-generating and only a few locations are presently acidic.

5. Conclusions

This study highlights an innovative approach for the spatially referenced geochemical characterization of inactive tailings storage facilities using geographical information systems. The multi-technique characterization was undertaken using a systematic sampling approach. This allowed for the production of georeferenced databases that were used to advance two main objectives: (i) To integrate and correlate the geochemical properties of the oxidized tailings at the Joutel mine site, and (ii) to identify the most appropriate reclamation scenario of the TSF by taking into account the spatial variabilities of the site's geochemical properties.

The GIS-based approach adds a spatial component to traditional geochemical analyses, which is not usually incorporated when laboratory-based tests are performed alone, and allows for a better understanding of in situ conditions. For example, on-site observations of the hardpan thickness revealed variations from a few millimetres to more than 20 cm across the entire site. This spatial variability means that the water balance within the oxidized material will change depending on the location on the site. Additionally, acidic areas of the site were found to occur in the eastern portion of the southern zone and in the western portion of the northern zone. These areas appear to be limited to a few locations where the carbonates are almost completely depleted as confirmed by correlations between the spatial maps generated for paste pH and carbonates content.

Supplementary Materials: The following are available online at http://www.mdpi.com/2075-163X/9/3/180/s1. Figure S1: Graph comparing the NP calculated based on carbonate contents and NP analyzed using the Sobek method (1978) [36] as modified by Lawrence et Wang (1996) [67]. Figure S2: conceptual model illustrating the geochemical evolution of the Joutel tailings (Reproduced with permission from Elghali et al 2019) [19]; Table S1: Sampling depth of the collected samples.

Author Contributions: Conceptualization, A.E., M.B., B.B., T.G.; formal analysis and investigation, AE., T.G.; writing—original draft preparation, A.E., M.B.; writing review and editing, A.E., M.B., B.B., T.G.; writing—review and editing, A.B., M.B.; project administration, M.B., B.B.

Funding: Funding for this study was provided by Mitacs (https://www.mitacs.ca/fr) and the NSERC-UQAT Industrial Research Chair on Mine Site Reclamation and its partners.

Acknowledgments: The authors thank Chris Kennedy, Josée Noel, and Jean Cayouette for their assistance with this manuscript. The authors also thank the URSTM staff for their support with laboratory and field work.

Conflicts of Interest: The authors declare no conflict of interest.

References

1. Bussière, B. Hydro-geotechnical properties of hard rock tailing from metal mines and emerging geo-environmental disposal approaches. *Can. Geotech. J.* **2007**, *44*, 1019–1052. [CrossRef]
2. Blowes, D.; Ptacek, C.; Jambor, J.; Weisener, C.; Paktunc, D.; Gould, W.; Johnson, D. *The Geochemistry of Acid Mine Drainage*; Elsevier: Amsterdam, The Netherlands, 2014.
3. Benzaazoua, M.; Belem, T.; Bussière, B. Chemical factors that influence the performance of mine sulphidic paste backfill. *Cem. Concr. Res.* **2002**, *32*, 1133–1144. [CrossRef]
4. Benzaazoua, M.; Bussière, B.; Demers, I.; Aubertin, M.; Fried, É.; Blier, A. Integrated mine tailings management by combining environmental desulphurization and cemented paste backfill: Application to mine Doyon, Quebec, Canada. *Miner. Eng.* **2008**, *21*, 330–340. [CrossRef]
5. Belem, T.; Benzaazoua, M.; Bussière, B. Mechanical behaviour of cemented paste backfill. In Proceedings of the 53rd Candadian Geotechnical Conference, Montreal, QC, Canada, 15–18 October 2000; pp. 373–380.
6. Nicholson, R.V.; Gillham, R.W.; Reardon, E.J. Pyrite oxidation in carbonate-buffered solution: 1. Experimental kinetics. *Geochim. Cosmochim. Acta* **1988**, *52*, 1077–1085. [CrossRef]
7. Plante, B.; Bussière, B.; Benzaazoua, M. Static tests response on 5 Canadian hard rock mine tailings with low net acid-generating potentials. *J. Geochem. Explor.* **2012**, *114*, 57–69. [CrossRef]
8. Jamieson, H.E.; Walker, S.R.; Parsons, M.B. Mineralogical characterization of mine waste. *Appl. Geochem.* **2015**, *57*, 85–105. [CrossRef]
9. Moncur, M.C.; Ptacek, C.J.; Blowes, D.W.; Jambor, J.L. Release, transport and attenuation of metals from an old tailings impoundment. *Appl. Geochem.* **2005**, *20*, 639–659. [CrossRef]
10. McGregor, R.; Blowes, D.; Jambor, J.; Robertson, W. Mobilization and attenuation of heavy metals within a nickel mine tailings impoundment near Sudbury, Ontario, Canada. *Environ. Geol.* **1998**, *36*, 305–319. [CrossRef]
11. Benzaazoua, M.; Bussière, B.; Dagenais, A.-M.; Archambault, M. Kinetic tests comparison and interpretation for prediction of the Joutel tailings acid generation potential. *Environ. Geol.* **2004**, *46*, 1086–1101. [CrossRef]
12. Plante, B.; Benzaazoua, M.; Bussière, B. Predicting geochemical behaviour of waste rock with low acid generating potential using laboratory kinetic tests. *Mine Water Environ.* **2011**, *30*, 2–21. [CrossRef]
13. Bussière, B.; Aubertin, M.; Benzaazoua, M.; Gagnon, D. Modèle d'estimation des coûts de restauration de sites miniers générateurs de DMA. In *Séminaire Mines écologiques présentés dans le cadre du congrès APGGQ*; Les sciences de la terre: Rouyn-Noranda, QC, Canada, 1999.
14. Cornell, R.M.; Schwertmann, U. Environmental Significance. In *The Iron Oxide*; Wiley-VCH Verlag GmbH & Co. KGaA: Weinheim, Germany, 2004; pp. 541–551. [CrossRef]

15. Cravotta, C.A., III. Dissolved metals and associated constituents in abandoned coal-mine discharges, Pennsylvania, USA. Part 1: Constituent quantities and correlations. *Appl. Geochem.* **2008**, *23*, 166–202. [CrossRef]
16. Cravotta, C.A., III. Dissolved metals and associated constituents in abandoned coal-mine discharges, Pennsylvania, USA. Part 2: Geochemical controls on constituent concentrations. *Appl. Geochem.* **2008**, *23*, 203–226. [CrossRef]
17. Dold, B. Acid rock drainage prediction: A critical review. *J. Geochem. Explor.* **2017**, *172*, 120–132. [CrossRef]
18. Blowes, D.W.; Reardon, E.J.; Jambor, J.L.; Cherry, J.A. The formation and potential importance of cemented layers in inactive sulfide mine tailings. *Geochim.Cosmochim. Acta* **1991**, *55*, 965–978. [CrossRef]
19. Elghali, A.; Benzaazoua, M.; Bussière, B.; Kennedy, C.; Parwani, R.; Graham, S. The role of hardpan formation on the reactivity of sulfidic mine tailings: A case study at Joutel mine (Québec). *Sci. Total Environ.* **2019**, *654*, 118–128. [CrossRef] [PubMed]
20. Chermak, J.A.; Runnells, D.D. Self-sealing hardpan barriers to minimize infiltration of water into sulfide-bearing overburden, ore, and tailings piles. Proceedings of Tailings and mine waste, Fort Collins, CO, USA, 16–19 January 1996; pp. 265–273.
21. Holmström, H.; Öhlander, B. Layers rich in Fe- and Mn-oxyhydroxides formed at the tailings-pond water interface, a possible trap for trace metals in flooded mine tailings. *J. Geochem. Explor.* **2001**, *74*, 189–203. [CrossRef]
22. Kohfahl, C.; Graupner, T.; Fetzer, C.; Holzbecher, E.; Pekdeger, A. The impact of hardpans and cemented layers on oxygen diffusivity in mining waste heaps: Diffusion experiments and modelling studies. *Sci. Total Environ.* **2011**, *409*, 3197–3205. [CrossRef]
23. McGregor, R.; Blowes, D. The physical, chemical and mineralogical properties of three cemented layers within sulfide-bearing mine tailings. *J. Geochem. Explor.* **2002**, *76*, 195–207. [CrossRef]
24. Blowes, D.W.; Jambor, J.L.; Hanton-Fong, C.J.; Lortie, L.; Gould, W.D. Geochemical, mineralogical and microbiological characterization of a sulphide-bearing carbonate-rich gold-mine tailings impoundment, Joutel, Québec. *Appl. Geochem.* **1998**, *13*, 687–705. [CrossRef]
25. Barnett, E.; Hutchinson, R.; Adamcik, A.; Barnett, R. Geology of the Agnico-Eagle gold deposit, Quebec. *Precambrian Sulphide Depos. Geol. Assoc. Can. Spec. Pap.* **1982**, *25*, 403–426.
26. Bussière, B.; Benzaazoua, M.; Aubertin, M.; Mbonimpa, M. A laboratory study of covers made of low-sulphide tailings to prevent acid mine drainage. *Environ. Geol.* **2004**, *45*, 609–622. [CrossRef]
27. Chopard, A.; Benzaazoua, M.; Plante, B.; Bouzahzah, H.; Marion, P. Kinetic tests to evaluate the relative oxidation rates of various sulfides and sulfosalts. In Proceedings of the 10th International Conference on Acid Rock Drainage and IMWA Annual Conference, Santiago, Chile, 21–24 April 2015.
28. Allen, T. *Powder Sampling and Particle Size Determination*; Elsevier: Amsterdam, The Netherlands, 2003.
29. Atkinson, P.M.; Lloyd, C.D. Mapping precipitation in Switzerland with ordinary and indicator kriging. Special issue: Spatial Interpolation Comparison 97. *J. Geogr. Inf. Decis. Anal.* **1998**, *2*, 72–86.
30. Carter, M.R. *Soil Sampling and Methods of Analysis*; CRC Press: Boca Raton, FL, USA, 1993.
31. Poduri, S.; Rao, R. *Sampling Methodology with Applications*; Chapman and Hall: New York, NY, USA, 2000.
32. Popek, E.P. *Sampling and Analysis of Environmental Chemical Pollutants: A Complete Guide*; Academic Press: Cambridge, MA, USA, 2003.
33. Sampath, S. *Sampling Theory and Methods*; CRC press: Boca Raton, FL, USA, 2001.
34. Elghali, A.; Benzaazoua, M.; Bouzahzah, H.; Bussière, B.; Villarraga-Gómez, H. Determination of the available acid-generating potential of waste rock, part I: Mineralogical approach. *Appl. Geochem.* **2018**, *99*, 31–41. [CrossRef]
35. Elghali, A.; Benzaazoua, M.; Bussière, B.; Bouzahzah, H. Determination of the available acid-generating potential of waste rock, part II: Waste management involvement. *Appl. Geochem.* **2019**, *100*, 316–325. [CrossRef]
36. Sobek, A.A.; Schuller, W.; Freeman, J.; Smith, R. *Field and Laboratory Methods Applicable to Overburdens and Minesoils, 1978*; US Environmental Protection Agency: Cincinnati, OH, USA, 1978; Volume 45268, pp. 47–50.
37. Bouzahzah, H.; Benzaazoua, M.; Plante, B.; Bussiere, B. A quantitative approach for the estimation of the "fizz rating" parameter in the acid-base accounting tests: A new adaptations of the Sobek test. *J. Geochem. Explor.* **2015**, *153*, 53–65. [CrossRef]

38. Adam, K.; Kourtis, A.; Gazea, B.; Kontopoulos, A. Evaluation of static tests used to predict the potential for acid drainage generation at sulphide mine sites. *Min. Tech.* **1997**, *16*, A1–A8.
39. Chotpantarat, S. A review of static tests and recent studies. *Am. J. Appl. Sci.* **2011**, *8*, 400. [CrossRef]
40. Jambor, J.; Dutrizac, J.; Groat, L.; Raudsepp, M. Static tests of neutralization potentials of silicate and aluminosilicate minerals. *Environ. Geol.* **2002**, *43*, 1–17.
41. Paktunc, A.; Leaver, M.; Salley, J.; Wilson, J. A new standard material for acid base accounting tests. In Proceedings of the Securing the Future, International Conference on Mining and the Environment, Skelleftea, Sweden, 25 June–1 July 2001; pp. 644–652.
42. Price, W.A. Prediction manual for drainage chemistry from sulphidic geologic materials. *MEND Rep.* **2009**, *1*, 579.
43. Sapsford, D.; Bowell, R.; Dey, M.; Williams, C.; Williams, K. *A Comparison of Kinetic NAG Tests with Static and Humidity Cell Tests for the Prediction of ARD*; Mine Water and the Environment: Ostrava, Czech Republic, 2008; pp. 325–328.
44. Miller, S.; Robertson, A.; Donahue, T. Advances in acid drainage prediction using the net acid generation (NAG) test. In Proceedings of the 4th International Conference on Acid Rock Drainage, Vancouver, BC, Canada, 31 May–6 June 1997; pp. 0533–0549.
45. Smart, R.; Skinner, W.; Levay, G.; Gerson, A.; Thomas, J.; Sobieraj, H.; Schumann, R.; Weisener, C.; Weber, P.; Miller, S. *ARD Test Handbook: Project P387A, Prediction and Kinetic Control of Acid Mine Drainage*; AMIRA International Ltd.: Melbourne, Australia, 2002.
46. Elberling, B.; Nicholson, R.V. Field determination of sulphide oxidation rates in mine tailings. *Water Resour. Res.* **1996**, *32*, 1773–1784. [CrossRef]
47. Mbonimpa, M.; Aubertin, M.; Aachib, M.; Bussière, B. Diffusion and consumption of oxygen in unsaturated cover materials. *Can. Geotech. J.* **2003**, *40*, 916–932. [CrossRef]
48. Mbonimpa, M.; Aubertin, M.; Bussière, B. Oxygen consumption test to evaluate the diffusive flux into reactive tailings: Interpretation and numerical assessment. *Can. Geotech. J.* **2011**, *48*, 878–890. [CrossRef]
49. Cressie, N. Spatial prediction and ordinary kriging. *Math. Geol.* **1988**, *20*, 405–421. [CrossRef]
50. Gratton, Y. Le krigeage: La méthode optimale d'interpolation spatiale. *Les articles de l'Institut d'Analyse Géographique* **2002**, *1*, 4.
51. Oliver, M.A.; Webster, R. Kriging: A method of interpolation for geographical information systems. *Int. J. Geogr. Inf. Syst.* **1990**, *4*, 313–332. [CrossRef]
52. Singh, C.K.; Kumar, A.; Shashtri, S.; Kumar, A.; Kumar, P.; Mallick, J. Multivariate statistical analysis and geochemical modeling for geochemical assessment of groundwater of Delhi, India. *J. Geochem. Explor.* **2017**, *175*, 59–71. [CrossRef]
53. Paktunc, A.; Davé, N. Mineralogy of pyritic waste rock leached by column experiments and prediction of acid mine drainage. In *Applied Mineralogy in Research, Economy, Technology, Ecology and Culture, Proceedings of the 6th International Congress on Applied Mineralogy*; Balkema: Avereest, The Netherlands, 2000; pp. 621–623.
54. Lawrence, R.W.; Wang, Y. Determination of neutralization potential in the prediction of acid rock drainage. In Proceedings of the Fourth International Conference on Acid Rock Drainage, Vancouver, BC, Canada, 31 May–6 June 1997; pp. 451–464.
55. Jambor, J.; Dutrizac, J.; Raudsepp, M.; Groat, L. Effect of peroxide on neutralization-potential values of siderite and other carbonate minerals. *J. Environ. Qual.* **2003**, *32*, 2373–2378. [CrossRef]
56. Weber, P.A.; Thomas, J.E.; Skinner, W.M.; Smart, R.S.C. Improved acid neutralisation capacity assessment of iron carbonates by titration and theoretical calculation. *Appl. Geochem.* **2004**, *19*, 687–694. [CrossRef]
57. Paktunc, A.D. Characterization of Mine Wastes for Prediction of Acid Mine Drainage. In *Environmental Impacts of Mining Activities: Emphasis on Mitigation and Remedial Measures*; Azcue, J.M., Ed.; Springer: Berlin/Heidelberg, Germany, 1999; pp. 19–40. [CrossRef]
58. Elghali, A.; Benzaazoua, M.; Bussière, B.; Schaumann, D.; Graham, S.; Genty, T.; Noel, J.; Kenedy, C.; Cayouette, J. Investigation of the role of hardpans on the geochemical behavior of the Joutel mine tailings. In Proceedings of the Twenty-First International Conference on Tailings and Mine Waste (TMW'17), Banff, AB, Canada, 5–8 November 2017; pp. 193–202.
59. Lapakko, K.A.; Engstrom, J.N.; Antonson, D.A. Effects of particle size on drainage quality from three lithologies. In Proceedings of the 7th International Conference on Acid Rock Drainage (ICARD), St. Louis, MI, USA, 26–30 March 2006; pp. 1026–1050.

60. Elberling, B.; Nicholson, R.; Reardon, E.; Tibble, R. Evaluation of sulphide oxidation rates: A laboratory study comparing oxygen fluxes and rates of oxidation product release. *Can. Geotech. J.* **1994**, *31*, 375–383. [CrossRef]
61. Tibble, P.; Nicholson, R. Oxygen consumption on sulphide tailings and tailings covers: Measured rates and applications. In Proceedings of the 4th International Conference on Acid Rock Drainage (ICARD), Vancouver, BC, Canada, 31 May–6 June 1997; pp. 647–661.
62. Coulombe, V. Performance de recouvrements isolants partiels pour contrôler l'oxydation de résidus miniers sulfureux. Master's Thesis, Université du Québec en Abitibi-Témiscamingue, Rouyn-Noranda, QC, Canada, 2012.
63. Pabst, T.; Aubertin, M.; Bussière, B.; Molson, J. Analysis of monolayer covers for the reclamation of acidgenerating tailings–Column testing and interpretation. In Proceedings of the 63rd Canadian Geotechnical Conference and 1st Joint CGS/CNC-IPA Permafrost Specialty Conference, Calgary, AB, Canada, 12–16 September 2010; pp. 1119–1127.
64. Talmon, A.M.; van Kesteren, W.G.; Sittoni, L.; Hedblom, E.P. Shear cell tests for quantification of tailings segregation. *Can. J. Chem. Eng.* **2014**, *92*, 362–373. [CrossRef]
65. Weber, P.A.; Hughes, J.B.; Conner, L.B.; Lindsay, P.; Smart, R. Short-Term Acid Rock Drainage Characteristics Determined by Paste pH and Kinetic NAG Testing: Cypress, Prospect, New Zealand. In Proceedings of the 7th International Conference on Acid Rock Drainage (ICARD), St. Louis, MO, USA, 26–30 March 2006.
66. Parbhakar-Fox, A.; Fox, N.; Hill, R.; Ferguson, T.; Maynard, B. Improved mine waste characterisation through static blended test work. *Miner. Eng.* **2018**, *116*, 132–142. [CrossRef]
67. Lawrence, R.; Wang, Y. Determinations of neutralization potential for acid rock drainage prediction. *MEND Proj.* **1996**, *1*, 38.

© 2019 by the authors. Licensee MDPI, Basel, Switzerland. This article is an open access article distributed under the terms and conditions of the Creative Commons Attribution (CC BY) license (http://creativecommons.org/licenses/by/4.0/).

Article

Desulfurization of the Old Tailings at the Au-Ag-Cu Tiouit Mine (Anti-Atlas Morocco)

Abdelkrim Nadeif [1], Yassine Taha [2,*], Hassan Bouzahzah [3], Rachid Hakkou [1,2,*] and Mostafa Benzaazoua [4,5]

1. LCME, Faculté des Sciences et Techniques, Université Cadi Ayyad, Marrakech 40000, Morocco
2. Materials Science and Nano-engineering Department, Mohammed VI Polytechnic University, Ben Guerir 43150, Morocco
3. GeMMe Laboratory, Argenco Department, Université de Liège, 4000 Liège, Belgium
4. Institut de Recherche en Mines et en Environnement, Université du Québec en Abitibi Témiscamingue, Rouyn-Noranda, QC J9X 5E4, Canada
5. Geology and Sustainable Mining Department, Mohammed VI Polytechnic University, Ben Guerir 43150, Morocco
* Correspondence: yassine.taha@um6p.ma (Y.T.); r.hakkou@uca.ma (R.H.)

Received: 30 April 2019; Accepted: 26 June 2019; Published: 30 June 2019

Abstract: Tailings from the abandoned Tiouit mine site in Morocco are mainly composed of sulfides, hematite, and quartz. They contain 0.06–1.50 wt % sulfur, mostly in the form of pyrite, pyrrhotite, and chalcopyrite. The tailings also contain gold (3.36–5.00 ppm), silver (24–37 ppm), and copper (0.06–0.08 wt %). Flotation tests were conducted to reprocess the tailings for Au, Ag, and Cu recovery, and at the same time to prevent acid mine drainage (AMD) generation through the oxidation of sulfide minerals, including pyrite, sphalerite, arsenopyrite, chalcopyrite, galena. The flotation results confirmed that environmental desulfurization is effective at reducing the overall sulfide content in the tailings. The recovery of sulfides was between 69% and 75%, while Au recovery weight-yield was between 2.8% and 4.7%. The test that showed the best sulfur recovery rate and weight-yield was carried out with 100 g/t $CuSO_4$ (sulfide activator) and 50 g/t of amyl xanthate (collector). The goal of this study was also to assess the remaining acid-generating potential (AP) and acid-neutralizing potential (NP) of the desulfurized tailing. The geochemical behavior of the initial tailings sample was compared to that of the desulfurized tailings using kinetic weathering cell tests. The leachates from the desulfurized tailings showed higher pH values than those from the initial tailings, which were clearly acid-generating. The residual acidity produced by the desulfurized tailings was most likely caused by the hydrolysis of Fe-oxyhydroxides.

Keywords: acid mine drainage; mine tailings; sulfides; gold reprocessing; desulfurization; flotation; kinetic test

1. Introduction

Acid mine drainage (AMD) generated by mine wastes and the subsequent contamination of surface waters and soils through leached metals is one of the most significant environmental problems facing the mining industry [1,2]. Higher operating costs result when AMD needs to be treated [3], and acid-generating wastes need to be reclaimed [4]. AMD is produced as a result of abiotic and biotic reactions involving water, air, and the sulfide minerals present in mine wastes (e.g., pyrite, pyrrhotite) [1,5,6]. These reactions produce acidic effluents that tend to be loaded with various heavy metals and metalloids (e.g., Fe, Mn, Cu, As, Hg) [1,2,7].

The need to manage sulfidic waste rocks and tailings is becoming an increasingly urgent issue. Safety issues (e.g., tailings dam failures), air pollution (e.g., dust generation), and water contamination (e.g., mine drainage) constitute the main concerns [8–10]. Many researchers have worked on the development of sustainable and eco-friendly solutions to eliminate or minimize the risks related to the disposal of mine tailings. Current waste management methods are site-specific and depend on local conditions. Tailings are generally stored in engineered dams and impoundments, dewatered, dried, and/or surface landfilled, stabilized with binders for open-pit or underground backfilling purposes [11,12]. Many challenges are faced such as seismicity risks, rainfall potential, land availability, etc.

Due to the depletion of metals ores and as these tailings could contain some valuable minerals, many studies were employed to investigate alternative technologies for a sustainable and integrated tailings management. Froth flotation has previously been used to recover Cu, Co, and other metals from tailings [13,14]. Prior works investigated the particle size of the tailings, the behavior of the froth during flotation, the age of the tailings, and the nature of the metal-bearing minerals (e.g., sulfides, oxides) as the main parameters affecting metal recovery [15–18]. Several recent studies also investigated bioleaching as a potential technology for removing heavy metals and sulfides from mine tailings [19–21]. Other innovative technologies for metal extraction from mine tailings were developed [22,23]. In addition to the reprocessing of mine tailings, researchers have also developed techniques for recycling mine tailings. For example, tailings could be put to use in products such as ceramics [24,25], and construction and building materials [26,27] among other applications [28].

Environmental desulfurization of tailings is one technology that has gained popularity in the last two decades. This method consists of separating sulfide minerals from mine tailings using froth flotation. The resulting desulfurized fraction can be managed separately [29,30] or backfilled in underground mines [31]. Desulfurization has been demonstrated as an economically and environmentally effective technique to decrease the acid-generation potential of mine tailings, in particular for operating mines and new mine projects [32–34].

Several old mine sites in Morocco have been abandoned without the implementation of proper closure plans, and most could still cause serious environmental damage [35]. The Tiouit Mine, which is located in the eastern Anti-Atlas Mountains, is one of these abandoned sites. From 1950 to 1996, the mine's Au, Ag, and Cu ore was estimated at approximately 1,743,000 tons [36,37]. Gold mineralization occurs exclusively as electrum as microscopic grains within various mineral phases. The gangue minerals consist of quartz, chlorite, and various feldspars. Approximately 743,000 tons of tailings were generated by the end of the mine's life and were landfilled in three tailings dams around the site [38]. The tailings facilities were abandoned without any restoration plans. The mine tailings contain 0.06–1.50 wt % sulfur, mostly in the sulfide minerals pyrite, pyrrhotite, and chalcopyrite [38]. AMD produced from these tailings has never been treated, resulting in the contamination of both surface and subsurface waters as well as the severe degradation of the surrounding ecosystems.

The main objective of this study was to investigate the feasibility of environmental desulfurization as a technique for reducing the acid-generating potential of the Tiouit mine tailings, while also recovering residual Au. The tailings samples were characterized for their main chemical and physical characteristics, as well as their acid-generating potential. A number of laboratory-scale flotation tests were conducted, using Denver flotation cells, to investigate precious metal recovery. Additionally, kinetic tests were performed to study the geochemical behavior of the desulfurized tailings, and consequently, determine the efficiency of the desulfurization process.

2. Materials and Methods

2.1. Tiouit Mine Site Description

Tiouit mine, discovered in 1947, is located in the eastern part of Jebel Saghro in the Anti-Atlas Mountains, Morocco, about 40 km south east of the city of Boumalne Dades. The Tiouit deposits were exploited intermittently by different companies: "Comansour" (1950–1956), "Westfield" (1959–1964), and finally the "SODECAT company" (1975–1996) [36]. The mineralization was identified as intrusive granite consisting of native gold associated with chalcopyrite ($CuFeS_2$), pyrite (FeS_2), sphalerite (ZnS), galena (PbS), magnetite (Fe_3O_4), arsenopyrite (FeAsS), bornite (Cu_5FeS_4), luzonite (Cu_3AsS_4), gold (Au), and gray copper ($X_{12}Y_4S_{13}$, with X: Ag, Cu, Fe, Hg, Zn and Y: As, Sb, Te). The ore processing involved crushing and milling followed by a bulk froth flotation process [37,39,40]. Over the duration of the entire mining operation, more than 740,000 tons of mine tailings (168,000 tons in the northern tailings pond, 346,000 tons in the central tailings pond, and 229,000 tons in the southern tailings pond) were deposited at the surface without any reclamation measures (Figure 1). Larger volumes of mine wastes were discharged directly into the Tiouit river. Many rainwater ravines were created in the tailings deposit, which caused their dispersion along the Tiouit river.

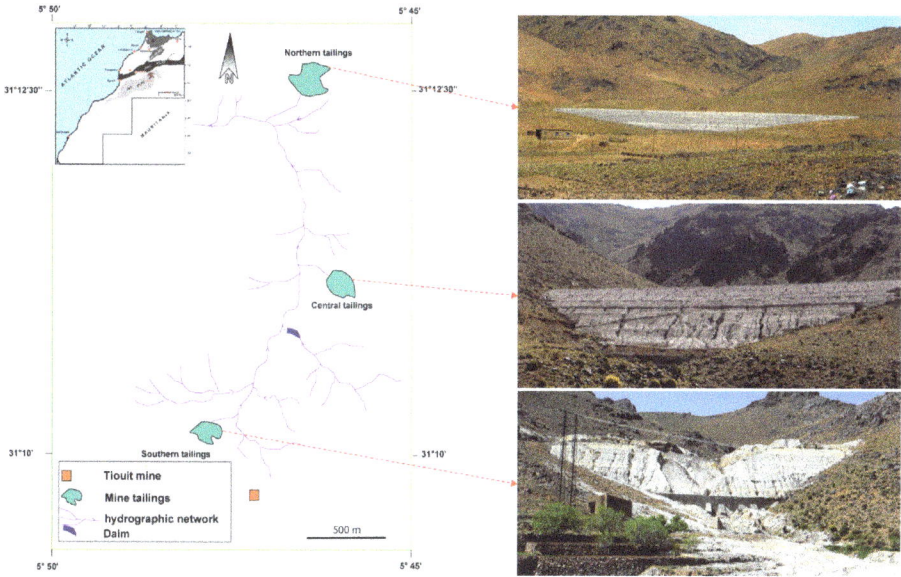

Figure 1. Tiouit mine site location and photographs of northern, central, and southern tailing ponds.

2.2. Methodology

Tailings samples were collected after excavating trenches in the tailings ponds. Sampling sites were selected, based on field observations, to obtain fresh, non-oxidized tailings. Tailings samples labeled ST, CT, and NT were collected from the southern, central, and northern tailings ponds, respectively. Samples were collected at depths between 90 and 110 cm. All samples were carefully conditioned, and immediately stored in double-sealed plastic bags from which the air was evacuated to prevent contact with oxygen. A small, representative sample from each pond was stored at low temperature (5 °C) and at their original saturation (without drying) for kinetic testing. The methodology followed in this study is compiled in Figure 2. The aim was to desulfurize the sampled tailings to reduce their environmental impact. The floated sulfide-rich tailings (SRT) needed to be properly managed due to their high concentration of sulfides.

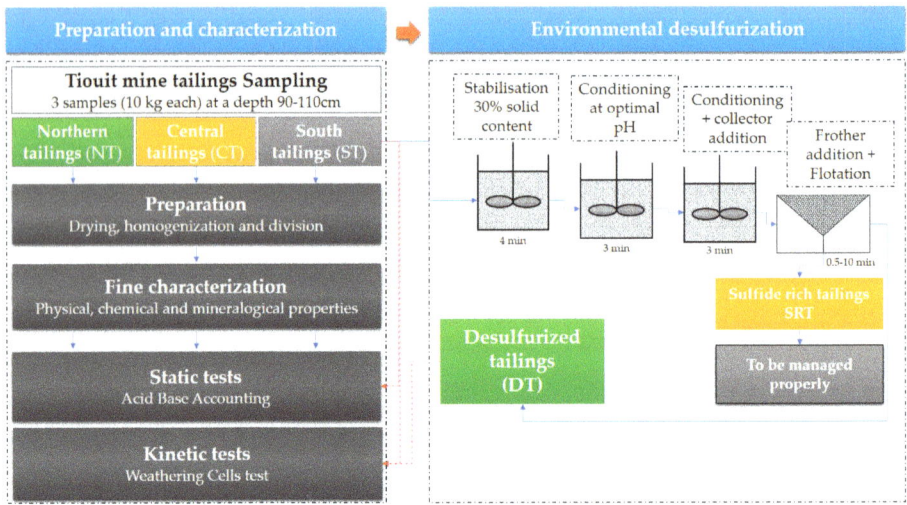

Figure 2. Schematic illustration of the methodology followed in this study.

2.2.1. Analytical Methods

Chemical compositions of the tailings samples were obtained by inductively coupled plasma atomic emission spectroscopy (ICP-AES; Perkin Elmer Optima 3100 RL, Waltham, MA, USA) following digestion with $HNO_3/Br_2/HF/HCl$. Dilute HCl was used to extract sulfate from the bulk solids; concentrations were determined by ICP-AES. Particle-size distribution was determined using a laser particle size analyzer (Malvern Mastersizer; ISO-13320, Panalytical, Almelo, The Netherlands). Specific gravity (GS) was measured using a gas pycnometer (Micromeritics Accupyc 1330, Micromeritics, Norcross, GA, USA).

The mineralogical compositions of the raw tailings samples were determined by X-ray diffraction (XRD) using a Bruker AXS Advance D8 (Bruker, Billerica, MA, USA) equipped with a copper anticathode, scanning over a diffraction angle (2θ) range from 5 to 60. Scan settings were 0.005° 2θ step size and 1 s counting time per step. The DiffracPlus EVA software (v. 9.0 rel. 2003) was used to identify mineral species. The TOPAS software (v. 2.1; Bruker, Billerica, MA, USA) was used to quantify the abundance of all identified mineral species implementing a Rietveld refinement. The absolute precision of this quantification method is ±0.5–1 wt % [41,42]. Before XRD analysis, samples were pulverized in isopropyl alcohol using a micronizing mill with corundum grinding media for 15 min to obtain ≈90 wt % <10 µm.

Sample mineralogy was further investigated by optical and electronic microscopy observations. Polished sections of bulk samples as well as the flotation products were prepared using an epoxy resin for reflected light microscopy. Scanning electron microscope (SEM) observations on polished sections using backscattered electrons (BSE) were performed using a Hitachi S-3500N microscope (Silicon Drift Detector Bruker, Bruker, Billerica, MA, USA) equipped with an X-ray energy dispersive spectrometer (EDS; Silicon drift spectrometer X-Max 20 mm^2) operated by INCA software (450 Energy). The operating conditions were 20 keV, ≈100 µA and 15 mm working distance.

To explain the Au deportment, the pyrite and chalcopyrite minerals in both samples (sulfide concentrates and tailings) were analyzed using microprobe analyses (Electron Probe Micro-Analysis (EPMA), Cameca SX-100) [43]. An accelerating voltage of 15 kV, a beam current of 20 nA (counting time 15 s, background 0 s) were used for major elements (Fe, S, Cu, Co, Ni, Zn, and As) and 120 nA for Au (counting time 60 s, background 15 s) at the same resolution/beam size.

2.2.2. Static Tests

The most commonly used static tests to predict acid-generating potential is the acid–base accounting (ABA) test [44–46]. The ABA test measures the balance between the acid-generating potential (AP) and neutralizing-potential (NP) of a given sample. NP values were obtained using the Sobek method [47] as modified by Lawrence and Wang [45] and each NP measurement was performed in duplicate. AP was calculated using the sulfide sulfur fraction, which was obtained by subtracting the sulfate fraction from the total sulfur assay (AP = 31.25 × % sulfur). Both AP and NP are expressed in kg CaCO3/t. The net neutralization potential (NNP) was calculated by subtracting the AP value from the NP. NNP values < −20 kg CaCO3/t indicate an acid-producing material, whereas materials with NNP > 20 kg CaCO3/t are considered acid-consuming. An uncertainty zone exists between the range 20 > NNP > −20 kg CaCO3/t [48]. Another useful tool to evaluate AMD production potential is the NP to AP ratio. Typically, a material is considered non-acid generating if the NP/AP > 2.5, uncertain if 2.5 > NP/AP > 1, and acid-generating if NP/AP < 1 [49].

2.2.3. Environmental Desulfurization by Froth Flotation

All flotation tests were carried out with a Denver D-12 lab flotation machine (2.5 L cell volume). All flotation tests were conducted on the southern tailings samples, which were generally characterized by high sulfur contents (Table 1). The pH during flotation was ~5 because the tailings' pore water was already acidic and the targeted solid percentage was around 30 solid wt % for all tests. The conditioning time was 3 min after the addition of a potassium amyl xanthate (PAX) collector (at various concentrations) and a frother (Methyl Isobutyl Carbinol (MIBC); 50 µg/kg of tailings). An activator of copper sulfate (CuSO$_4$) was used to reactivate the oxidized sulfide surfaces. The speed of the rotor-stator was adjusted to 1200 rpm with free airflow injection. To obtain consistent results, the same operator manually removed the froths with a spatula for all of the flotation tests. The pH was measured and adjusted by adding diluted NaOH solution to increase the pH. A summary of the flotation tests is presented in Table 1.

Table 1. Reagents used in the flotation tests.

Test Nbr	CuSO$_4$ (g/t)	KAX (g/t)	MIBC (g/t)
1	100	100	50
2	100	50	50
3	100	25	50
4	0	100	50
5	0	50	50
6	0	25	50

2.2.4. Kinetic Tests (Weathering Cells) Procedure

Representative samples of raw tailing (ST) and desulfurized tailings (DT), were selected for kinetic tests with weathering cells similar to the ones used by Cruz et al. [50]. The method uses a thin layer of sample and more frequent flushing–drying cycles compared to the ASTM D5744-96 humidity cell test [51]. The main advantage of this weathering cell test is its rapidity and the small volume of material required [52,53].

Approximately 67 g (dry weight) of tailings were placed in a 100 mm diameter Buchner funnel equipped with a glass-fiber filter. A seven-day cycle consisted of two days of exposure to ambient air, leaching on the third day, three days of exposure to air, and finally flushing during the seventh day. The flushes consisted of adding 50 mL of deionized water to the top of the Buchner funnel. The leachate was recovered by applying a slight suction on a filtering flask after 3 h of contact with the sample. The total duration of the experiments was 27 cycles (13 weeks).

The leachates recovered after each cycle were filtered using a 0.45 µm nylon filter and analyzed for several geochemical parameters to understand the tailings' reactivity, oxidation kinetics, metal solubility, and the overall leaching behavior of the materials. For each sample, pH, Eh, electrical conductivity, metal concentrations, acidity, and alkalinity were analyzed. These data were compiled as instantaneous and cumulative loads, as well as elemental depletion curves that were based on the volume and composition of the leachates and the initial geochemistry of the solid samples. Filtered leachates were acidified with 2% HNO_3 to avoid metal precipitation. The resulting solutions were analyzed by ICP-AES to determine metal and sulfate concentrations.

3. Results

3.1. Sample Characterizations

The physical, chemical, and mineralogical properties of the different mine tailings are shown in Table 2. The specific gravity of all tailings samples was around 2.9 g/cm^3. In terms of particle-size distribution, the Tiouit mine tailings showed a silty characteristic, with a D80 in the range of 100–230 µm. The NT and CT showed very low sulfur contents (0.4 and 0.06 wt %, respectively), while the ST contained up to 1.5 wt % sulfur. As the sulfur content was relatively low in the three samples related to low content of sulfide minerals, the high iron content (7.9–5 wt %) was related to the presence of iron-rich minerals such as hematite. Aluminum (~4–5 wt %), potassium (~2 wt %), manganese (~0.2 wt %), and sodium (~0.2 wt %) are related to silicate minerals. The low concentrations of Ca and Mg (~0.4 wt %) in the tailings suggest a negligible carbonate content. The Cu and Zn contents were around 0.1 wt %, while Pb and As concentrations were lower than 0.05 wt %. A fire assay of the selected samples from the Tiouit mine site showed significant concentrations of precious metals, including 3–5 g/t of Au and 23–37 g/t of Ag.

Table 2. Physical, chemical and mineralogical composition of the Tiouit tailings.

Characteristics	Northern Tailings (NT)	Central Tailings (CT)	South Tailings (ST)	Desulfurized Tailings (DT)	Sulfide Rich Tailings (SRT)
Physical properties					
SG (g/cm^3)	2.85	2.90	2.86	2.82	-
D_{80} (µm)	155	222	108	106	104
D_{50} (µm)	64	108	49	45	53
D_{10} (µm)	7	12	7	6	6
Chemical composition					
S_{total} (wt %)	0.37	0.061	1.5	0.51	21.57
$S_{sulfate}$ (wt %)	0.08	0.055	0.17	0.38	0.64
$S_{sulfide}$ (wt %)	0.29	0.006	1.33	0.13	20.93
Si (wt %)	38.24	37.68	34.12	33.6	18.7
Al (wt %)	4.31	5.19	4.26	4.89	4.04
Ca (wt %)	0.39	0.46	0.29	0.29	0.75
Fe (wt %)	8.29	8.54	7.95	8.52	22.52
K (wt %)	2	2.5	2	2.2	1.8
Mg (wt %)	0.42	0.464	0.369	0.44	0.61
Na (wt %)	0.1	0.22	0.21	0.22	0.63
Au (ppm)	5	3.95	3.36	-	-
Ag (ppm)	24	37	23	20	87
As (ppm)	300	190	846	860	860
Cu (ppm)	770	1140	924	870	2390
Pb (ppm)	290	110	296	280	740
Ti (ppm)	1190	1370	1680	1750	840
Zn (ppm)	1240	1130	1230	1140	3550
Mineralogical composition (%)					
Quartz SiO_2	62.1	61.2	63.8	62.2	31.8
Muscovite $KAl_2(Si_3Al)O_{10}(OH,F)_2$	19.2	18.5	16.2	20.1	15.2
Clinochlore $(Mg,Fe)_6(Si,Al)_4O_{10}(OH)_8$	7.8	10.2	9.8	6.6	5.7
Hematite Fe_2O_3	10.2	9.6	7.6	10.8	5.8
Pyrite FeS_2	0.3	0.1	2.1	0.1	38.2
Sphalerite ZnS	0.2	0.2	0.2	0.1	0.6
Arsenopyrite FeAsS	0.1	0.1	0.2	0.2	0.2
Chalcopyrite $CuFeS2$	0.2	0.3	0.2	0.2	0.7
Galena PbS	0.1	0.0	0.0	0.0	0.8

The XRD results showed that the only sulfide detected in the ST was pyrite (1.7 wt %), which is in accordance with the S assays. Quartz was the most abundant mineral (73–77 wt %), with no carbonates detected (absence of neutralizing minerals) and low muscovite (5–10 wt %), clinochlore (4–5 wt %), and hematite (9–13 wt %) contents. Although analyzed, Cd, Co, Mo, Ni, and Ta were all below the detection limit of the ICP-AES and are not presented.

3.2. Flotation Test Results

Flotation experiments were carried out in order to select the optimal collector mix and to assess the effect of adding an activation reagent. As the ST had the highest sulfur content (in the form of sulfide minerals), these tailings were chosen for the desulfurization tests. Desulfurization was performed through sulfide bulk flotation tests. The sulfur recoveries obtained for various flotation test conditions are summarized in Figure 3.

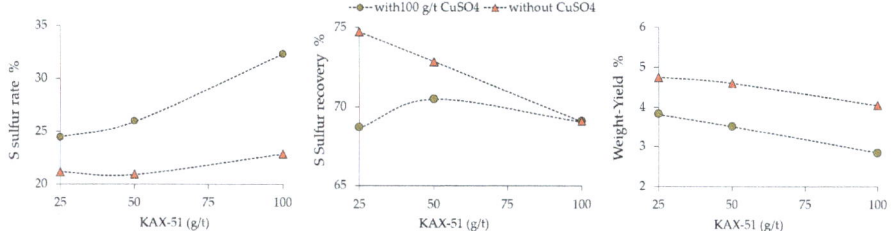

Figure 3. Yield weight and sulfur recovery during flotation tests (desulfurization).

The recovery yield of sulfur-bearing minerals was relatively high (68–74 wt %) with very negligible amounts of residual sulfides. Achieving a maximum recovery of around 73 wt % was possible with collector concentrations as low as 50 g/t and without any activators. Concentrate weight proportions were between 2.8% (for a collector concentration of 100 g/t and 100 g/t $CuSO_4$) and 4.7% (for a collector concentration of 25 g/t and without $CuSO_4$).

Results achieved using the optimal flotation conditions are presented in Figure 4. This experiment showed that the residual sulfur content in the tailings was <24% after 1 min and 13% after 10 min, which corresponds to a recovery of approximately 76% after 1 min and 87% after 10 min. The weight percentage of the concentrate was 2.9% after 1 min and 5.3% after 10 min.

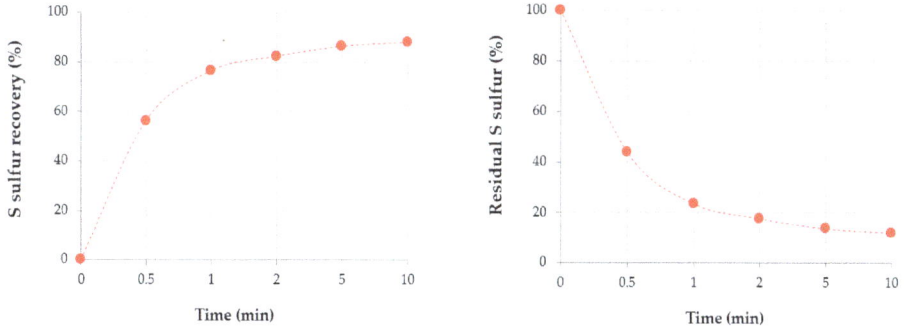

Figure 4. Flotation kinetic of the tailing: (**a**) % of $S_{sulfide}$ recovery vs. time; (**b**) residual % S_{sulfur} vs. time.

The concentrate recovered from the first 30 s of the flotation test and from the desulfurized tailings were analyzed for their Au content. The results, which are displayed in Table 3, show that only around 4% of the initial gold in the ST sample was recovered and most of the gold remained within the desulfurized tailings, which could indicate unliberated gold within silicates.

Table 3. Au recovery within concentrate of flotation test.

Samples	Weight (g)	Au (g/t)	Au Distribution (%)
Initial sample (ST)	832	3.36	
Flotation concentrate	16.1	6.07	3.99
Flotation tailings	770.9	3.05	96.01

3.3. Mineralogy of Samples Obtained by Flotation Test and Gold Deportment

The sulfide concentrate sample contained mainly pyrite occurring in two different forms: (1) euhedral pyrite with flat faces and sharp edges (Figure 5a,c); and (2) anhedral pyrite represented by grains that show no well-formed crystal faces (xenomorphic) and grains overlapping one another (Figure 5a). Chalcopyrite, pyrrhotite, sphalerite, and galena were also present in trace amounts (Figure 5c,e). Chalcopyrite was often associated with pyrite as inclusions (Figure 5c).

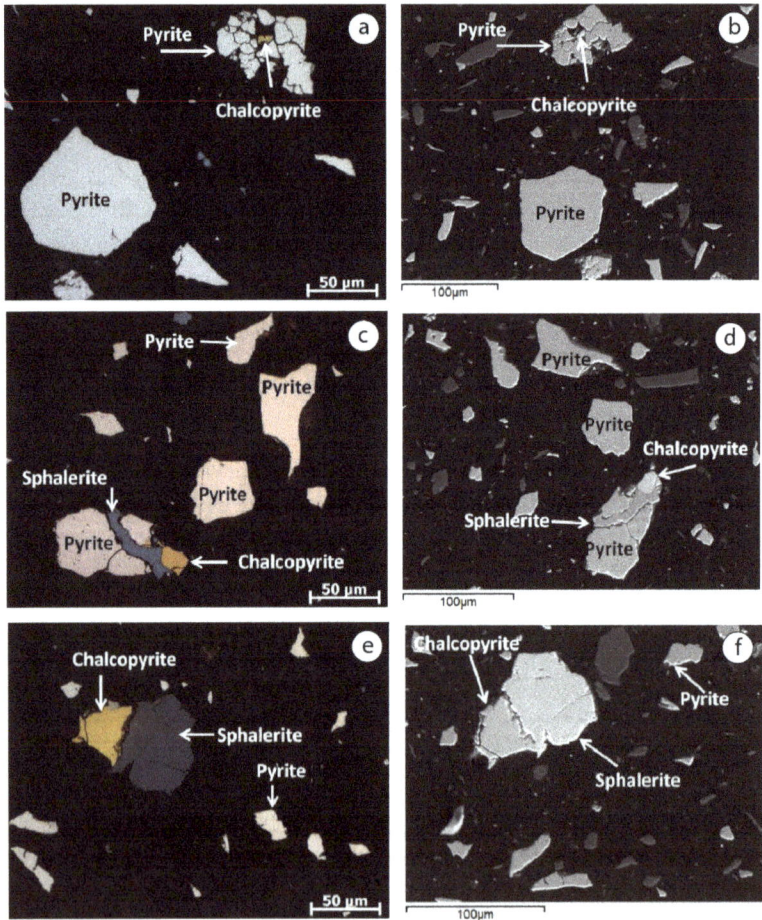

Figure 5. The identified sulfide minerals within the concentrate obtained from the flotation test; optical microscopy (a,c,e) and scanning electron microscopy (SEM) (b,d,f).

The tailings sample from the flotation test contained traces of chalcopyrite, sphalerite, pyrite and pyrrhotite with sizes generally lower than 20 μm. Chalcopyrite and pyrite were almost encapsulated

(Figure 6a,e) or attached to the gangue minerals (Figure 6c). The sample contained significant hematite and amorphous iron oxides as a result of pyrite oxidation (Figure 6g).

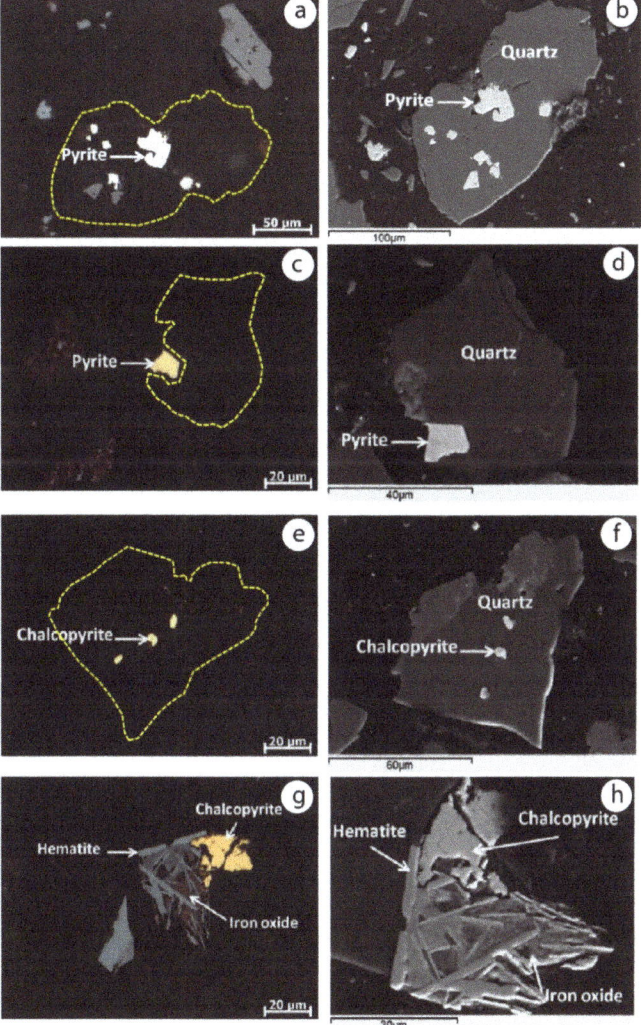

Figure 6. The identified sulfide minerals within the desulfurized tailings obtained from the flotation test; optical microscopy (**a**,**c**,**e**,**g**) and SEM (**b**,**d**,**f**,**h**).

Results from the EPMA (Table 4) show that 100% of Au in the concentrate was associated within the pyrite lattice. However, in the desulfurized tailings, only 0.4% of the total Au (3.05 g/t) was related to the pyrite. This could potentially be due to (1) insufficient statics in the EPMA measurements in the tailings; only 35 grains were analyzed instead of 104 in the concentrate, and/or (2) the presence of sub-micrometric gold locked in sulfides, iron-oxides, or gangue minerals (quartz) that the microscopic observations could not capture. Since the amount of the structural Au in the desulfurized sample was very low (0.4% of 3.05 g/t), and the sample contained substantial amounts of hematite and amorphous iron oxides resulting from pyrite oxidation, the Au was most probably associated with hematite and Fe-oxides as sub-micrometric particles as already demonstrated by Benzaazoua et al. [43]. This suggests that, prior to weathering, the Au was refractory within the Tiouit ore body and explains its losses at such high amounts [38]. On the other hand, exploratory results from three independent laboratories showed up to 86% Au recovery from the Tiouit tailings can be attained by regrinding and cyanide-in-leach or cyanide-in-pulp processing, thus confirming the non-refractory behavior of Au.

Table 4. Compilation of the Electron Probe Micro-Analysis (EPMA) analyses of pyrite and chalcopyrite for the structural Au quantification.

Au Occurrence	Flotation Concentrate	Flotation Tailings
Total Au in the samples (by fire-assay, g/t)	6.07	3.05
Structural Au in Pyrite and chalcopyrite (g/t)	6.14	0.012
Structural Au in Pyrite (g/t)	6.04	0.01
Structural Au in Chalcopyrite (g/t)	0.06	0.002
% of structural Au compared to total Au	100%	0.4%

3.4. Static Test Prediction

The ABA results, based on the Sobek et al. [47] test modified by Lawrence and Wang [45], are shown in Table 5. The Tiouit tailings showed a low NP (4–5 kg CaCO$_3$/t) and an AP of 0.1–9 kg CaCO$_3$/t. The ST showed the highest AP (41 kg CaCO$_3$/t) due to their high sulfide content (1.3 wt % S). According to the classification criteria proposed by Miller et al. [48], only the ST of Tiouit is potentially acid-generating. In contrast, the acid-generating potentials of the NT and CT are uncertain. Based on the NP/AP ratio [50], the ST and NT are acid-generating, while the CT is non-acid generating. Importantly, this acid-generating potential does not take into account the possible existence of Fe and sulfate secondary minerals that could contribute to the materials' acidities [54,55].

Table 5. Static test results of the Tiouit tailings.

Characteristics	Northern Tailings Pond (NT)	Central Tailings Pond (CT)	South Tailings Pond (ST)
AP (kg CaCO$_3$/t)	9.1	0.2	41.6
NP (kg CaCO$_3$/t)	5	5.4	4.2
NNP (kg CaCO$_3$/t)	−4	5.3	−37.4
NP/AP	0.6	29.0	0.1

3.5. Kinetic Test Results

For the initial tailings, the pH oscillated around 4 (2.8–5), and for the desulfurized tailings, the pH was slightly high and fluctuated around 5 (2.8–6.6) due to its low sulfide content after desulfurization (S$_{sulfide}$ = 0.13 wt %). Both the initial and desulfurized tailings generated leachates with low pH values, however, this was only partially due to sulfide oxidation. Most likely the majority of the acid generation was related to the solubilization of secondary oxy-hydroxide minerals (Figure 6g). As the tailings were free of carbonates, they had no significant neutralizing potential. The solubilized Fe^{2+} might oxidize into Fe^{3+} under acidic conditions (Equation (1)), with subsequent hydrolysis generating acidity (Equation (2)). In addition to oxygen, acidity could also provide from the biological processes that could happen within these kind of conditions (*Acidithiobacillus*).

$$4Fe^{2+} + O_2 + H^+ \rightarrow 4Fe^{3+} + 2H_2O, \tag{1}$$

$$Fe^{3+} + 3H_2O \rightarrow Fe(OH)_3 + 3H^+. \tag{2}$$

Eh values were slightly higher for the initial ST compared to the desulfurized tailings, indicating an oxidizing environment in agreement with their high sulfur content (152–296 mV and 91–251 mV, respectively for initial tailing and desulfurized tailing samples) (Figure 7). For electrical conductivity (EC) and all dissolved elements, the measured values in the first nine recovered leachates showed the highest values compared to the subsequent rinsing cycles for both samples. The highest EC values (6270 µS/cm for the two samples) and dissolved elements (SO_4^{2-}, Fe, Cu, Zn, Ca) were recorded at the beginning of the tests, presumably due to the dissolution of oxidation products generated by in situ weathering. In fact, EC and dissolved elements resulting from sulfide oxidation (SO_4^{2-}, Fe, Cu, Zn) and neutralization reactions (Ca, Mg) decreased and stabilized after the 9th cycle of the test, reaching a steady state (Figure 7). The EC and dissolved ions associated with oxidation/neutralization reactions were higher in the leachates recovered from the initial tailings sample than those of the desulfurized tailings (Figure 7), indicating the effectiveness of the desulfurization process at improving the environmental behavior of the tested mine tailings [56,57].

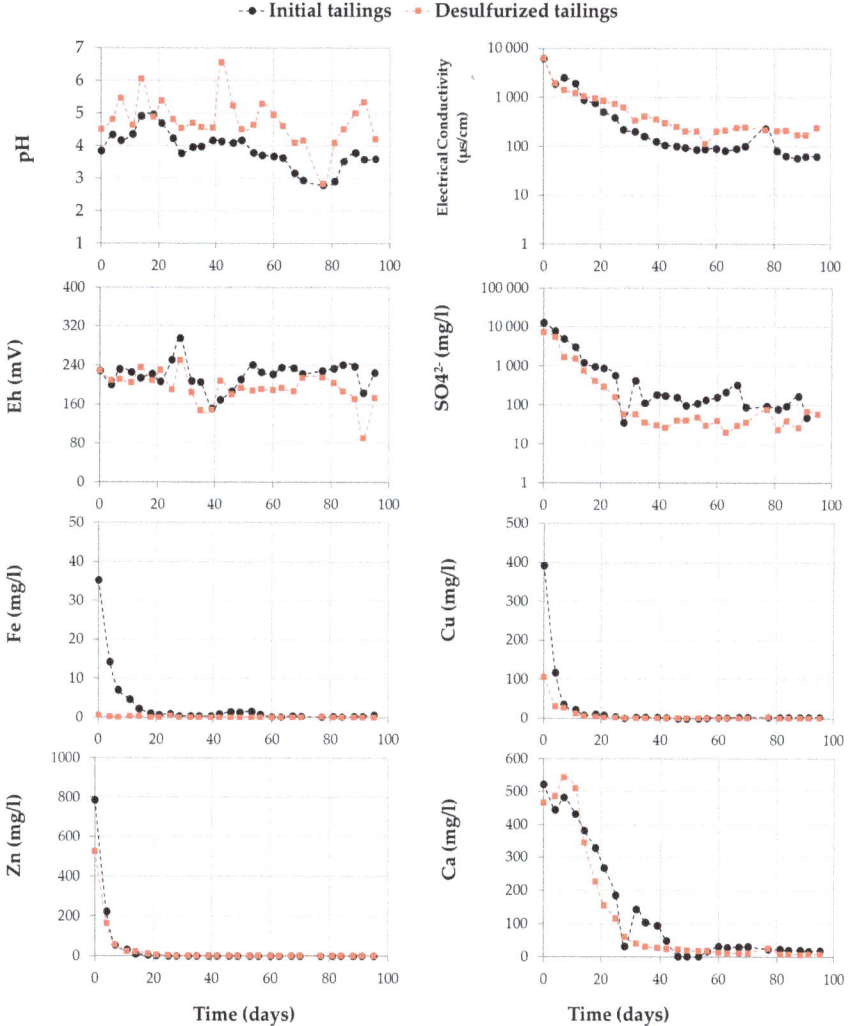

Figure 7. Evolution of pH, Eh, electrical conductivity (EC), and dissolved SO_4, Fe, Cu, Zn, and Ca in leachates after 27 kinetic test cycles.

4. Conclusions

Approximately 743,000 tons of tailings were generated at the Tiouit mine between 1950 and 1996 and are now stored in three tailings ponds around the mine site. The tailings are characterized by low sulfur contents (0.06–1.3 wt %), with sulfur present mostly in pyrite, as well as in trace amounts of chalcopyrite. These tailings, which are acid-generating according to static tests, have been exposed to atmospheric weathering and erosion for 60 years with no reclamation strategy in place. The Tiouit tailings are, however, also a potential secondary resource due to the presence of Au (3.36–5 ppm) and Ag (24–37 ppm). The residual grades of Au and Ag were originally refractory (inserted within sulfides), but the transformation of sulfides to oxy-hydroxides (due to natural weathering) has allowed for their recovery. Desulfurization of the southern tailings was achieved through flotation as a technique for controlling the production of AMD. The acid-generating potential of the desulfurized sample was assessed by kinetic testing. Geochemical data showed that desulfurization allows for good but not completely effective sulfide recovery. Furthermore, the desulfurized tailings produced leachates with slightly higher pH values than the initial tailings and were less loaded in terms of sulfate and dissolved metals.

Gold recovery from the flotation concentrate sample was only 4%, which suggests that the Au is no longer associated with sulfides. The flotation results helped to determine an adequate amount of activator and collector (50 g/t) to achieve sulfide recovery with the lowest weight-yield. As the remaining sulfides are almost entirely encapsulated, they are likely unreactive and, therefore, there was no need for additional desulfurization. Meanwhile, it is important to confirm these results with other laboratory tests on other samples and at pilot scale during the possible reopening of the flotation plant. In a wider context, it is always important to characterize the polluting potential of tailings in closed mines as well as the potential value of any residual metals. This may allow economically feasible mine waste-processing strategies (already milled ore) to be implemented, instead of very costly mitigation/reclamation of mine sites.

Author Contributions: A.N. conducted the majority of the flotation tests aside from the physical and chemical characterizations. The mineralogical characterization was conducted by H.B. The interpretation of results and writing were done by A.N. and Y.T. under the supervision of M.B. and R.H.

Funding: This work was supported through the International Research Chairs Initiative, a program funded by the International Development Research Centre and supported by the Canadian Research Chairs program, Canada.

Acknowledgments: Our acknowledgements go to the International Development Research Centre for funding and to the Unité de Recherche et de Service en Technologie Minérale (UQAT) for their technical support.

Conflicts of Interest: The authors declare no conflict of interest.

Abbreviations

AP	Acid-generating potential
CT	Central tailings
EC	Electrical conductivity
Eh	Redox potential
DT	Desulfurized tailings
EDS	Energy dispersive spectroscopy
SG	Specific gravity
MIBC	Methyl Isobutyl Carbinol
NT	Northern tailings
NNP	Net neutralization potential
NP	Acid neutralizing potential
ST	South tailings
SEM	Scanning electron microscopy
XRD	X-ray diffraction

References

1. Simate, G.S.; Ndlovu, S. Acid mine drainage: Challenges and opportunities. *J. Environ. Chem. Eng.* **2014**, *2*, 1785–1803. [CrossRef]
2. Kefeni, K.K.; Msagati, T.A.; Mamba, B.B. Acid mine drainage: Prevention, treatment options, and resource recovery: A review. *J. Clean. Prod.* **2017**, *151*, 475–493. [CrossRef]
3. Neculita, C.-M.; Zagury, G.J.; Bussière, B. Passive treatment of acid mine drainage in bioreactors using sulfate-reducing bacteria. *J. Environ. Qual.* **2007**, *36*, 1–16. [CrossRef] [PubMed]
4. Bussière, B.; Aubertin, M.; Zagury, G.J.; Potvin, P.; Benzaazoua, M. Principaux défis et pistes de solution pour la restauration des aires d'entreposage de rejets miniers abandonnées. In Proceedings of the Symposium 2005 sur l'environnement et les mines, Rouyn-Noranda, QC, Canada, 15–18 May 2005.
5. Aubertin, M.; Bernier, L.; Bussière, B. *Environnement et Gestion des Rejets Miniers [Ressource Électronique]: Manuel sur Cédérom*; Presses internationales Polytechnique: Mont-Royal, QC, Canada, 2002.
6. Parbhakar-Fox, A.; Lottermoser, B.G. A critical review of acid rock drainage prediction methods and practices. *Miner. Eng.* **2015**, *82*, 107–124. [CrossRef]
7. Hakkou, R.; Benzaazoua, M.; Bussière, B. Acid mine drainage at the abandoned Kettara mine (Morocco): 2. Mine waste geochemical behavior. *Mine Water Environ.* **2008**, *27*, 160–170. [CrossRef]
8. Ilankoon, I.M.S.K.; Tang, Y.; Ghorbani, Y.; Northey, S.; Yellishetty, M.; Deng, X.; McBride, D. The current state and future directions of percolation leaching in the Chinese mining industry: Challenges and opportunities. *Miner. Eng.* **2018**, *125*, 206–222. [CrossRef]
9. Parbhakar-Fox, A.; Lottermoser, B.; Bradshaw, D. Evaluating waste rock mineralogy and microtexture during kinetic testing for improved acid rock drainage prediction. *Miner. Eng.* **2013**, *52*, 111–124. [CrossRef]
10. Bouzahzah, H.; Benzaazoua, M.; Bussiere, B.; Plante, B. Prediction of acid mine drainage: Importance of mineralogy and the test protocols for static and kinetic tests. *Mine Water Environ.* **2014**, *33*, 54–65. [CrossRef]
11. Hudson-Edwards, K. Tackling mine wastes. *Science* **2016**, *352*, 288–290. [CrossRef]
12. Ritcey, G.M. *Tailings Management: Problems and Solutions in the Mining Industry*; Elsevier Science Publishers: Amsterdam, The Netherlands, 1989.
13. Mackay, I.; Mendez, E.; Molina, I.; Videla, A.R.; Cilliers, J.J.; Brito-Parada, P.R. Dynamic froth stability of copper flotation tailings. *Miner. Eng.* **2018**, *124*, 103–107. [CrossRef]
14. Lutandula, M.S.; Maloba, B. Recovery of cobalt and copper through reprocessing of tailings from flotation of oxidised ores. *J. Environ. Chem. Eng.* **2013**, *1*, 1085–1090. [CrossRef]
15. Edraki, M.; Baumgartl, T.; Manlapig, E.; Bradshaw, D.; Franks, D.M.; Moran, C.J. Designing mine tailings for better environmental, social and economic outcomes: A review of alternative approaches. *J. Clean. Prod.* **2014**, *84*, 411–420. [CrossRef]
16. Falagán, C.; Grail, B.M.; Johnson, D.B. New approaches for extracting and recovering metals from mine tailings. *Miner. Eng.* **2017**, *106*, 71–78. [CrossRef]
17. Alcalde, J.; Kelm, U.; Vergara, D. Historical assessment of metal recovery potential from old mine tailings: A study case for porphyry copper tailings, Chile. *Miner. Eng.* **2018**, *127*, 334–338. [CrossRef]
18. Yin, Z.; Sun, W.; Hu, Y.; Zhang, C.; Guan, Q.; Wu, K. Evaluation of the possibility of copper recovery from tailings by flotation through bench-scale, commissioning, and industrial tests. *J. Clean. Prod.* **2018**, *171*, 1039–1048. [CrossRef]
19. Park, J.; Han, Y.; Lee, E.; Choi, U.; Yoo, K.; Song, Y.; Kim, H. Bioleaching of highly concentrated arsenic mine tailings by Acidithiobacillus ferrooxidans. *Sep. Purif. Technol.* **2014**, *133*, 291–296. [CrossRef]
20. Lee, E.; Han, Y.; Park, J.; Hong, J.; Silva, R.A.; Kim, S.; Kim, H. Bioleaching of arsenic from highly contaminated mine tailings using Acidithiobacillus thiooxidans. *J. Environ. Manag.* **2015**, *147*, 124–131. [CrossRef] [PubMed]
21. Borja, D.; Lee, E.; Silva, R.A.; Kim, H.; Park, J.H.; Kim, H. Column bioleaching of arsenic from mine tailings using a mixed acidophilic culture: A technical feasibility assessment. *J. Korean Inst. Resour. Recycl.* **2015**, *24*, 69–77.
22. Chen, T.; Lei, C.; Yan, B.; Xiao, X. Metal recovery from the copper sulfide tailing with leaching and fractional precipitation technology. *Hydrometallurgy* **2014**, *147*, 178–182. [CrossRef]
23. Golik, V.; Komashchenko, V.; Morkun, V. Innovative technologies of metal extraction from the ore processing mill tailings and their integrated use. *Metall. Min. Ind.* **2015**, *7*, 49–52.

24. Taha, Y.; Benzaazoua, M.; Mansori, M.; Yvon, J.; Kanari, N.; Hakkou, R. Manufacturing of ceramic products using calamine hydrometallurgical processing wastes. *J. Clean. Prod.* **2016**, *127*, 500–510. [CrossRef]
25. Kinnunen, P.; Ismailov, A.; Solismaa, S.; Sreenivasan, H.; Räisänen, M.-L.; Levänen, E.; Illikainen, M. Recycling mine tailings in chemically bonded ceramics—A review. *J. Clean. Prod.* **2018**, *174*, 634–649. [CrossRef]
26. Qi, C.; Fourie, A.; Chen, Q.; Zhang, Q. A strength prediction model using artificial intelligence for recycling waste tailings as cemented paste backfill. *J. Clean. Prod.* **2018**, *183*, 566–578. [CrossRef]
27. Loutou, M.; Misrar, W.; Koudad, M.; Mansori, M.; Grase, L.; Favotto, C.; Taha, Y.; Hakkou, R. Phosphate Mine Tailing Recycling in Membrane Filter Manufacturing: Microstructure and Filtration Suitability. *Minerals* **2019**, *9*, 318. [CrossRef]
28. Brend, L.G. *Mine Wastes, Characterization, Treatment and Environmental Impacts*; Springer: Berlin, Germany, 2007.
29. Leppinen, J.; Salonsaari, P.; Palosaari, V. Flotation in acid mine drainage control: Beneficiation of concentrate. *Can. Metall. Q.* **1997**, *36*, 225–230. [CrossRef]
30. Benzaazoua, M.; Bussière, B.; Kongolo, M.; McLaughlin, J.; Marion, P. Environmental desulphurization of four Canadian mine tailings using froth flotation. *Int. J. Miner. Process.* **2000**, *60*, 57–74. [CrossRef]
31. Benzaazoua, M.; Kongolo, M. Physico-chemical properties of tailing slurries during environmental desulphurization by froth flotation. *Int. J. Miner. Process.* **2003**, *69*, 221–234. [CrossRef]
32. Bois, D.; Poirier, P.; Benzaazoua, M.; Bussière, B.; Kongolo, M. A feasibility study on the use of desulphurized tailings to control acid mine drainage. *Cim Bull.* **2005**, *98*, 1.
33. Mermillod-Blondin, R. Influence des Propriétés Superficielles de la Pyrite et des Minéraux Sulfurés Associés sur la Rétention de Molécules Organiques Soufrées et Aminées: Application à la Désulfuration Environnementale. Ph.D. Thesis, Institut National Polytechnique de Lorraine, Nancy, France, et École Polytechnique de Montréal, Montreal, QC, Canada, 2005.
34. Demers, I.; Bussière, B.; Benzaazoua, M.; Mbonimpa, M.; Blier, A. Column test investigation on the performance of monolayer covers made of desulphurized tailings to prevent acid mine drainage. *Miner. Eng.* **2008**, *21*, 317–329. [CrossRef]
35. Hakkou, R.; Benzaazoua, M.; Bussière, B. Acid mine drainage at the abandoned Kettara mine (Morocco): 1. Environmental characterization. *Mine Water Environ.* **2008**, *27*, 145–159. [CrossRef]
36. Alansari, A.; Mouguina, E.; Maacha, L. 1.5-Le gisement de Tiouit à Au-Cu-Ag (Massif néoprotérozoïque du J. Saghro). In *Les principales mines du Maroc*; Mouttaqi, A., Rjimati, E.C., Maacha, L., Michard, A., Soulaimani, A., Ibouh, H., Eds.; Editions du Service géologique du Maroc: Rabat, Morocco, 2011; Volume 564, pp. 53–57.
37. Chaker, M. Geochimie et Metallogenie de la Mine d'or de Tiouit, Anti-Atlas Oriental, Sud du Maroc. Ph.D. Thesis, Université du Québec à Chicoutimi, Saguenay, QC, Canada, 1997.
38. Roche-Invest. *Oral Communication (Confidential Report)*; Roche-Invest: Basel, Switzerland, 2016.
39. Alansari, A. La Mine D'or de Tiouit: Un Exemple de Veines Aurifères Mésothermales, Associées à Une Granodiorite d'âge Protérozoïque Supérieur (Massif panafrican du Jbel Saghro, Anti-Atlas, Maroc). Ph.D. Thesis, University Cadi Ayyad, Marrakech, Morocco, 1997; p. 284.
40. Alansari, A.; Sagon, J.P. Le gisement d'or de Tiouit (Jbel Saghro, Anti-Atlas, Maroc), un système mésothérmal polyphasé à sulfures-or et hématite-or dans une granodiorite potassique d'âge protérozoique supérieur. *Chron. Rech. Min.* **1997**, *527*, 3–25.
41. Pirard, E. Modal analysis of mineralogical blends using optical image analysis versus X-ray diffraction and ICP. In Proceedings of the 9th International Congress for Applied Mineralogy (ICAM), Brisbane, Australia, 8–10 September 2008; p. 673679.
42. Raudsepp, M.; Pani, E. Application of Rietveld analysis to environmental mineralogy. *Environ. Asp. Mine Wastes* **2003**, *31*, 165–180.
43. Benzaazoua, M.; Marion, P.; Robaut, F.; Pinto, A. Gold-bearing arsenopyrite and pyrite in refractory ores: Analytical refinements and new understanding of gold mineralogy. *Mineral. Mag.* **2007**, *71*, 123–142. [CrossRef]
44. Lawrence, R. Prediction of the behavior of mining and processing wastes in the environment. In *Proc. Western Regional Symposium on Mining and Mineral Processing Wastes*; Doyle, F., Ed.; Society for Mining, Metallurgy, and Exploration, Inc.: Littleton, CO, USA, 1990.
45. Lawrence, R.W.; Wang, Y. Determination of neutralization potential in the prediction of acid rock drainage. In Proceedings of the Fourth International Conference on Acid Rock Drainage, Vancouver, BC, Canada, 31 May–6 June 1997; pp. 451–464.

46. Lawrence, R.W.; Marchant, P.M. *Acid Rock Drainage Prediction Manual*; MEND/NEDEM Report 1.16.1b; Canadian Centre for Mineral and Energy Technology: Ottawa, ON, Canada, 1991.
47. Sobek, A.A.; Schuller, W.; Freeman, J.; Smith, R. *Field and Laboratory Methods Applicable to Overburdens and Minesoils*; US Environmental Protection Agency: Cincinnati, OH, USA, 1978; Volume 45268, pp. 47–50.
48. Miller, S.; Jeffery, J.; Wong, J. Use and misuse of the acid base account for "AMD" prediction. In Proceedings of the 2nd International Conference on the Abatement of Acidic Drainage, Montréal, QC, Canada, 16–18 September 1991; pp. 16–18.
49. Adam, K.; Kourtis, A.; Gazea, B.; Kontopoulos, A. Evaluation of static tests used to predict the potential for acid drainage generation at sulphide mine sites. *Trans. Inst. Min. Metall. Sect. A Min. Ind.* **1997**, *106*, A1.
50. Cruz, R.; Bertrand, V.; Monroy, M.; González, I. Effect of sulfide impurities on the reactivity of pyrite and pyritic concentrates: A multi-tool approach. *Appl. Geochem.* **2001**, *16*, 803–819. [CrossRef]
51. Bouzahzah, H.; Benzaazoua, M.; Bussière, B. A modified protocol of the ASTM normalized humidity cell test as laboratory weathering method of concentrator tailings. In Proceedings of the International Mine Water and the Environment (IMWA), Mine Water and Innovative Thinking, Sydney, NS, Canada, 5–9 September 2010; pp. 15–18.
52. Bouzahzah, H.; Benzaazoua, M.; Bussière, B.; Plante, B. ASTM Normalized Humidity Cell Kinetic Test: Protocol Improvements for Optimal Sulfide Tailings Reactivity. *Mine Water Environ.* **2015**, *34*, 242–257. [CrossRef]
53. Villeneuve, M. *Évaluation du Comportement Géochimique à Long Terme de Rejets Miniers à Faible Potentiel de Génération D'acide à L'aide D'essais Cinétiques*; Université de Montréal—Ecole Polytechnique Montréal: Montreal, QC, Canada, 2004.
54. Elghali, A.; Benzaazoua, M.; Bussière, B.; Bouzahzah, H. Determination of the available acid-generating potential of waste rock, part II: Waste management involvement. *Appl. Geochem.* **2019**, *100*, 316–325. [CrossRef]
55. Elghali, A.; Benzaazoua, M.; Bussière, B.; Genty, T. Spatial Mapping of Acidity and Geochemical Properties of Oxidized Tailings within the Former Eagle/Telbel Mine Site. *Minerals* **2019**, *9*, 180. [CrossRef]
56. Benzaazoua, M.; Bouzahzah, H.; Taha, Y.; Kormos, L.; Kabombo, D.; Lessard, F.; Bussière, B.; Demers, I.; Kongolo, M. Integrated environmental management of pyrrhotite tailings at Raglan Mine: Part 1 challenges of desulphurization process and reactivity prediction. *J. Clean. Prod.* **2018**, *162*, 86–95. [CrossRef]
57. Elghali, A.; Benzaazoua, M.; Bouzahzah, H.; Bussière, B.; Villarraga-Gómez, H. Determination of the available acid-generating potential of waste rock, part I: Mineralogical approach. *Appl. Geochem.* **2018**, *99*, 31–41. [CrossRef]

© 2019 by the authors. Licensee MDPI, Basel, Switzerland. This article is an open access article distributed under the terms and conditions of the Creative Commons Attribution (CC BY) license (http://creativecommons.org/licenses/by/4.0/).

Article

Recovery of Residual Silver-Bearing Minerals from Low-Grade Tailings by Froth Flotation: The Case of Zgounder Mine, Morocco

Boujemaa Drif [1,2], Yassine Taha [3,*], Rachid Hakkou [1,3,*] and Mostafa Benzaazoua [4,5]

1. Laboratoire de Chimie des Matériaux et de l'Environnement, Faculté des Sciences et Techniques, Université Cadi Ayyad, Marrakech BP 549/40000, Morocco; boj.adrif@gmail.com
2. Zgounder Millenium Silver Mining Company, 3 Rue de l'Epargne, 1st Floor Racine, Casablanca 20100, Morocco
3. Materials Science and Nano-engineering Department, Mohammed VI Polytechnic University, Lot 660, Hay Moulay Rachid, Ben Guerir 43150, Morocco
4. Institut de Recherche en Mines et en Environnement, Université du Québec en Abitibi-Témiscamingue, 445 Boulevard de l'Université, Rouyn-Noranda, QC J9X5E4, Canada; mostafa.benzaazoua@uqat.ca
5. Geology and Sustainable Mining Department, Mohammed VI Polytechnic University, Lot 660, Hay Moulay Rachid, Ben Guerir 43150, Morocco
* Correspondence: yassine.taha@um6p.ma (Y.T.); r.hakkou@uca.ma (R.H.)

Received: 23 May 2018; Accepted: 25 June 2018; Published: 27 June 2018

Abstract: The need to explore more complex and low-grade silver ores and to develop novel and cost-effective processes to recover silver from waste is becoming an important challenge. This paper aims to characterize old, low-grade, silver tailings generated by the former Zgounder silver mine, located in Morocco. Understanding the mineralogical composition, particularly the silver deportment, was critical to allow the recovery of silver from these tailings. More than 88 samples of low grade tailings were sampled and characterized using chemical and mineralogical techniques. Froth flotation was used to recover silver bearing minerals using a combination of different collectors (dithiophosphate, dialkyl dithiophosphinates, Aero 7518, Aero 7640, alkyl dithiophosphates and potassium butyl-xanthate). The main goal was to optimize the flotation process at a laboratory scale through the testing of different parameters, such as collectors and frother types and dosage, activators and sulphidizing agents, and pH conditions. The characterization results showed that silver content varied between 30 and 440 ppm with an overall average content of 148 ppm. Silver occurs mainly in the form of native silver as well as in association with sulphides, such as acanthite and pyrite. Minor amounts of sphalerite, chalcopyrite, arsenopyrite, and hematite were identified. The flotation results showed the following optimum conditions: particle size of 63 µm, conditioning pH of 8.5, a combination of butyl-xanthate and dithiophosphate as collectors at a dosage of 80 g/t each, a concentration of 200 g/t of the activating agent ($CuSO_4$), 30 g/t of methyl isobutyl carbonyl (MIBC) frother and a duration time of 8 min with slow kinetics. With these optimal conditions, it was possible to achieve a maximum silver recovery yield of 84% with 1745 ppm Ag grade to be cyanided. Moreover, the environmental behavior of the final clean tailings was demonstrated to be inert using Toxicity characteristic leaching procedure (TCLP) leaching tests.

Keywords: Zgounder mine; silver; tailings; low-grade silver ores; flotation; mineralogical characterization

1. Introduction

Recently, the demand for silver has increased exponentially due to the growing development of green technologies and nanotechnologies where silver is greatly required. Paradoxically to this high demand, the price of silver has increased, and world reserves are becoming scarcer. Worldwide silver reserves are significantly decreasing, and their price is continuously growing [1,2]. Silver price has more than tripled in the last two decades [3], and in the near future, silver will potentially be more expensive than gold. Therefore, the processing of complex low-grade ores has become a critical issue which the mining industry is facing. Many challenges and opportunities need to be addressed. Further research should be carried out in order to develop processing methods that take into account the effects of mineralogy, pre-processing methods, surface chemistry in regard to flotation rates, the activation and passivation of gangue minerals, the liberation degree of valuable minerals, etc. These aspects have been widely studied for gold, lead, zinc, and copper minerals. However, the literature shows that only few studies have dealt with the recovery of silver bearing minerals by froth flotation.

Regarding mineralogy, silver might be associated with hundreds of different silver bearing minerals with a wide range of compositions. This natural issue makes the recovery of silver bearing minerals a more complex and challenging task. Researchers have classified silver bearing minerals into eight different groups: silver alloys [4,5], sulphides [6], selenides [7,8], antimonides [9], tellurides [8,10], sulphosalts [11,12], halides [13], and solid solution [14]. Acanthite and argentite are the main silver sulphide minerals reported in the literature [14–16]. Silver may be associated with halide group minerals and sulphosalts. In some sulphosalts, Ag does not appear in the chemical formula because it occurs in the lattice sites. Ag can also be associated with telluride group minerals. It is highly associated with Au and sulphides. The naumannite mineral is the main Ag bearing mineral in the group of selenides. It is generally found in association with quartz and carbonates in hydrothermal veins. Silver can also take the form of a solid solution when some sulphides, such as chalcopyrite, sphalerite, galena and pyrite, are present in the silver ore. Special attention should be given to these minerals, because silver can replace iron, zinc, palladium, nickel, and cobalt [14].

Concerning silver beneficiation, different metallurgical processing techniques have been developed in order to recover silver from complex ores. Two main parameters need to be understood before selecting the appropriate recovery process: (i) silver deportment among Ag minerals (sulfosalts) and as a trace substitution within other minerals (pyrite, galena, etc.); and (ii) silver occurrence in terms of particle size distribution. These parameters can help to better choose the appropriate treatment method to recover silver bearing minerals of interest.

The flotation technique has generally been used to selectively separate valuable minerals from gangue non-valuable materials using the hydrophobicity properties [17–22]. It has been widely used to recover silver minerals from different ore deposits [23–26]. Silver minerals can be recovered either by selective or bulk flotation depending on the amount of minerals of interest within the ore [11]. Various types of reagents could be used to recover silver, based on the surface properties and the nature of present minerals. Activators, pH modifiers, depressants, collectors, frothers, and dispersants in different commercial nomenclatures are commonly used. Dorr and Bosqui [25] were among the first researchers interested in the effects of different reagents on the efficiency of silver mineral recovery. They described the reagents used in Mochito mine collectors (Aerofloat 25/31, amyl and butyl xanthates, Aerofloat 208 and 404); pH modifiers (sodium carbonate); activators (copper sulphate); depressants (starch); dispersants (sodium silicate); and sulphidising reagents. A study performed by Thompson [27] investigated the most appropriate reagents to use for the recovery of silver from an ore containing 350 ppm of Ag. Silver was associated with sulphosalts (freibergite, pyrargyrite), sulphides (proustite, argentite), and in solid solution with pyrite. It was shown that it is possible to float 82% of silver by using a combination of collectors (thionocarbamate with ethyl xanthate) to produce a concentrate containing 17,500 ppm of Ag. In another study, xanthates and dithiophosphates as collectors with MIBC, pine oil, or polypropylene glycol as frothers were used to recover Cu–Ag, Co–Ag, or Cu–Ag–Bi bearing minerals [6]. Lime was used to achieve the neutral pH. The flotation recovery yields might also depend on the grain particle size. Indeed, it was demonstrated in the literature that low flotation rates are obtained when particles are finely ground due to the low collision efficiency between particles and bubbles [28–30]. Therefore, multiple flotation processes have been developed in order to enhance the bubble–particle collision efficiency, either by decreasing the bubble size or by increasing the apparent particle size [31].

However, it is important to mention that, to the best of knowledge of the current paper's authors, no study related to the recovery of silver from old mine tailings using flotation has yet been reported. Pioneer studies should be carried out in order to understand the reactional mechanisms affecting the flotation of silver bearing minerals. The recovery of silver from old, weathered tailings may present a challenge (since mineral surfaces are oxidized), and at the same time, could be a profitable opportunity if it is economically feasible and environmentally viable.

Morocco is one of the main world producers of silver, classified 16th according to 2014 statistics [32]. In this study, complex low-grade and weathered tailings from the Zgounder mine, one of the oldest silver mines in Morocco, are tested for the recovery of residual silver bearing minerals using froth flotation as a pretreatment before cyanidation. The unexpectedly high silver values in the Zgounder old tailings storage facility (TSF) are believed to be due to insufficient residence time in the reaction tanks; the presence of coarse silver particles; and an insufficiently fine grind during previous operations. Therefore, this study aimed to recover silver from Zgounder low grade tailings using froth flotation. The effect of several parameters on the silver recovery rate, such as particle size, pH, types and amounts of collectors, frothers, activation, and sulphidization were investigated. The main objective was to optimize the parameters affecting silver recovery conditions.

2. Materials and Methods

2.1. Sampling Campaign

Zgounder mine is located in the Western part of the central Anti-Atlas Mountain Range, 150 km south of Marrakech, Morocco. Silver in Zgounder ores is enriched at narrow mineralized veins of chlorite, quartz, siderite, and disseminated ore minerals [33]. It generally occurs in the form of native silver, mercurian silver, silver bearing sulphides and sulphosalts. Silver ore was processed by crushing, milling and cyanidation. The mine had undergone two mine exploitation phases. In the first phase, silver was extracted from 1982 to 1990 by crushing and milling, followed by cyanidation. The site was closed in 1990 and the mine was reopened again in July 2014 by Zgounder Millenium Silver Mining company. The north tailings pond of the Zgounder mine was selected in this study due to its significant volume of tailings produced between 1982 and 1990 (more than 400,000 tons) and its content of Ag (an average grade of 125 g/t silver) [34]. A quantity of 88 samples (approximately 6 kg for each sample) was taken from the tailings pond at a depth of 1.6 m, as presented in Figure 1. The samples were homogenized and stored in sealed bottles to ensure minimal contact with air. All of the samples were dried at a low temperature, disaggregated, homogenized, separated, and sealed in plastic bags until testing.

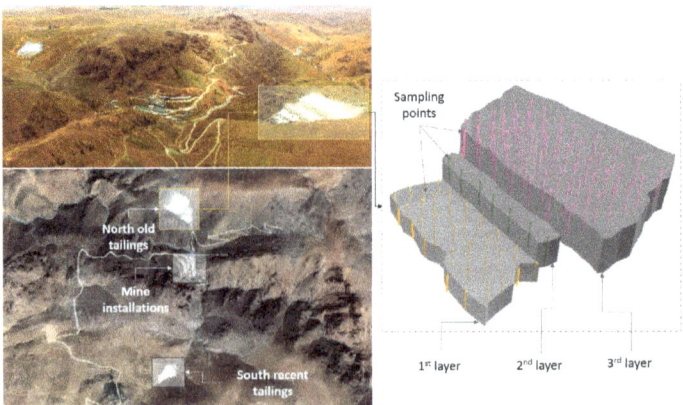

Figure 1. Location of the Zgounder mine tailings storage facilities and the used sampling procedure.

2.2. Tailing Characterization

The volumetric particle size distribution of the tailings was determined using a laser analyzer (Malvern Mastersizer; ISO-13320, Panalytical, Almelo, The Netherlands). The specific gravity (Gs) measurement was performed with a helium gas pycnometer (Micromeritics Accupyc 1330, Micromeritics, Norcross, GA, USA). The chemical composition of the tailings was determined with atomic absorption equipment (Thermo Fisher Scientific, Waltham, MA, USA) following HNO_3/HCl digestion. The inorganic carbon content (Cin) was determined using a LECO furnace with a ±0.05 to 0.1 wt % precision. The crystalline minerals performed only for the composite sample were determined by X-ray diffraction (Bruker AXS Advance D8, Bruker, Billerica, MA, USA), using CuKα radiation with a fixed counting time of 4 s in the range from 5° to 70° with steps of 0.02° in 2θ. The quantitative phase analysis of the tailings was performed using the TOPAS 4.2 program (Bruker, Billerica, MA, USA), based on the Rietveld analysis and data reconciliation against chemical analysis. Scanning electron microscopy (SEM) observations using backscattered electrons (BSE) were performed on polished sections prepared with the bulk samples and an epoxy resin using a Hitachi S-3500N microscope equipped with an X-ray energy dispersive spectrometer (Silicon Drift Detector Bruker, Bruker, Billerica, MA, USA) with ESPRIT 2® software (Bruker, Billerica, MA, USA). The static tests were also performed to evaluate the environmental behavior of Zgounder mine tailing (ZMT) waste. This test was conducted using acid-base accounting, according to the Sobek method modified by Lawrence and Scheske [35].

2.3. Flotation Tests

Before flotation, the samples were tested as raw tailings or were reground at two levels (less 105 and 63 microns) in order to liberate silver and refresh mineral surfaces. Numerous flotation tests were performed in triplicate on Zgounder mine tailings (ZMT) using a Denver D-12 lab flotation machine (911 Metallurgist Corporation, Kamloops, BC, Canada) with a cell volume of 4 L. All of the samples were dried, mixed, homogenized, and divided into representative samples for the flotation tests. The objective was to assess the effects of the main parameters on silver recovery and grade. Many reagents were used. It is known that the reagent types and their dosages highly affect the recovery of silver bearing minerals. The effectiveness of these reagents depends mainly on their functionalities and the minerals' particle compositions and surfaces [36,37]. The synthesized methodology and the reagents used in this study are illustrated in Figure 2.

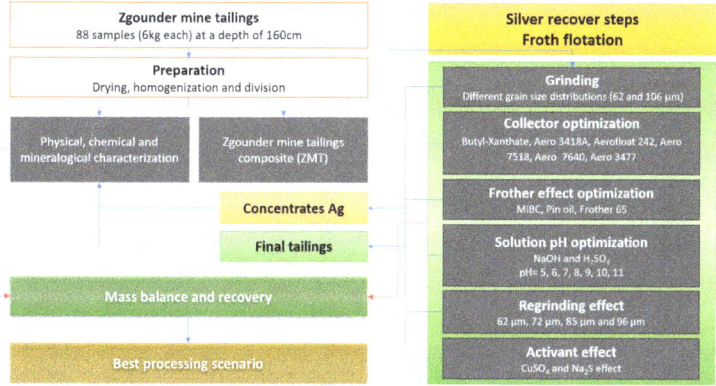

Figure 2. The methodology scheme used in this study.

Figure 3 presents the flotation setup used to recover silver bearing minerals. During all rougher and scavenging flotation experiments, the solid content was maintained between 30% and 32%, as commonly used in the industry. The speed of the rotor-stator was adjusted to 1500 rpm, and airflow varied between at 2 and 3 L·min^{-1}. Combinations of different families of collectors were used, such as dithiophosphate (Aerofloat 242), dialkyl dithiophosphinates (Aero 3418A), alkyl dithiophosphates (Aero 3477) and potassium butyl-xanthate (PBX), as well as other special formulations, such as Aero 7518 and Aero 7640 [38]. Sodium sulfide (Na$_2$S) was used as a sulphidising reagent, while CuSO$_4$ was used as an activator. Sodium hydroxide (NaOH) and sulphuric acid (H$_2$SO$_4$) were used as pH modifiers. Methyl isobutyl carbonyl (MIBC), pin oil and polypropylene glycol frother (Aerofroth 65) were used as frothers. The test flotation procedure involved a series of tests with two roughing stages followed by one stage of cleaning and scavenging. Flotation optimization was based on testing different parameters classified in a priority order as well as using industrial experience. The flotation kinetics; effect of pH; effect of grinding; and the effects of several types of collectors and frothers were verified in the study (Table 1).

Table 1. Synthesis of all the studied parameters and related conditions.

1-	Effect of pH and Type of Collector								
	pH	D90	Methyl Isobutyl Carbonyl	Butyl-Xanthate	Dialkyl Dithiophosphinates	Dithiophosphate	Aero 7518	Aero 7640	Alkyl Dithiophosphates
Test-1					100 g/t				
Test-2						100 g/t			
Test-3	7.5	62 μm	30 g/t	60 g/t			100 g/t		
Test-4								100 g/t	
Test-5									100 g/t
Test-6					100 g/t				
Test-7						100 g/t			
Test-8	9	62 μm	30 g/t	60 g/t			100 g/t		
Test-9								100 g/t	
Test-10									100 g/t
Test-11					100 g/t				
Test-12						100 g/t			
Test-13	10.5	62 μm	30 g/t	60 g/t			100 g/t		
Test-14								100 g/t	
Test-15									100 g/t
2-	Effect of pH and Type of Frother								
Parameter	pH	D90	Methyl Isobutyl Carbonyl	Butyl-Xanthate	Polypropy-lene Glycol	Dithiophosphate		Pin Oil	
Test-16	7.5		30 g/t		30 g/t	100 g/t		30 g/t	
Test-17	8.5	62 μm	30 g/t	60 g/t	30 g/t			30 g/t	
Test-18	9.5		30 g/t		30 g/t			30 g/t	
Test-19	10.5		30 g/t		30 g/t			30 g/t	
3-	Effect of Optimized Collector Dosage								
	pH		D90		Methyl Isobutyl Carbonyl		Butyl-Xanthate		Dithiophosphate
Test-20							40 g/t		60 g/t
Test-21	8.5		62 μm		30 g/t		60 g/t		40 g/t
Test-22							60 g/t		60 g/t

Table 1. Cont.

Test-23					80 g/t	60 g/t
Test-24					80 g/t	80 g/t
Test-25					80 g/t	100 g/t
Test-26					100 g/t	100 g/t
4-	Effect of Solution pH					
	pH	D90	Methyl Isobutyl Carbonyl		Butyl-Xanthate	Dithiophosphate
Test-27	5					
Test-28	6					
Test-29	7					
Test-30	8	62 µm	30 g/t		80 g/t	80 g/t
Test-31	9					
Test-32	10					
Test-33	11					
5-	Effect of Grinding					
	pH	D90	Methyl Isobutyl Carbonyl		Butyl-Xanthate	Dithiophosphate
Test-34		62 µm				
Test-35	8.5	72 µm	30 g/t		80 g/t	80 g/t
Test-36		85 µm				
Test-37		96 µm				
6-	Effect of Activation					
	pH	D90	Methyl Isobutyl Carbonyl	Butyl-xanthate	Dithiophosphate	CuSO₄
Test-38						0 g/t
Test-39	8.5	62 µm	30 g/t	80 g/t	80 g/t	100 g/t
Test-40						200 g/t
Test-41						300 g/t
7-	Effect of Sulphidization					
	pH	D90	Methyl Isobutyl Carbonyl	Butyl-Xanthate	Dithiophosphate	Na₂S
Test-42						0 g/t
Test-43	8.5	62 µm	30 g/t	80 g/t	80 g/t	300 g/t
Test-44						500 g/t

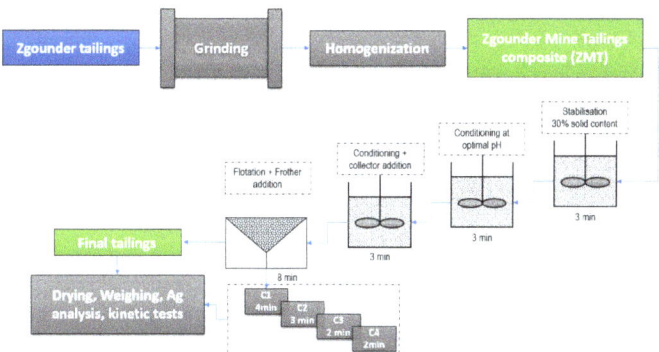

Figure 3. Zgounder mine tailing (ZMT) preparation and flotation test scheme.

3. Results and Discussion

3.1. Tailing Characterization Results

The chemical properties obtained by atomic absorption spectrometry (AAS) of all of the samples are presented in Figure 4. The results showed that the content of silver in old Zgounder mine tailings varied between 30 and 440 ppm. The general average of the samples was about 142 ppm. After a composite sample was prepared from the different samples, the chemical analysis showed that silver content in these samples was about 148 ppm.

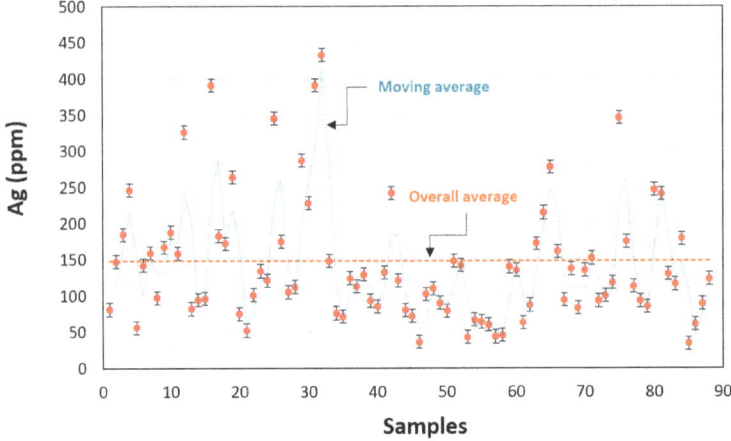

Figure 4. Variation of silver grade in the different samples taken from Zgounder mine tailings (TSF).

Table 2 presents the physical, chemical, mineralogical, and environmental characteristics of the prepared ZMT composite sample. The particle size distribution results show the fine-grained particle size distribution of the tailings with a D80 of 150 µm. Some traces of heavy metals (Pb, Zn, Cu) and metalloids (As and Sb) were also detected. The sulphur content was about 0.58% in the ZMT composite. The results of the static tests showed that the net neutralization potential (NNP) of ZMT was around -9.8 kg $CaCO_3$/t, thus classifying the ZMT as having an uncertain acid generation potential (-20 kg $CaCO_3$/t < NNP < +20 kg $CaCO_3$/t). This result was expected due to the mineralogical composition of Zgounder tailings mainly comprising non-sulfidic gangue minerals (more than 95 wt %). The XRD results showed that the ZMT composite is mainly composed of gangue minerals, muscovite, quartz, albite, actinolite, and orthoclase. Minor amounts of pyrite, sphalerite, chalcopyrite, arsenopyrite, and hematite were identified by SEM observations. Figure 5 shows some particles of these minerals and their associations.

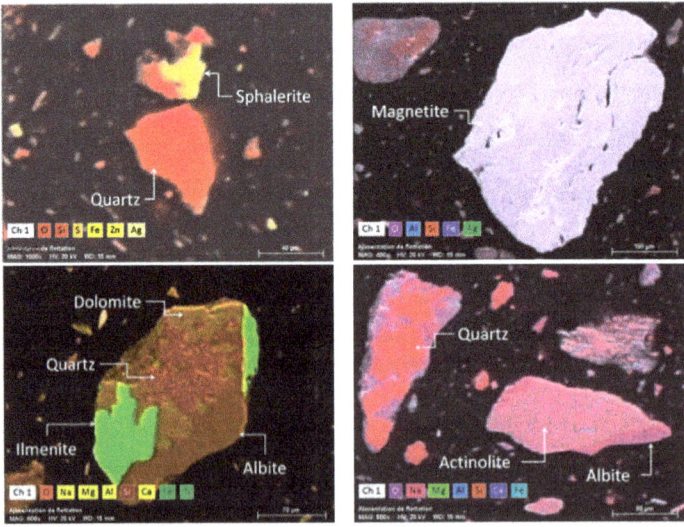

Figure 5. Combined BSE SEM and X-mapping images of some minerals in the ZMT composite sample.

Table 2. Physical, chemical, and mineralogical characteristics of the ZMT composite sample.

Physical Properties		Zgounder Mine Tailings (ZMT)
D_{20} (μm)		43
D_{50} (μm)		74
D_{80} (μm)		150
Gs (g/cm^3)		1.6
Chemical Composition		
Ag	ppm	148
Au	ppm	<0.1
As	ppm	0.04
Sb	ppm	86
Pb	%	0.24
Zn	%	0.37
Fe	%	5.41
Cu	%	0.08
S	%	0.58
C	%	0.10
Static Test		
AP (kg CaCO$_3$/t)		18.1
NP (kg CaCO$_3$/t)		8.3
NNP (kg CaCO$_3$/t)		−9.8
Acid generation		Uncertain
Mineralogical Composition (%)		
Quartz	SiO_2	14.8
Muscovite	$KAl2(Si_3Al)O_{10}(OHF)_2$	16.2
Albite	$NaAlSi_3O_8$	31.0
Chlorite	$(Fe,Mg,Al)_6(SiAl)_4O_{10}(OH)_8$	13.6
Actinolite	$Ca_2(Mg,Fe)_5Si_8O_{22}(OH)_2$	14.7
Orthoclase	$KAlSi_3O_8$	5.4
Rutile	TiO_2	2.1
Sphalerite	ZnS	0.5
Pyrite	FeS_2	0.5
Chalcopyrite	$CuFeS_2$	0.1
Arsenopyrite	FeAsS	0.1
Galena	PbS	0.3
Dolomite	$CaMg(CO_3)_2$	0.1
Butlerite	$Fe(SO_4)(OH)_2(H_2O)$	0.5

Abbreviation: AP—Acidity potential; NP—Neutralization potential; NNP—Net neutralization potential.

3.2. Size-by-Size Chemistry Analysis

The Zgounder mine tailing sample was homogenized and sieved into different grain size fractions: +300 μm, 212–300 μm, 150–212 μm, 106–150 μm, 75 + 106 μm, 45–75 μm, and 0–45 μm. The objective was to define the fraction that is richest in silver. The distributions and grades of silver in these fractions were determined and are presented in Figure 6. The measured silver content in the ZMT composite sample was 148 ppm. An Ag analysis on a size-by-size basis indicated that silver varied between 100 and 200 ppm in the finer and intermediate size fractions (0–300 μm), while the coarse fraction recorded the highest silver content: 540 ppm.

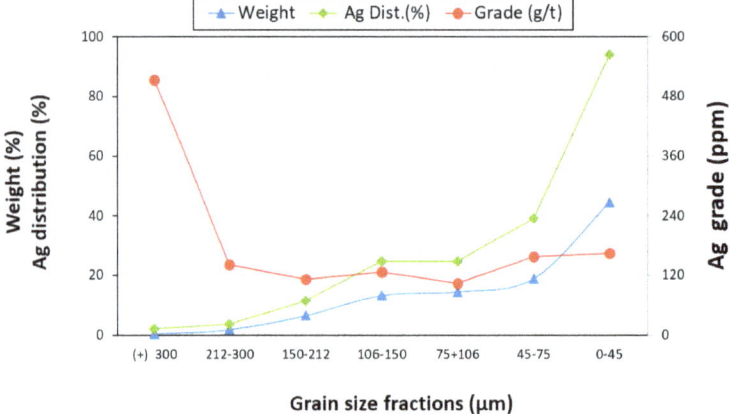

Figure 6. Distributions and grades of silver in different grain size fractions.

3.3. Flotation Kinetics

Preliminary tests were conducted in order to determine the optimal conditions of flotation in terms of flotation residence time and grain size distribution. Two rougher kinetic flotation tests were conducted on the composite following the conditions described below. The ZMT composite sample was ground to prepare distinct samples with D_{90} of 62 µm and 106 µm to be floated as well as the raw ZMT sample. Butyl-xanthate (60 g/t) and dithiophosphate (100 g/t) were used as collectors, MIBC (35 g/t) as collector. This combination of reagents is widely recognized as being a strong and selective Ag bearing mineral collector in the industry. The rougher kinetics of silver (Ag) flotation at two different grinding levels are graphically shown in Figure 7.

The results of the rougher testing proved that the recovery of Ag is higher when the particle size distribution is finer. A rougher recovery rate of nearly 52 wt % was achieved when a D90 of 62 µm was used, while only 38 wt % and 32 wt % (106 µm and 150 µm, respectively) were achieved for coarser tailings. The maximum mass recovery was also obtained with the fine grinded material (31 wt %). The silver recovery reached a recovery plateau after a flotation duration of around 8 min. Under these conditions, to obtain a greater Ag recovery yield, particular attention should be given to the effects of reagents (type and dosage). Therefore, the following tests were all carried out with a particle size of D90 = 62 µm and a flotation time of 8 min.

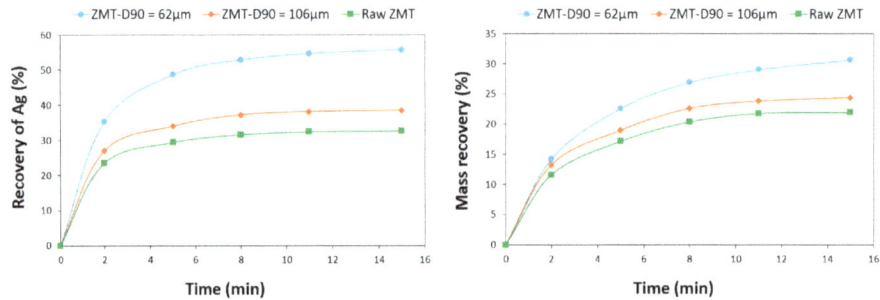

Figure 7. Flotation kinetics of mineral silver at two grain sizes: (**a**) silver recovery and (**b**) mass recovery as a function of flotation residence time.

3.4. Selection of Collector and Optimal pH

The next flotation experiments were carried out in order to select the optimal collector mix and conditioning pH. Combinations of butyl-xanthate and five other collectors were tested. A summary of the conducted tests is highlighted in Table 1-1. The dosage of each collector was 60 g/t for collector 1 (butyl-xanthate) and 100 g/t for collector 2 (dialkyl dithiophosphinates (Aero 3418A), dithiophosphate (Aerofloat 242), Aero7518, Aer7640 and alkyl dithiophosphates (Aero 3477)). The results presented in Figure 8 show that the best Ag recovery yields were obtained when butyl-xanthate was mixed with dithiophosphate or dialkyl dithiophosphinates. It was possible to reach a silver recovery greater than 50 wt % with these collectors, while a recovery yield of 30 wt % was achieved with the other collector combinations at each of the three pHs (7.5, 9 and 10.5). At a pH of 7.5, recovery yields and silver grades of 56 wt % and 337 ppm and 53 wt % and 295 ppm were achieved, respectively, with dithiophosphate and dialkyl dithiophosphinates. It was also observed that the best recovery yield was obtained with pH values of 7.5 and 9. Even though an increase in silver grade was recorded at a pH of 10 for the same collector combinations, recovery yields decreased significantly to less than 50 wt %. Out of these two combinations, butyl-xanthate/dithiophosphate gave the best rates. Also, from a price/efficiency ratio perspectives, dithiophosphate was chosen for the rest of experiments in this study.

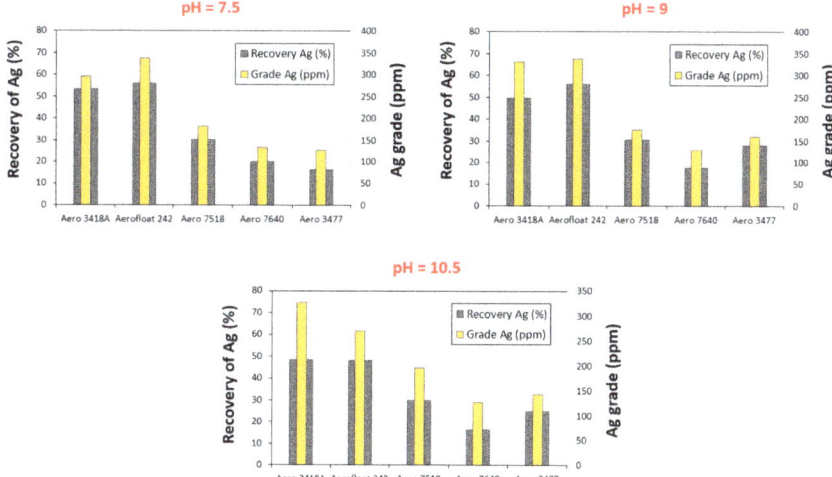

Figure 8. Grade and recovery of silver for different collector combinations and under different pH conditions.

3.5. Selection of the Frother

Several frothers were tested using the previous optimized conditions in terms of grain size, residence time, and collector combinations. More details about the tests that were carried out are given in Table 1-2. The results are graphically highlighted in Figure 9. The best silver recovery yield and grade were obtained with MIBC at the different pH conditions. The overall yield for MIBC at the four pH levels was greater than 40 wt %. Therefore, the MIBC frother was selected for the other tests due the froth quality, silver recovery, and concentrate quality.

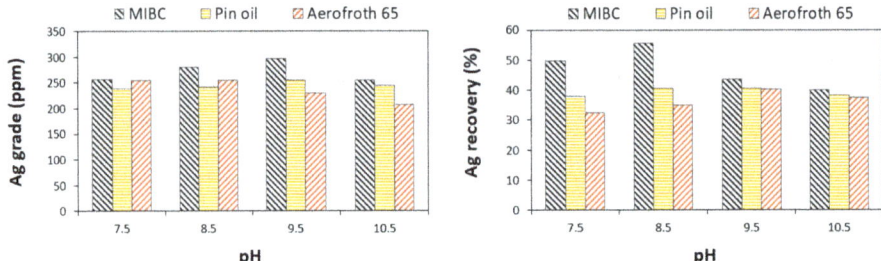

Figure 9. Grade and recovery of silver for different types of frothers and under different pH levels.

3.6. Effect of Collector Dosage

During this stage, the ZMT composite sample was floated at the previously determined optimal (Table 1-3). A series of flotation tests with varying doses of butyl-xanthate and dithiophosphate were carried out under a pH of 8.5. These tests were conducted in three stages: roughing, cleaning, and scavenging. The results (Figure 10) show that an increase in the collector dosage led to an increase in the silver recovery yield and a decrease in the silver grade. Based on the obtained results, it could be concluded that the best flotation can be obtained with a dosage of 80 g/t of each collector (butyl-xanthate and dithiophosphate).

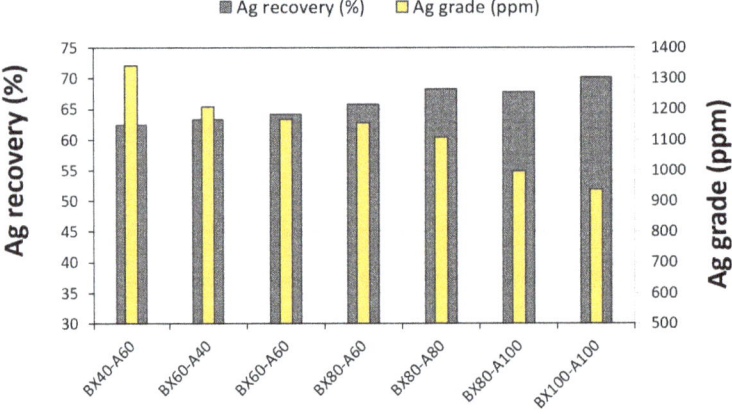

Figure 10. The evolution of silver grade and recovery as a function of the dosage of collector combinations given; BX = butyl-xanthate and A = Dithiophosphate (Aerofloat-242).

3.7. Effects of Other Parameters

Multiple flotation tests were performed under various pH levels: 5, 6, 7, 8, 9, 10 and 11. The previously optimized conditions were used in these tests (Table 1-4). The pH prior to flotation was measured, and sulphuric acid or lime was used as a regulator to set the desired pH. Figure 11a highlights the obtained results. The silver recovery reached its maximum (64 to 67 wt %) at a pH interval of 6 to 8. The best silver grade was obtained at a pH of 8 (Ag grade concentrate = 1210 ppm).

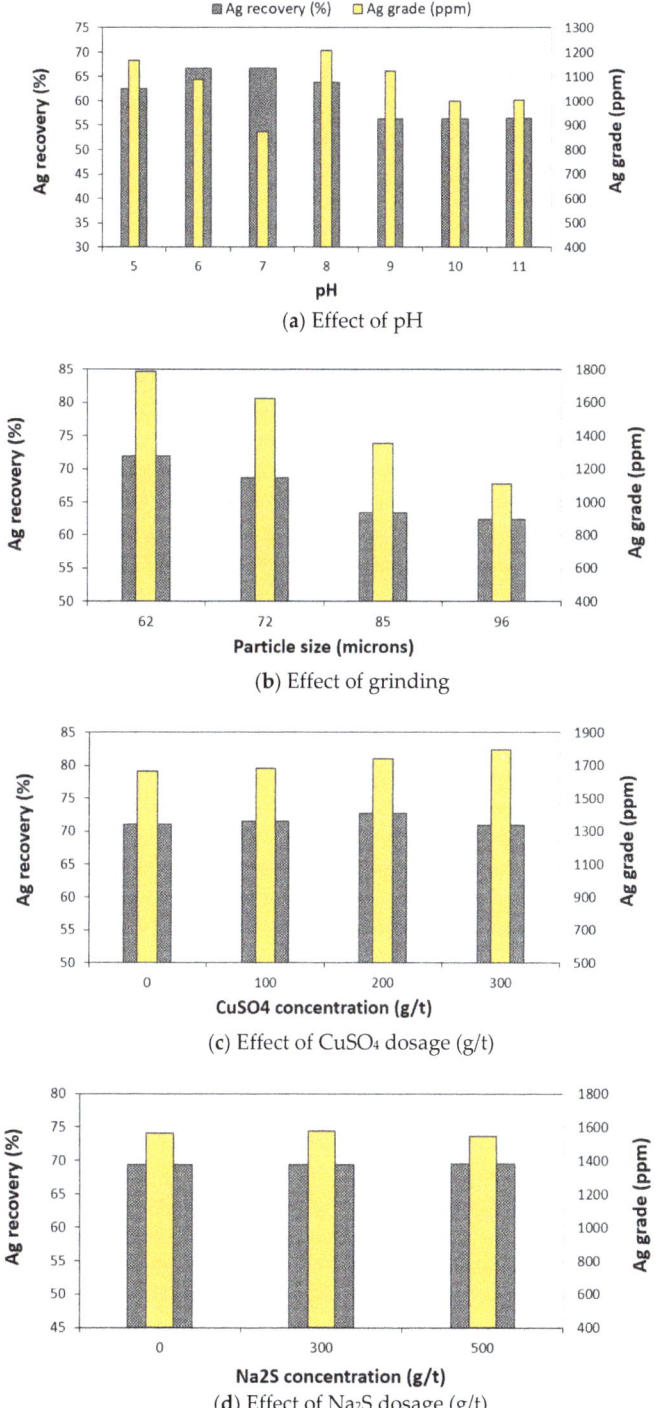

Figure 11. Evolution of silver grade and recovery as a function of: (**a**) pH; (**b**) grain size fraction (D90); (**c**) $CuSO_4$ dosage (g/t) and (**d**) Na_2S dosage (g/t).

In order to enhance the recovery of silver, the effect of grinding on mineral liberation and Ag recovery was re-evaluated. The goal was to reduce the grinding time and consequently, reduce the treatment costs to achieve the best particle size distribution. Also, it is well known that very fine particles present a big challenge due to their difficult flotation and entrainment abilities. Therefore, four different grain size distributions were evaluated (62, 72, 85 and 96 µm), as illustrated in Table 1-5. The overall recovery yield and silver grade are highlighted in Figure 11b. These results indicate that as fineness increased, the recovery yield and silver grade increased. An increase of 38 wt % of silver grade was reached when the ZMT sample was ground from a D90 of 96 µm (1110 ppm) to 62 µm (1784 ppm). An overall silver recovery yield of 72 wt % was achieved.

Due to the possible association of silver with other sulphide minerals and their probable surface alterations, it was important to evaluate the effects of activating agents on silver recovery yield, and grade. $CuSO_4$ was used to activate sulphides, in particular, the sphalerite mineral. A series of tests was completed in three stages: roughing, cleaning, and scavenging, and the $CuSO_4$ dosage was from 0 to 300 g/t (Table 1-6). The results are graphically highlighted in Figure 11c. It was observed that $CuSO_4$ significantly affected the recovery and grade of silver. The silver grade was increased from 1660 ppm to 1800 ppm when a CuSO4 dosage of 300 g/t was used. Recovery yields varied within the range of 71–73 wt %. Based on these results, the optimal dosage of $CuSO_4$ is 200 g/t. the silver grade is around 1742 g/t and the overall recovery yield is around 72.8 wt %.

As it is possible to have some oxidized minerals in the ZMT tailings, it was important to use sulphidizing agents to recover them. Sulphidization enables the recovery of this type of mineral by flotation. In this study, sodium sulphide (Na_2S) was used to recover the oxidized sulphide minerals. Before doing so, all of the sulphide minerals were first floated, and then the sulphidizing agent was added. Figure 11d presents the effects of the addition of Na_2S on the recovery and grade of silver. It was observed that the addition of the sulphidizing agent did not lead to any increase or decrease in silver grade or recovery yield.

3.8. Flotation Tests with Optimized Parameters (Flotation Open Circuit Tests)

Three flotation tests (triplicates) were conducted using the optimized conditions for each parameter (Table 3). The results are graphically highlighted in Figure 12. Under these conditions and with two rounds of roughing, three rounds of scavenging and one round of concentration, it was possible to reach a recovery yield of 83 wt % with an average silver grade of 1745 ppm. Figure 13 shows some SEM images of silver bearing minerals found in the final concentrate. It was observed that silver was found mainly in three forms: free silver, Ag associated with acanthite, and Ag as a substitution within pyrite (refractory silver, 1200 ppm Ag in FeS_2). It was also observed that acanthite was associated with pyrite particles.

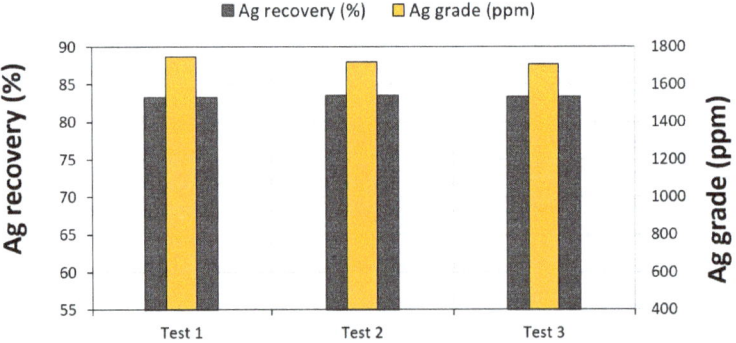

Figure 12. Silver recovery and grade under the optimized conditions.

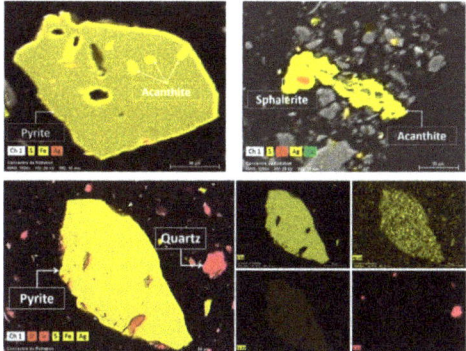

Figure 13. Combined BSE SEM and X-mapping images of the main silver bearing minerals.

Table 3. Final flotation tests under the optimized conditions.

	pH	Butyl-Xanthate	Dithiophosphate	MIBC	D90	CuSO$_4$
Test-45						
Test-46	8.5	80 g/t	80 g/t	30 g/t	62 µm	200 g/t
Test-47						

3.9. Environmental Behavior of Cleaned Tailings

The environmental behavior of cleaned tailings was inspected using two tests: a static test and a toxicity characteristic leaching procedure test. The results presented in Table 4 show that the flotation allowed the acid generation potential of tailings to be reduced from 18.1 to 2.6 kg CaCO$_3$/t and the neutralization potential was increased from 8.3 to 16.7 kg CaCO$_3$/t. The net neutralization potential therefore increased from −9.8 (ZMT) to 14.1 CaCO$_3$/t. according to the TCLP test results. It was observed that pollutants, such as As, Ba, Cd, Cr, Mo, and Pb, were in accordance with United States Environmental Protection Agency (US-EPA) limits, while the Zn concentration was superior to the requested limit. The presence of Zn in cleaned tailings is associated with sphalerite mineral. In general, the released concentrations of Cu and Zn may present an ecological risk to the environment. It is unsafe to deposit tailings that contain these elements unless implemented techniques aim at controlling the mine drainage generation. The use of covers over these tailings should protect them from reactivity (sulfide oxidation). In addition, the reprocessed tailings may be valorized/recycled during the production of ceramic materials.

Table 4. Static and toxicity characteristic leaching procedure tests results.

Static Test	Zgounder Tailings	Cleaned Tailings
AP (kg CaCO$_3$/t)	18.1	2.6
NP (kg CaCO$_3$/t)	8.3	16.7
NNP (kg CaCO$_3$/t)	−9.8	14.1
Acid generation	Uncertain	Uncertain
Toxicity Characteristic Leaching Procedure Test	**Cleaned Tailings (µg/L)**	**Regulation Limit (µg/L) ***
As	<DL	5000
Ba	121	100,000
Cd	<DL	1000
Cr	<DL	5000
Cu	416	-
Mo	<DL	-
Pb	20	5000
Zn	6445	2000

DL: detection limit; * United States Environmental Protection Agency (US-EPA) limit.

4. Conclusions

The aim of this paper was to optimize the recovery of silver from old, low-grade tailings based on laboratory regrinding and flotation tests. Tailings were sampled in the TSF of Zgounder mine, taking into account the need for a representative sample. Mineralogical characterization results obtained by a combination of SEM-EDS examination and XRD analysis showed that silver is mainly associated with acanthite (Ag_2S) and within pyrite as a trace element. The chemical results showed that the overall grade of silver in tailings (88 samples) was about 148 ppm. Flotation tests occurred with different parameters being optimized, namely, particle size distribution (regrinding), collector type and dosage, pH, D80, activator, and sulphidizing agents, to achieve a silver concentrate grade of 1745 ppm with a silver recovery grade of 84 wt % and a mass recovery of around 40 wt %. These were obtained with two rounds of roughing, three rounds of scavenging and one round of cleaning. The particle grain size was reduced to a D90 of 63 microns, the conditioning pH was fixed at 8.5, a combination of butyl-xanthate and dithiophosphate was used at a dosage of 80 g/t each, the dosage of the activating agent ($CuSO_4$) was fixed at 200 g/t and the frother (MIBC) dosage was maintained at 30 g/t. Our perspective (paper in progress) is that the obtained concentrate should be submitted to cyanidation for further silver concentration, requiring a full technico-economical and feasibility study to be carried out. The cleaned tailings could be valorized as secondary raw materials for ceramic manufacturing. This could lead to a potential stabilization of released elements, such as Zn and Cu.

Author Contributions: B.D. conducted all flotation tests besides the physical and chemical characterizations. The interpretation of results and paper writing were done by B.D. and T.Y. under the supervision of M.B. and R.H.

Acknowledgments: The authors greatly acknowledge the Zgounder Millenium Silver Mining for the great help concerning the sampling, flotation tests, and chemical analyses to valorize their wastes. The authors are also grateful to the staff of URSTM at the University of Quebec in Abitibi Temiscamingue for their valuable contribution to the solid samples analysis.

Conflicts of Interest: The authors declare no conflict of interest.

Acronyms and Abbreviations

AAS	Atomic absorption spectroscopy
AP	Acidity potential
BSE	Back Scattered Electrons
Cin	Inorganic carbon content
EDS	Energy dispersive spectroscopy
Gs	Specific gravity
MIB	Methyl isobutyl carbinol
NNP	Net neutralization potential
NP	Neutralization potential
SEM	Scanning electron microscopy
TCLP	Toxicity characteristic leaching procedure
TSF	Tailings storage facility
XRD	X-ray diffraction
ZMT	Zgounder Mine Tailings

References

1. Schweikert, K. Are gold and silver cointegrated? New evidence from quantile cointegrating regressions. *J. Bank. Financ.* **2018**, *88*, 44–51. [CrossRef]
2. Lucey, M.E.; O'Connor, F.A. Mind the gap: Psychological barriers in gold and silver prices. *Financ. Res. Lett.* **2016**, *17*, 135–140. [CrossRef]
3. Eryiğit, M. Short-term and long-term relationships between gold prices and precious metal (palladium, silver and platinum) and energy (crude oil and gasoline) prices. *Econ. Res.* **2017**, *30*, 499–510. [CrossRef]
4. Wallace, T.C.; Barton, M.; Wilson, W.E. Silver & silver-bearing minerals. *Rocks Miner.* **1994**, *69*, 16–38.

5. Allan, G.; Woodcock, J. A review of the flotation of native gold and electrum. *Miner. Eng.* **2001**, *14*, 931–962. [CrossRef]
6. Quinteros, J.; Wightman, E.; Bradshaw, D. Applying process mineralogy to complex low-grade silver ores. In Proceedings of the XXVII International Mineral Processing Congress-IMPC, Santiago, Chile, 20–24 October 2014.
7. Wiegers, G. Crystal-structure of low-temperature form of silver selenide. *Am. Mineral.* **1971**, *56*, 1882.
8. Cook, N.J.; Ciobanu, C.L.; Spry, P.G.; Voudouris, P. Understanding gold-(silver)-telluride-(selenide) mineral deposits. *Episodes* **2009**, *32*, 249–263.
9. Celep, O.; Alp, İ.; Deveci, H. Improved gold and silver extraction from a refractory antimony ore by pretreatment with alkaline sulphide leach. *Hydrometallurgy* **2011**, *105*, 234–239. [CrossRef]
10. Mueller, A.G.; Muhling, J.R. Silver-rich telluride mineralization at Mount Charlotte and Au–Ag zonation in the giant Golden Mile deposit, Kalgoorlie, Western Australia. *Miner. Depos.* **2013**, *48*, 295–311. [CrossRef]
11. Woodcock, J.T.; Henley, K.; Cathro, K. *Metallurgy of Gold and Silver with Reference to Other Precious Metals*; Course Notes for an Australian Mineral Foundation Workshop Course; Australian Mineral Foundation: Glenside, Australia, 1976.
12. Sack, R.; Brackebusch, F. Fahlore as an indicator of mineralization temperature and gold fineness. *Can. Min. Metall. Bull.* **2004**, *97*, 78–83.
13. Viñals, J.; Roca, A.; Cruells, M.; Núñez, C. Characterization and cyanidation of Rio Tinto gossan ores. *Can. Metall. Q.* **1995**, *34*, 115–122. [CrossRef]
14. Gasparrini, C. *Gold and Other Precious Metals: From Ore to Market*; Springer Science & Business Media: Berlin, Germany, 2012.
15. Klein, C.; Hurlbut, C.S.; Dana, J.D. *The 22nd Edition of the Manual of Mineral Science: (After James D. Dana)*; Wiley: New York, NY, USA, 2002.
16. Fleischer, M.; Mandarino, J.A. *Glossary of Mineral Species, 1995*; Mineralogical Record Incorporated: Tucson, AZ, USA, 1995.
17. Horwood, E.J. Process of Treating and Subsequently Separating Sulfid Ores, &C. U.S. Patent US1020353A, 12 March 1912.
18. Hoover, T.J. *Concentrating Ores by Flotation*; Mining Magazine: London, UK, 1914.
19. Gaudin, A. *Flotation*; McGraw-Hill Book Co., Inc.: New York, NY, USA, 1957; Volume 463, p. 31.
20. Fuerstenau, D.W. Fine Particle Flotation. In Proceedings of the International Symposium on Fine Particles Processing, Las Vegas, NV, USA, 24–28 February 1980; pp. 669–705.
21. King, R. *Modeling & Simulation of Mineral Processing Systems Butterworth*; Heinemann: Oxford, UK, 2001; 403p.
22. Wills, B.A.; Napier-Munn, T. *Wills' Mineral Processing Technology: An Introduction to the Practical Aspects of Ore Treatment and Mineral Recovery*; Elsevier Science and Technology: New York, NY, USA, 2006.
23. Leaver, E.; Woolf, J. *Flotation of Silver Minerals*; Research Investigation 3436; US Bureau of Mines: Washington, DC, USA, 1939.
24. Taggart, A.F. *Handbook of Mineral Dressing*; Wiley: New York, NY, USA, 1945.
25. Dorr, J.; Bosqui, F. *Cyanidation and Concentration of Gold and Silver Ores*, 2nd ed.; McGraw Hill: New York, NY, USA, 1950.
26. Malhotra, D.; Harris, L. Review of plant practice of flotation of gold and silver ores. In *Advances in Flotation Technology*; Parekh, B., Miller, J., Eds.; SME: Southfield, MI, USA, 1999; pp. 167–181.
27. Thompson, P. The selection of flotation reagents via batch flotation test. In *Mineral Processing Plant Design, Practice, and Control: Proceedings*; SME: Southfield, MI, USA, 2002; Volume 1.
28. Trahar, W.; Warren, L. The flotability of very fine particles—A review. *Int. J. Min. Process.* **1976**, *3*, 103–131. [CrossRef]
29. Yoon, R.; Luttrell, G. The effect of bubble size on fine particle flotation. *Min. Process. Extr. Metall. Rev.* **1989**, *5*, 101–122. [CrossRef]
30. Dai, Z.; Fornasiero, D.; Ralston, J. Particle–bubble collision models—A review. *Adv. Colloid Interface Sci.* **2000**, *85*, 231–256. [CrossRef]
31. Miettinen, T.; Ralston, J.; Fornasiero, D. The limits of fine particle flotation. *Min. Eng.* **2010**, *23*, 420–437. [CrossRef]
32. Institute, S. *World Silver Survey 2014*; The Silver Institute: Washington, DC, USA, 2014.
33. Petruk, W. Mineralogy and geology of the Zgounder silver deposit in Morocco. *Can. Mineral.* **1975**, *13*, 43–54.

34. El Adnani, M.; Plante, B.; Benzaazoua, M.; Hakkou, R.; Bouzahzah, H. Tailings Weathering and Arsenic Mobility at the Abandoned Zgounder Silver Mine, Morocco. *Mine Water Environ.* **2016**, *35*, 508–524. [CrossRef]
35. Lawrence, R.W.; Scheske, M. A method to calculate the neutralization potential of mining wastes. *Environ. Geol.* **1997**, *32*, 100–106. [CrossRef]
36. Welsby, S.D.D. On the Interpretation of Floatability Using the Bubble Load. Ph.D. Thesis, The University of Queensland, Brisbane, Australia, 2009.
37. Vianna, S.M.S. The Effect of Particle Size Collector Coverage and Liberation on the Floatability of Galena Particles in an Ore. Ph.D. Thesis, The University of Queensland, Brisbane, Australia, 2004.
38. Thomas, W. *Mining Chemical Handbook*; Cytec Industries Inc.: Woodland Park, NJ, USA, 2010.

© 2018 by the authors. Licensee MDPI, Basel, Switzerland. This article is an open access article distributed under the terms and conditions of the Creative Commons Attribution (CC BY) license (http://creativecommons.org/licenses/by/4.0/).

Article

Towards Greener Lixiviants in Value Recovery from Mine Wastes: Efficacy of Organic Acids for the Dissolution of Copper and Arsenic from Legacy Mine Tailings

Richard A. Crane [1],* and Devin J. Sapsford [2]

1. Camborne School of Mines, College of Engineering, Mathematics and Physical Sciences, University of Exeter, Penryn Campus, Penryn TR10 9FE, Cornwall, UK
2. School of Engineering, Cardiff University, Queen's Buildings, The Parade, Cardiff CF24 3AA, UK; sapsforddj@cf.ac.uk
* Correspondence: r.crane@exeter.ac.uk; Tel.: +44-(0)1326-214370

Received: 20 June 2018; Accepted: 31 July 2018; Published: 3 September 2018

Abstract: In many cases, it may be possible to recover value (e.g., metals, land) from legacy mine wastes and tailings when applying leaching-based remediation such as dump/heap leaching or in-vessel soil washing. However, if the lixiviant used has the potential to cause environmental damage upon leakage, then this approach will have limited practicability due to actual or perceived risk. This study focused on comparing the efficacy of organic acids, namely methanesulfonic (CH_3SO_3H) and citric ($C_6H_8O_7$) acid, with mineral acids, namely sulfuric (H_2SO_4) and hydrochloric (HCl) acid, for the dissolution of Cu and As from mine tailings. The advantage of the former acid type is the fact that its conjugate base is readily biodegradable which should thereby limit the environmental impact of accidental spill/leakage (particularly in non-carbonate terrain) and might also be directly useful in capture/recovery systems coupled with percolation leaching (e.g., as an electron donor in sulphate-reducing bioreactors). The operational factors acid concentration, leaching time, mixing intensity and solid–liquid ratio, were tested in order to determine the optimum conditions for metal dissolution. HCl, H_2SO_4, and CH_3SO_3H typically exhibited a relatively similar leaching ability for As despite their different pKa values, with dissolutions of 58%, 56%, 55%, and 44% recorded for H_2SO_4, HCl, CH_3SO_3H, and $C_6H_8O_7$, respectively, after 48 h when using 1 M concentrations and a 10:1 L:S ratio. For the same conditions, H_2SO_4 was generally the most effective acid type for Cu removal with 38% compared to 32%, 29% and 22% for HCl, CH_3SO_3H and $C_6H_8O_7$. As such, CH_3SO_3H and $C_6H_8O_7$ demonstrated similar performances to strong mineral acids and, as such, hold great promise as environmentally compatible alternatives to conventional mineral acids for metal recovery from ores and waste.

Keywords: valorisation; mine waste; soil washing; heap leaching; dump leaching; mine drainage; remediation

---

1. Introduction

Legacy mining wastes have polluted and continue to pollute the environment on decadal to millennial timescales [1–3]. For example, in the UK, one of the most significant metal pollution contributors to fresh waters is legacy non-ferrous metalliferous mines [4,5]. Several thousand mines are known to be discharging environmentally deleterious quantities of metal/metalloid pollutants into surrounding watercourses [4]. It has been estimated that nine percent of rivers in England and Wales, and two percent in Scotland are at risk of failing to meet their European Water Framework Directive targets of chemical and ecological quality because of legacy mines [4]. These rivers carry some of the largest quantities of contaminant metals, such as As, Cu, Cd, Pb, and Zn, into the seas surrounding the UK each year. Similar situations exist in most other locations with a metal mining legacy worldwide. Some examples include the USA and Canada, which have approximately 35,000 and 10,000 legacy metal mine sites, respectively, Japan with approximately 5500 legacy metal mines, and Sweden with around 1000 legacy metal mines [6].

As well as causing environmental degradation, mine sites can also represent opportunities for resource recovery [7]. Resources can be present in many forms, including as metals that can be leached and recovered; decontaminated residue that can be reused (for example as aggregate); or as a landscape resource. This latter resource value is recognised by planning designations based on its cultural or ecological value and the increase in heritage tourism [8]. These unique geological, ecological and cultural designations would act as significant constraints to mine waste remediation and site reclamation if the existence of these features was to be adversely affected by such activities [5]. The metal resources in these historical deposits are often not insubstantial but are generally not sufficient to present (planning issues aside) a justification for intervention for resource recovery alone [8]. However, if designed correctly, remediation of such sites could be implemented where economically valuable metals are also recovered and used to offset the costs of such remediation activity.

Leaching-based technology can be applied to achieve the removal of metals and metalloids (hereafter, "metals") for the purpose of remediation and/or recovery of resource(s), either through ex situ in-vessel soil washing or via in situ percolation leaching, with the latter technology being preferable to reduce capital and operational costs (e.g., of material handling). Percolation leaching could be carried out on the waste in situ (dump leaching), or the waste could be excavated, agglomerated, and placed on an engineered liner for leaching (i.e., heap leaching).

However, if the lixiviant used has the potential to cause environmental damage through leakage or run-off from residues, then heap leaching and soil washing might have limited practicability due to actual or perceived risk. Towards the goal of leaching with more environmentally acceptable ("greener" see for example (e.g., [9–11]) lixiviants, this study focused on the efficacy of organic acids in remediation/value recovery (e.g., reference [12]) because of the advantage that the conjungate base is biodegradable which should limit the environmental impact of accidental spill/leakage (e.g., references [13–15]) and might be directly useful in capture/recovery systems (e.g., to alleviate metal toxicity and act as an electron donor in sulphate-reducing bioreactors (e.g., references [16,17]) or microbial electrochemical systems.

There are several reasons why a biodegradable lixiviant is preferable for in-vessel soil washing, in situ and/or in heap leach scenarios. For soil washing, whilst the main process is contained within tanks, leaks/spills can occur, but more perhaps more importantly, when waste is replaced, there is the risk of environmental degradation from residual acidic leachate run-off. With in situ leaching of legacy mine wastes, it is highly unlikely that there is an engineered liner under the waste. Thus, for in situ dump leaching of waste piles, either there has to be a high-level of confidence that the underlying rock is impermeable, or a liner needs to be retrospectively installed. This is possible through the injection of cement-based grouts and chemical grouts or jet grouting. However, given that the use of such a technique for creating an impermeable barrier should ideally be limited to homogenous soils [18] and that even well engineered liner systems can leak, the use of biodegradable lixiviant would add confidence that should leakage occur, biodegradation will retard the transport of the escaping plume. This is particularly the case in areas of non-carbonate rocks where the effect of acid escape is far more serious because of the lack of acid-neutralisation. Furthermore, even if the mine waste is placed on to a highly engineered liner, leakage can still occur through pinholes and shrinkage cracks in geomembrane and clay liners, respectively. The strong mineral acids hydrochloric (HCl) and sulfuric acid (H_2SO_4) are conventional hydrometallurgical reagents, and the latter, in particular, is extensively used as a lixiviant in heap leaching of Cu and U ores. Environmental concerns have arisen from the failure to contain process solutions within the heap leach circuit, which is compounded by large surface areas, the use of open drainage trenches (rather than enclosed in pipework) [19]. Its strong acidity and environmental persistence also dictates that even after leach pad decommissioning, the leachate from the spent ore (ripios) has to be carefully managed for many years until residual acidity has been fully flushed by meteoric water.

A further disadvantage of using sulfuric acid as a lixiviant is the subsequent formation of sulfate precipiates (e.g., jarosite and gypsum) that consume sulfuric acid whilst simultaneously resulting in unwanted permeability loss.

Strong organic acids, such as some sulfonic acids, are a class of strong acids with readily (bio)degradable conjugate bases that have been demonstrated as generally less environmentally persistent than mineral acids, such as sulfuric acid. Methanesulfonic acid, CH_3SO_3H (pKa = −1.9), has been demonstrated as being highly efficient for the dissolution of a number of different heavy metals via the formation of soluble methanesulfonate complexes [20]. Furthermore, the properties of methanesulfonic acid, such as its high conductivity, stability against volatilisation and hydrolysis, and low corrosiveness [20] are advantageous for its widespread use in hydrometallurgy and chemical engineering. Another advantageous property of methanesulfonic acid is the stability of reduced metal ions in methanesulfonic acid solutions, which is best known in the Sn^{2+}/Sn^{4+} system. Furthermore, methanesulfonic acid has a comparable conductivity to hydrochloric and sulfuric acid (299.6, 346.1 and 444.9 S cm^2 mol^{-1}, respectively) allowing for efficient recovery of metals from solution using electrowinning. Despite these attributes, the application of methanesulfonic acid to metal recovery from tailings has not been widely explored, aside from leaching of rare earth elements from bauxite residue [21].

As a weak acid, citric acid ($C_6H_8O_7$) partially dissociates in water to form hydrogen ions and its conjugate base, citrate, which is readily biodegradable in aerobic and anaerobic environments [12,14,15] and is commonly used as a raw material in the manufacturing/food and beverage industry and thus, is readily available as a bulk commodity and may be public acceptable due to common use in food products. The role of citric acid in metal mobilisation from tailings has been investigated more widely than methanesulfonic acid, for example, by Burckhard et al. [22], and the efficacy of citric acid in remedial application has been previously investigated in relation to the leaching of metal-contaminated soils, e.g., in references [12,23], and for hydrometallurgical application in Ni-bearing lateritic/saprolitic ores [24–26].

Here, we present a preliminary study that assesses the comparative efficacy of citric, hydrochloric, methanesulfonic and sulfuric acids for the recovery of economically valuable and/or contaminant metals from mine tailings waste taken from a legacy Cu/As mine in the southwest England. The work was established in order to demonstrate the feasibility of using such lixiviants for the recovery of toxic and/or economically valuable metals from metalliferous mine waste.

2. Methodology

2.1. Site Description

The Devon Great Consols (DGC) mine is a disused Cu/As mine located in Devon, England (50°32′16″ N, 04°13′17″ W). It was selected for study here because as a member of the UK mine water directive sites, it is known to release significant quantities of heavy metals into the surrounding environment each year [4,27]. The principal minerals extracted at the DGC mine were chalcopyrite ($CuFeS_2$) and arsenopyrite (AsFeS) and at its peak in the late 19th century it was the largest producer of both Cu and As in the world [28]. Its output for the period spanning 1844–1902 is estimated to have been approximately 736,200 tons of Cu ore and 72,300 tons of refined As [29]. The on site processing of both Cu and As ores has resulted in the accumulation of large quantities of mine tailings which are currently predominantly located in two piles with an estimated total volume of approximately 258,600 m^3 [8].

2.2. Mine Tailing Collection Procedure

This study focuses on two major mine tailing piles at the DGC site. The larger Northern pile (50°32′16″ N, 4°13′14″ W) is predominantly composed of sand and silt sized particles, and the smaller Southern pile (50°32′13″ N, 4°13′09″ W) is finer grained, because it was reworked during the period from 1902–1925 to extract As. A total of 18 mine tailing samples were collected (15 from the Northern heap and 3 from the Southern heap) following the methodology of ASTM D6009-12 [30]. Samples were collected using a stainless-steel trowel at equal distances around the base of each tailing pile at a depth of 0.3 m. Each sample had a volume of approximately 5 L with a mass typically between 6 and 8 kg (depending on bulk density). Once collected, the samples were heated at 105 °C for 24 h in order to remove any moisture present.

2.3. Physical and Chemical Characterisation of the Mine Tailings

A composite sample was created by riffling each sample 6 times and then thoroughly mixing each final subsample together using a mixing pad. Each composite sample was then riffled to yield an appropriate mass for each analysis technique. All analysis methods were performed using duplicate samples with the average taken. Particle size distribution (PSD) measurements were performed via dry sieving and sedimentation (ISO 11277:2009) [31] using approximately 200 g of the composite. Uncompacted aggregate bulk density measurements were performed in accordance with BS 812: 1995 [32] using a cylinder of 1876 mL in volume and a tamping rod of 16 mm in diameter. Paste pH measurements were preformed via ASTM D4972-13 [33] using 40 g from each composite and 40 mL of Milli-Q water (resistivity > 18.2 MΩ cm). Samples were prepared for X-ray diffraction (XRD), inductively coupled plasma optical emission spectroscopy (ICP-OES), total organic carbon (TOC) analysis, and total inorganic carbon (TIC) analysis by crushing (to particle size <75 μm) a riffled (approximately 200 g) subsample of the composite sample using a Labtech Essa LM1-P puck mill crusher at 935 RPM for 120 seconds. The XRD analysis was performed using a Phillips Xpert Pro diffractometer with a CuKα radiation source (λ = 1.5406 A°; generator voltage of 40 keV; tube current of 30 mA). Spectra were acquired between 2θ angles of 5–90° with a step size of 0.02° and a 2 s dwell time. The sample was prepared by packing approximately 2 g of the material into an aluminium XRD stub. ICP-OES analysis was performed using a Perkin Elmer Optima 2100 DV ICP-OES. The sample was prepared for analysis via the 4 acid digest method (EPA 3052-12) [34]. Firstly, 0.01 g was placed in a PTFE lined microwave digest cell, and 3 mL of analytical grade 45.71% hydrofluoric acid (HF) was then added and left for 12 h. Six millilitres of aqua regia solution (1:1 ratio of analytical grade 32% hydrochloric acid (HCl) and 70% nitric acid (HNO_3)) was then added, and the container was then placed in a microwave digest oven (Anton Paar Multiwave 3000) and heated at 200 °C (1400 watts) for 30 min (after a 10 min up ramp time period) and then allowed to cool for 15 min. The resultant solution was then neutralised using 18 mL of analytical grade 4% Boric acid (H_3BO_3) at 150 °C (900 watts) for 20 min (after a 5 min up ramp time period) and then allowed to cool for 15 min. Total carbon (TC) measurements were performed using a Leco SC-144DR sulfur/carbon analyser. Samples of 0.35 g mass were loaded into the instrument and heated at 1350 °C in a pure O_2 (>99.9%) atmosphere. The concentration of CO_2 released by each sample was then measured using an infrared detection cell at a constant flow rate. Total inorganic carbon (TIC) measurements were performed using a Shimadzu (Kyoto, Japan) SSM-5000A using 99.9% O_2 at 500 mL/min and catalytically aided combustion oxidation performed at 900 °C. Total organic carbon (TOC) was calculated by subtracting each TIC measurement from the corresponding TC measurement for each sample. SEM-EDX maps of the mine tailings were acquired using a Zeiss Sigma HD Field Emission Gun SEM outfitted with 2*Oxford Instruments (Abingdon, UK) 150 mm^2 X-Max EDS detectors. The sample was first mounted onto an aluminium stub using an adhesive carbon tab. A 15 nm thick conductive carbon layer was applied by thermal evaporation using an Agar Turbo Carbon Coater to prevent charging effects. SEM-EDX data were obtained using a 20 kV accelerating voltage with a nominal beam current of 4.7 nA, a pixel dwell time of 10–20 ms, and a pixel size of 0.5 to 1.3 μm using Oxford Instruments Aztec acquisition software. The shortest process time (time constant) was used to maximise counts in each pixel. Background and peak overlap corrections were applied within Aztec to provide semi-quantitative results for the samples.

2.4. Hydrometallurgical Extraction

Batch hydrometallurgical extraction experiments were conducted using an acid strength of 1 M, a solid liquid ratio of 1:10, a mixing speed of 200 RPM (using a Stuart SSL1 orbital shaker table), a solution volume of 200 mL, and an equilibration time of 24 h for all experiments, unless specified elsewhere. The following variables were investigated: acid strength (0, 0.01, 0.1, 1, 2, and 4 M), solid–liquid (S/L) ratio (1:5, 1:10, 1:20, and 1:40) and mixing speed (0, 50, 100, and 150 RPM) using separate batch systems (sealed glass jars that were 250 mL in volume). Aqueous samples of 5 mL volume were extracted from each batch system using a 10 mL syringe and then filtered through a 0.45 µm PTFE membrane. Experiments were conducted at room temperature which was measured to be 21.0 °C ± 1.5 °C.

3. Results and Discussion

3.1. Physical and Chemical Characterisation of the Mine Tailings

Table 1 displays the volume, bulk density, paste pH, TOC, and TIC data for the mine tailing composite along with the calculated total mass of the tailing piles. It was noted that the paste pH was 3.33, which indicates that the tailings are acid-producing and therefore, likely to represent a source of metals into the surrounding environment when flushed by meteoric water. In addition, the TIC was recorded as <0.00%, which indicates that the tailings have no appreciable carbonate content; thus, the materials are not strongly acid-consuming and are thus amenable to acid leaching. Table 2 displays notable metal and metalloid concentration data for the composite sample. It can be seen that the Cu, Zn, and Pb levels were recorded as exceeding the screening levels for ecological health. As levels were above the guideline levels for human health, and Cr levels exceeded both parameters. In addition, relatively high concentrations of Al (4.60%) and Fe (9.99%) were recorded. Figure 1 displays the XRD spectra for the composite sample, which indicate quartz (α-SiO_2) as the major crystalline component present with a relatively minor contribution from muscovite ($H_2KAl_3(SiO_4)_3$). The original primary ore minerals, arsenopyrite and chalcopyrite, were not detected. Figure 2 displays SEM-EDX data for the mine tailings and indicates that As and Cu are both relatively widely distributed in the mine tailings but in varying concentrations. Moreover, where present, both elements are often located in discrete regions on each particle. In general, As, Cu, and Fe are distributed in relatively similar locations, whereas no clear correlation with S was observed. This suggests that As, Cu, and Fe are unlikely to be still present as the original ore minerals (chalcopyrite and arsenopyrite) with S likely to have been removed during the original ore processing and/or subsequent weathering of the tailings (possibly via dissolution). This result is in contrast to previous work where arsenopyrite was detected using SEM-EDX, but it was also commonly recorded as partially altered to iron oxyhydroxides [35].

Figure 3 displays the particle size as a function of percentage passing by mass for the composite mine tailing sample. It can be noted that the majority of the mass of particles was recorded within the size fraction range of sand (particle sizes 0.063–2 mm) with 78.82% recorded, compared to 18.78 and 2.41% recorded for gravel (particle sizes 2–64 mm), silt, and clay sized particles (particle sizes < 0.063 mm) respectively. The D_{50}, D_{10}, and D_{90} were 0.99, 0.18, and 2.80 mm, respectively, indicating that the median particle mass is approximately 1 mm in diameter. Figure 4 shows that, in general, the greatest concentrations of Cu and As are in the largest particle size fractions. In addition, a slight increase in both Cu and As concentrations was recorded for the finest size fractions of the tailings.

Table 2 displays the notable metal concentrations recorded for the composite mine tailings sample, Table 3 displays their values per ton and subsequent estimated total values in the tailings pile. It can be observed that Cu and As had relatively high concentrations, with 0.183 and 1.92 wt % recorded respectively. Although cut-off values are invariably specific to the ore, mine setting, and reserve tonnage, a survey of typical cut-off grades (percentage w/w) for a range of heavy metals indicated that Cu is economic at grades of approximately >0.5% [36], and Sn is economic at grades of >0.15% [37]. It is therefore unlikely that such metals would be considered to be suitable targets at present for extraction (i.e., when compared to conventional ore deposits). However, if combined with site remediation, then the value of such metals could be used to offset the remediation cost. In particular, As concentrations were recorded as being significantly high. Therefore, conceivably, a site remediation strategy could be designed whereby As is removed along with any economically valuable metals present which would enable clean-up to be conducted at a significantly lower cost.

Table 1. Volume, bulk density, paste pH, TOC, and TIC data for the composite mine tailings sample along with the total mass of the tailing piles.

Volume (m³)	Bulk Density (g/cm³)	Mass (Tonnes) [6]	Paste pH	TOC (wt %)	TIC (wt %)
198.923 [6]	1.30	258,600	3.33	0.16	0.00

Table 2. Notable metal and metalloid concentration data (wt %) for composite samples from all sites † indicates concentrations above screening levels for ecological risk (1); ‡ indicates those above guideline levels for human health risk (2,3).

Li	Na	Mg	Al	K	Ca	Ti	Cr	Mn
0.0135	0.4312	0.5295	4.6035	0.8871	1.1426	0.2207	0.0315 ‡†	0.0610
Fe	Ni1	Cu	Zn	As	Ag	Cd	Sn	Pb
9.9893	0.0019	0.1833 †	0.0101 †	1.9176 ‡	<DL	0.0012 †	0.0290	0.0067

¹ Proposed Soil Screening Values under the framework for Ecological Risk Assessment [38]; ² Category 4 Screening Values for public open space where there is considered to be a "negligible tracking back of soil" [39]; ³ Soil Guideline Value for commercial land use [40–42].

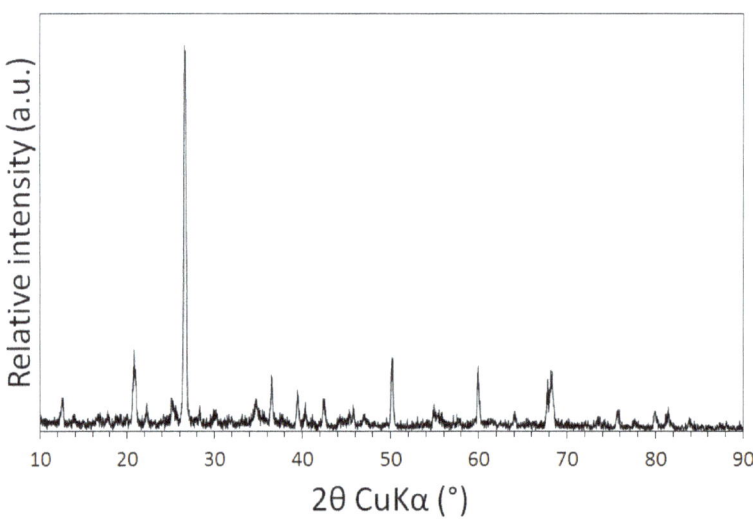

Figure 1. XRD spectra of the composite mine tailings sample.

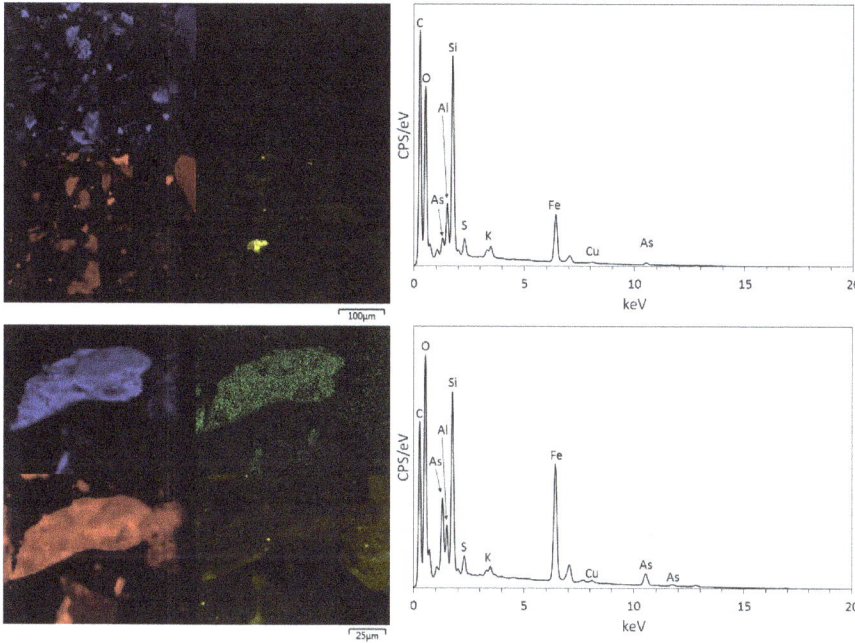

Figure 2. SEM-EDX maps for As (blue), Cu (green), Fe (yellow), and S (red) for the mine tailings along with corresponding spectra for CPS/eV detected for each element within the detection range of 0–20 keV.

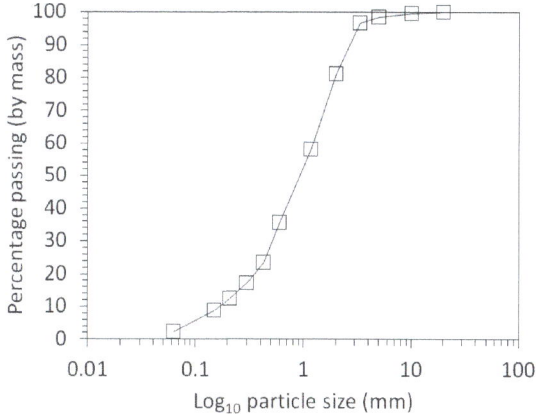

Figure 3. Particle size as a function of percentage passing (by mass) for the composite mine tailing sample.

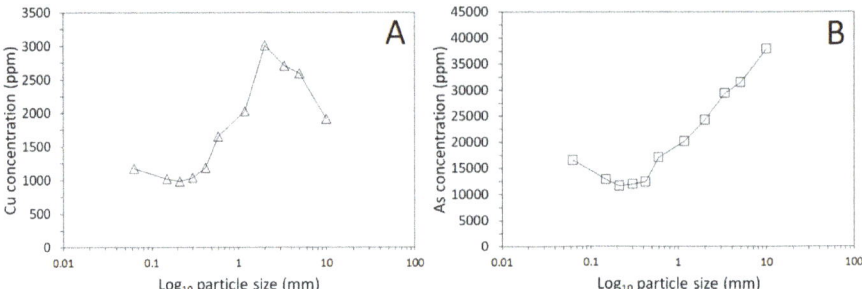

Figure 4. Cu (**A**) and As (**B**) concentrations as a function of particle size.

Table 3. Notable metal value per ton and subsequent estimated total value in the tailings pile. * Value per ton was calculated by multiplying the current metal price (26/07/2018) of each metal by their concentrations in the mine waste composite. The metal prices used were as follows: Cu = £4635/ton, Zn = £2010/ton, Sn = £14,683/ton, and Pb = £1601/ton. † The estimated value in the tailings pile was calculated by multiplying the metal value per ton by 258,600 (estimated total mass in tons of the tailings pile [8]) and rounded to the nearest thousand.

Element	Value Per Ton (£) *	Estimated Value in Tailings Pile †
Cu	8.50	1,511,000
Zn	0.20	36,000
Sn	4.26	757,000
Pb	0.11	19,000

3.2. Influence of Acid Concentration on Cu and As Dissolution

The influence of the acid concentration on the extent of metal leaching (after 48 h) is displayed in Figure 5. It can be observed that all acid types are relatively non-selective for metal recovery from the tailings, with simultaneous removal of a wide range of different metals. The quantity of leached metals was recorded as being dependent on the acid concentration at <1 M concentration with relatively limited additional metal recovery recorded for greater acid concentrations. The extent of Cu and As leaching as a function of time is shown in Figures 6 and 7 with Cu and As recovery after 48 h as a function of the acid concentration displayed in Figure 8. The recovery of Cu and As was shown to depend strongly on the acid concentration and increased as a function of time. The greatest Cu and As recovery was typically recorded for 4 M acid concentration after 48 h reaction time, however, relatively similar recovery was also often recorded for 1 M acid concentrations and a 24 h reaction time. The recovery of As was typically higher than Cu for all acid types with removal of 59.32%, 61.53%, 59.35%, and 43.71% of As recorded for 4 M solutions of H_2SO_4, HCl, CH_3SO_3H, and $C_6H_8O_7$, respectively, after a 48 h reaction, compared to 47.41%, 54.23%, 32.34%, and 23.24% removal of Cu. H_2SO_4 and HCl were generally shown to be the most effective acid types for Cu removal, and H_2SO_4, HCl and CH_3SO_3H were generally shown to be the most effective for As removal.

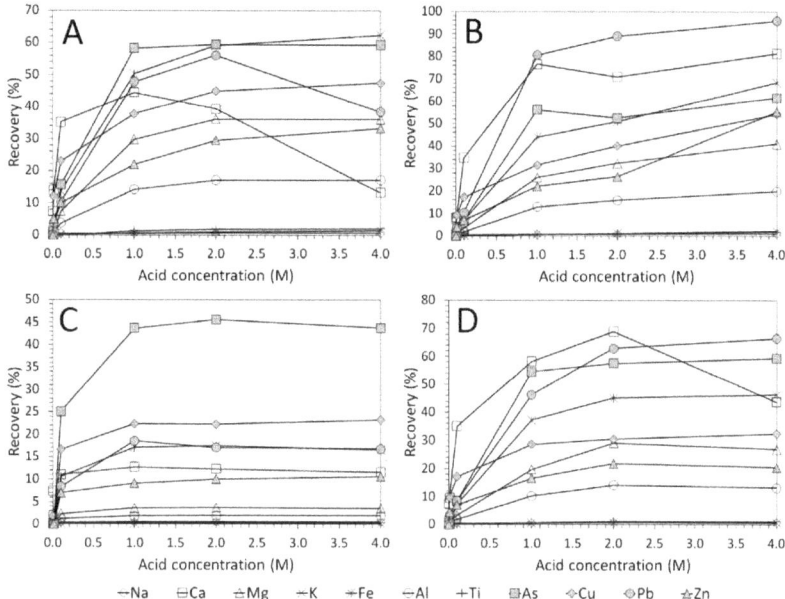

Figure 5. Metal recovery as a function of acid concentration after 48 h for each acid type: H_2SO_4 (**A**), HCl (**B**), $C_6H_8O_7$ (**C**) and CH_3SO_3H (**D**) A solid–liquid ratio of 0.1 and lixiviant volume of 200 mL were used for each experiment.

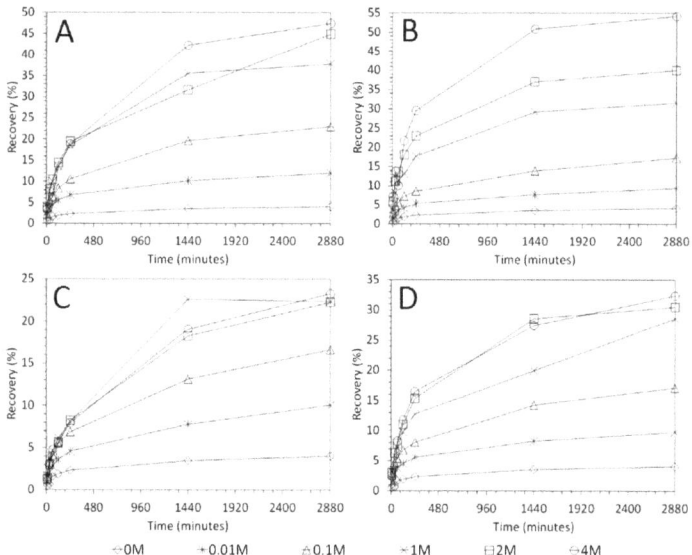

Figure 6. Cu recovery as a function of time and acid concentration for: H_2SO_4 (**A**), HCl (**B**), $C_6H_8O_7$ (**C**) and CH_3SO_3H (**D**). A solid–liquid ratio of 0.1 and lixiviant volume of 200 mL were used for each experiment.

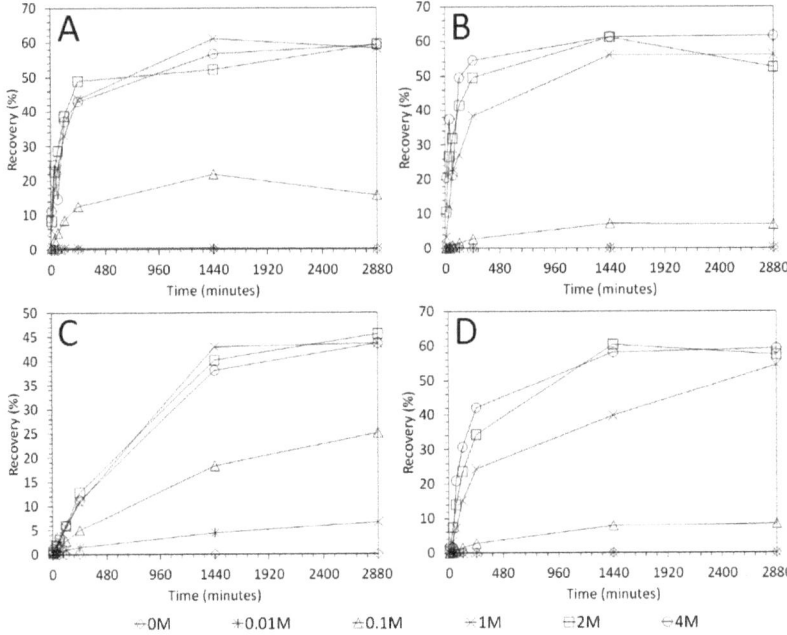

Figure 7. As recovery as a function of time for each acid concentration for each acid type: H_2SO_4 (**A**), HCl (**B**), $C_6H_8O_7$ (**C**) and CH_3SO_3H (**D**). A solid–liquid ratio of 0.1 and lixiviant volume of 200 mL were used for each experiment.

3.3. Influence of Solid–Liquid Ratio for Cu and As Dissolution

The influence of the S/L ratio on the leaching degrees of Cu and As is displayed in Figures 8 and 9. An inverse correlation can be observed between the S/L ratio and both Cu and As recovery, with 32.28%, 28.04%, 23.77%, and 14.96% Cu recovery recorded for H_2SO_4, HCl, CH_3SO_3H, and $C_6H_8O_7$, and a S/L ratio of 0.025, respectively, compared to 21.49%, 23.94%, 18.17%, and 12.93% respectively recorded for a S/L ratio of 0.2. Similarly, As recovery levels of 59.86%, 59.60%, 57.27% and 35.93% were recorded for an S/L ratio of 0.025 compared with 36.50%, 34.96%, 28.03%, and 20.37%, respectively, for an S/L ratio of 0.2. The results therefore demonstrate that the recovery efficacy of Cu and As can be improved by lowering the S/L ratio when within the range of 0.025–0.2, which is likely to be due to greater acid consumption (neutralisation) at higher S/L ratios.

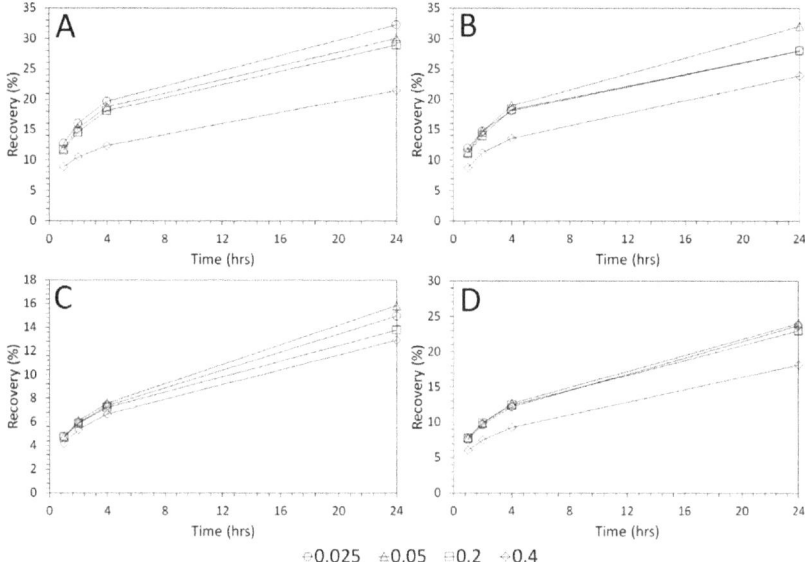

Figure 8. Cu recovery as a function of time for solid–liquid ratios of 0.025, 0.05, 0.2, and 0.4 using: H_2SO_4 (**A**), HCl (**B**), $C_6H_8O_7$ (**C**) and CH_3SO_3H (**D**). The acid concentrations were 1 M and lixiviant volumes were 200 mL.

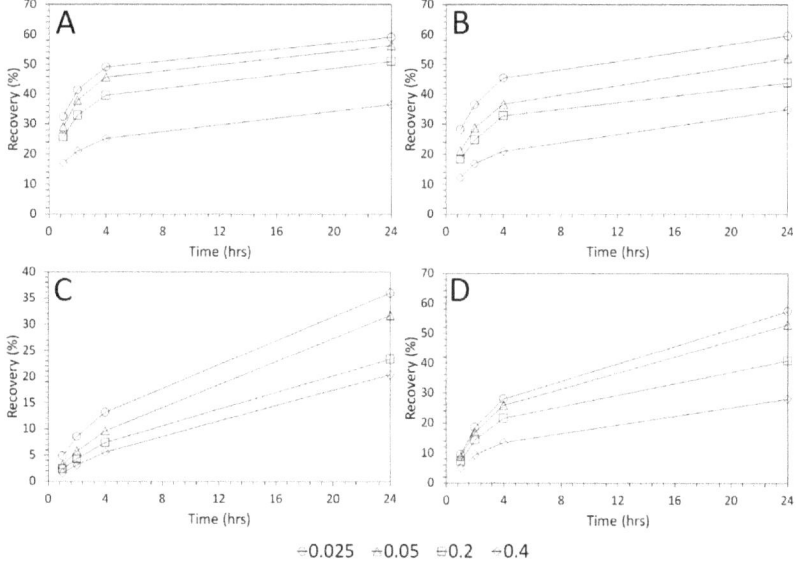

Figure 9. As recovery as a function of time for solid–liquid ratios of 0.025, 0.05, 0.2, and 0.4 using: H_2SO_4 (**A**), HCl (**B**), $C_6H_8O_7$ (**C**) and CH_3SO_3H (**D**). The acid concentrations were 1 M and the lixiviant volumes were 200 mL.

3.4. Influence of Mixing Speed on Cu and As Dissolution

The influence of the mixing speed (0 to 150 RPM) on the As and Cu leaching degrees are displayed in Figures 10 and 11. It can be observed that the stirring speed has a clear positive influence on both As and Cu dissolution with greater dissolution occurring at a greater stirring speed. This is attributed to a decrease in the thickness of the diffusion layer surrounding the mine tailing particles, which, in turn, increases the metal dissolution rate [43].

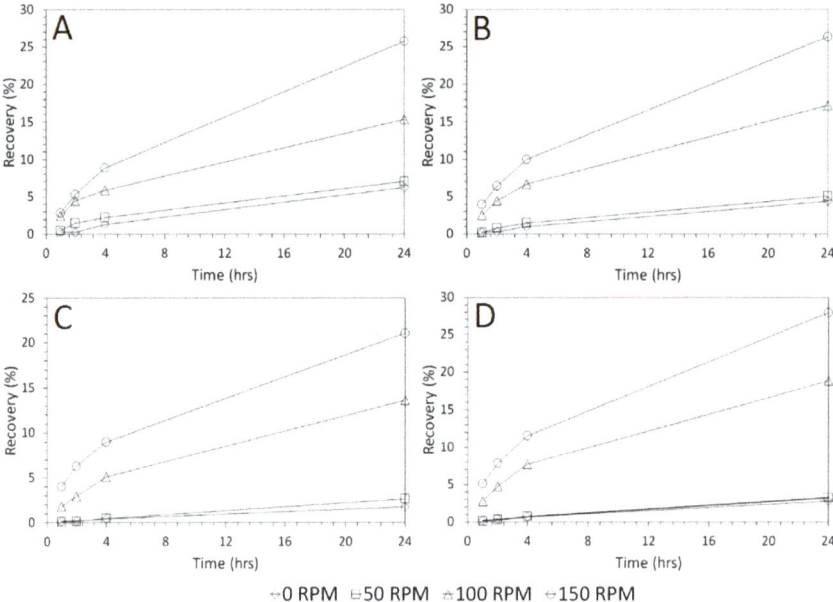

Figure 10. Cu recovery as a function of time for mixing speeds of 0, 50, 100, and 150 RPM using: H_2SO_4 (**A**), HCl (**B**), $C_6H_8O_7$ (**C**) and CH_3SO_3H (**D**). The acid concentrations were 1 M and the lixiviant volumes were 200 mL.

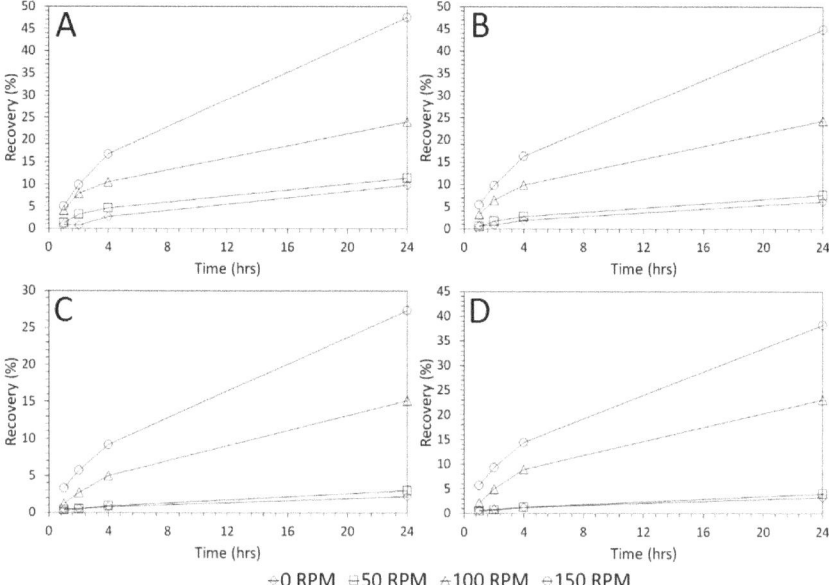

Figure 11. As recovery as a function of time for mixing speeds of 0, 50, 100, and 150 RPM using: H_2SO_4 (**A**), HCl (**B**), $C_6H_8O_7$ (**C**) and CH_3SO_3H (**D**). The acid concentrations were 1 M and the lixiviant volumes were 200 mL.

4. Discussion

The results demonstrate that CH_3SO_3H has relatively similar ability to remove Cu and As from mine waste as H_2SO_4 and HCl. As such, it is suggested that CH_3SO_3H is a suitable alternative for soil washing and heap leaching. $C_6H_8O_7$ is a relatively weak organic acid (pKa = 2.79) and complexing agent, and has been demonstrated as secondary to the other acids but is still effective for the dissolution of Cu. However, it was demonstrated as being able to exhibit relatively high removal of As from mine tailings. This result is of particular significance because weak organic acids, such as $C_6H_8O_7$, are an attractive alternative to stronger acids due to their lower acidity and subsequent lower environmental impact.

In soil washing and dump/heap leaching applications, the recovery of economically valuable and/or contaminant metals and metalloids is the primary objective. Such interventions must be conducted, however, whilst also ensuring that the treated residue is chemically stable (e.g., a low tendency to release aqueous contaminants). Certain mineral acids, such as HCl and H_2SO_4, which have strong acidity and persistence in the environment are therefore not ideal lixiviants for such applications. In contrast, the use of organic acids, such as CH_3SO_3H and $C_6H_8O_7$, which exhibit lower acidity, but also lower environmental persistence, could provide an alternative which would enable targeted metal recovery with a lower detrimental impact on the environment, particularly for non-carbonate lithologies, such as in this study, which are very susceptible to acidification. An assessment of the feasibility of any proposed percolation or in-vessel leaching process based on the lixiviants studied herein would have to be made in light of (i) the ability of the lixiviant to achieve the site-specific remedial target concentrations in the residue, and (ii) a site-specific environmental risk assessment.

5. Conclusions

This study has demonstrated that H_2SO_4, HCl, CH_3SO_3H, and $C_6H_8O_7$ solutions can be applied for the recovery of both Cu and As from mine tailings taken from the Devon Great Consols mine in southwest Devon, England. The leaching potential of each acid was investigated under changing operational factors—acid concentration, leaching time, mixing intensity, and solid–liquid ratio—in order to determine the optimum conditions for metal recovery. The following can be concluded:

(a) Cu and As dissolution rates were determined to typically increase with an increase in the acid concentration, mixing speed, and liquid to solid ratio.

(b) HCl, H_2SO_4 and CH_3SO_3H generally exhibited relatively similar leaching abilities for As despite their different pKa values, with removal percentages after 48 h of 58%, 56%, and 55% recorded for 1 M H_2SO_4, HCl and CH_3SO_3H respectively, compared to 44% exhibited by $C_6H_8O_7$.

(c) H_2SO_4 was generally shown to be the most effective acid type for Cu removal with 38% removal for 1 M solutions after 48 h, compared to 32%, 29%, and 22% recorded for HCl, CH_3SO_3H, and $C_6H_8O_7$ respectively.

(d) Overall the optimum leaching conditions was found to be 1 M acid concentration, 200 RPM mixing speed and a mixing time of 24 h, with only minor improvements in leaching efficacy recorded for concentrations greater than 1 M or time periods greater than 24 h.

The results therefore suggest that processes, such as in-vessel soil washing or percolation leaching, could be relatively low-cost, and in the case of percolation leaching, non-invasive metal recovery techniques that enable simultaneous contaminant and/or economically valuable metal recovery from mine tailing waste. In particular, the use of organic acids, such as CH_3SO_3H and $C_6H_8O_7$, could provide similar As and Cu removal efficacies to H_2SO_4 and HCl but with a potentially lower environmental impact, especially in dump and heap leaching applications for environmental remediation and particularly, in cases where host rocks have low neutralising potential, which are more susceptible to acidification by non-degrading mineral acids. This work also provides a foundation which substantiates further research into the relative environmental performances and other co-benefits of the application of organic acids in metal mine waste remediation and value recovery, particularly for methanesulfonic acid, which, to the authors' knowledge, has not previously been applied to metal recovery from mine tailings.

Author Contributions: R.A.C. and D.J.S. conceived and designed the experiments; R.A.C. performed the experiments; R.A.C. and D.J.S. analyzed the data and wrote the paper.

Acknowledgments: We would like to thank Jeff Rowlands and Marco Santonastaso from the School of Engineering, Cardiff University for performing the ICP-OES and TOC/TIC analyses, respectively. We would also like to thank Geoff Smart from the School of Engineering, Cardiff University for organising the mine waste sample collection and for helping with the sample analysis. We would like to thank Mark Snellgrove of Bakes Down Farm and James Squier from Bidwells LLP for facilitating access to Devon Great Consols. The work was financially supported by the Natural Environment Research Council (grant number: NE/L013908/1).

Conflicts of Interest: The authors declare no conflict of interest.

References

1. Tordoff, G.M.; Baker, A.J.M.; Willis, A.J. Current approaches to the revegetation and reclamation of metalliferous mine wastes. *Chemosphere* **2000**, *41*, 219–228. [CrossRef]
2. Hudson-Edwards, K.; Jamieson, H.; Lottermoser, B. Mine Wastes: Past, Present and Future. *Elements* **2011**, *7*, 375–380. [CrossRef]
3. Plumlee, G.; Morman, S. Mine Wastes and Human Health. *Elements* **2011**, *7*, 399–404. [CrossRef]
4. Environment Agency (EA). *Abandoned Mines and the Water Environment*; Science Project SC030136-41; Environment Agency: Bristol, UK, 2008; ISBN 978-1-84432-894-9.

5. Crane, R.A.; Sapsford, D.J. Selective formation of copper nanoparticles from acid mine drainage using nanoscale zerovalent iron particles. *J. Hazard. Mater.* **2018**, *347*, 252–265. [CrossRef] [PubMed]
6. Mayes, W.M.; Johnston, D.; Potter, H.A.; Jarvis, A.P. A national strategy for identification, prioritisation and management of pollution from abandoned non-coal mine sites in England and Wales. I. Methodology development and initial results. *Sci. Total Environ.* **2009**, *407*, 5435–5447. [CrossRef] [PubMed]
7. Crane, R.A.; Sapsford, D.J. Towards "Precision Mining" of wastewater: Selective recovery of Cu from acid mine drainage onto diatomite supported nanoscale zerovalent iron particles. *Chemosphere* **2018**, *202*, 339–348. [CrossRef] [PubMed]
8. Crane, R.A.; Sinnett, D.E.; Cleall, P.J.; Sapsford, D.J. Physicochemical composition of wastes and co-located environmental designations at legacy mine sites in the south west of England and Wales: Implications for their resource potential. *Res. Conserv. Recycl.* **2017**, *123*, 117–134. [CrossRef]
9. Lee, J.C.; Srivastava, R.R. Leaching of gold from the spent/end-of-life mobile phone-PCBs using "greener reagents". In *The Recovery of Gold from Secondary Sources*; Imperial College Press: London, UK, 2016; pp. 7–56.
10. Jadhao, P.; Chauhan, G.; Pant, K.K.; Nigam, K.D.P. Greener approach for the extraction of copper metal from electronic waste. *Waste Manag.* **2016**, *57*, 102–112. [CrossRef] [PubMed]
11. Li, L.; Ge, J.; Chen, R.; Wu, F.; Chen, S.; Zhang, X. Environmental friendly leaching reagent for cobalt and lithium recovery from spent lithium-ion batteries. *Waste Manag.* **2010**, *30*, 2615–2621. [CrossRef] [PubMed]
12. Wasay, S.A.; Barrington, S.; Tokunaga, S. Organic acids for the in situ remediation of soils polluted by heavy metals: Soil flushing in columns. *Water Air Soil Pollut.* **2001**, *127*, 301–314. [CrossRef]
13. Francis, A.J.; Dodge, C.J.; Gillow, J.B. Biodegradation of metal citrate complexes and implications for toxic-metal mobility. *Nature* **1992**, *356*, 140–142. [CrossRef]
14. Renella, G.; Landi, L.; Nannipieri, P. Degradation of low molecular weight organic acids complexed with heavy metals in soil. *Geoderma* **2004**, *122*, 311–315. [CrossRef]
15. Huang, F.Y.; Brady, P.V.; Lindgren, E.R.; Guerra, P. Biodegradation of Uranium–Citrate Complexes: Implications for Extraction of Uranium from Soils. *Environ. Sci. Technol.* **1998**, *32*, 379–382. [CrossRef]
16. Qian, J.; Zhu, X.; Tao, Y.; Zhou, Y.; He, X.; Li, D. Promotion of Ni^{2+} Removal by masking toxicity to sulfate-reducing bacteria: Addition of citrate. *Int. J. Mol. Sci.* **2015**, *16*, 7932–7943. [CrossRef] [PubMed]
17. Gámez, V.M.; Sierra-Alvarez, R.; Waltz, R.J.; Field, J.A. Anaerobic degradation of citrate under sulfate reducing and methanogenic conditions. *Biodegradation* **2009**, *20*, 499–510. [CrossRef] [PubMed]
18. Dwyer, B.P. *Feasibility of Permeation Grouting for Creating Subsurface Barriers*; SAND94-0786.UC-721; Sandia National Laboratories: Albuquerque, NM, USA, 1994.
19. Ghorbani, Y.; Franzidis, J.P.; Petersen, J. Heap leaching technology—Current state, innovations, and future directions: A. review. *Miner. Process. Extr. Metall. Rev.* **2016**, *37*, 73–119. [CrossRef]
20. Michael, D.G.; Min, W.; Thomas, B.; Patrick, J. Environmental benefits of methanesulfonic acid: Comparative properties and advantages. *Green Chem.* **1999**, *1*, 127–140.
21. Borra, C.R.; Pontikes, Y.; Binnemans, K.; Van Gerven, T. Leaching of rare earths from bauxite residue (red mud). *Miner. Eng.* **2015**, *76*, 20–27. [CrossRef]
22. Burckhard, S.R.; Schwab, A.P.; Banks, M.K. The effects of organic acids on the leaching of heavy metals from mine tailings. *J. Hazard. Mater.* **1995**, *41*, 135–145. [CrossRef]
23. Yuan, S.; Xi, Z.; Jiang, Y.; Wan, J.; Wu, C.; Zheng, Z.; Lu, X. Desorption of copper and cadmium from soils enhanced by organic acids. *Chemosphere* **2007**, *68*, 1289–1297. [CrossRef] [PubMed]
24. Tzeferis, P.G.; Agatzini-Leonardou, S. Leaching of nickel and iron from Greek non-sulphide nickeliferrous ores by organic acid. *Hydrometallurgy* **1994**, *36*, 345–360. [CrossRef]
25. Astuti, W.; Hirajima, T.; Sasaki, K.; Okibe, N. Comparison of effectiveness of citric acid and other acids in leaching of low-grade Indonesian saprolitic ores. *Miner. Eng.* **2016**, *85*, 1–16. [CrossRef]
26. Wanta, K.C.; Perdana, I.; Petrus, H.T.B.M. Evaluation of shrinking core model in leaching process of Pomalaa nickel laterite using citric acid as leachant at atmospheric conditions. In *IOP Conference Series: Materials Science and Engineering*; IOP Publishing: Bristol, UK, 2016; Volume 162, p. 012018.
27. Environment Agency (EA). Dissolved Metal Contamination from Mine Wastes–Risk Assessment and Quantification in the Tamar Catchment. Project: SC060095; 2012. Available online: https://www.gov.uk/government/uploads/system/uploads/attachment_data/file/290507/LIT_7320_4babcfpdf (accessed on 8 June 2016).

28. Dines, H.G. *The Metalliferous Mining Region of South-West England*; HMSO Publications: London, UK, 1956; Volume 1, p. 410.
29. World Heritage Cornwall. Available online: http://www.worldheritagecornwall.com/mines/devon-great-consols.htm (accessed on 21 July 2016).
30. ASTM D6009-12. *Standard Guide for Sampling Waste Piles*; ASTM International: West Conshohocken, PA, USA, 2012; Available online: www.astm.org (accessed on 3 September 2018).
31. *ISO 11277: Soil Quality—Determination of Particle Size Distribution in Mineral Soil Material—Method by Sieving and Sedimentation*; International Organization for Standardization: Geneva, Switzerland, 2009; ISBN 978 0 580 67636 9.
32. *BS 812: Testing Aggregates of Density, Part 2. Methods of Determination*; British Standards Institution: London, UK, 1995; ISBN 0 580 24257 9.
33. ASTM D4972-13. *Standard Test Method for pH of Soils*; ASTM International: West Conshohocken, PA, USA, 2013; Available online: www.astm.org (accessed on 3 September 2018).
34. EPA. EPA Method 3052-1. Microwave Assisted Acid Digestion of Siliceous and Organically Based Matrices. 1996. Available online: https://www.epa.gov/sites/production/files/2015-12/documents/3052.pdf (accessed on 8 June 2016).
35. Klink, B.; Palumbo, B.; Cave, M.; Wragg, J. *Arsenic Dispersal and Bioaccessibility in Mine Contaminated Soils: A Case Study from an Abandoned Arsenic Mine in Devon, UK*; British Geological Survey Research Report RR/04/003; British Geological Survey, NERC: Keyworth, Nottingham, UK, 2005; 52p.
36. Environment Agency. *Mitigation of Pollution from Abandoned Metal Mines, Part 2: Review of Resource Recovery Options from the Passive Remediation of Metal-Rich Mine Waters*; SC090024/R2; Environment Agency: Bristol, UK, 2012. Available online: https://www.gov.uk/government/uploads/system/uploadsattachment_data/file/291554/scho1111buvo-e-e.pdf (accessed on 8 June 2016).
37. Vesborg, P.C.; Jaramillo, T.F. Addressing the terawatt challenge: Scalability in the supply of chemical elements for renewable energy. *RSC Adv.* **2012**, *2*, 7933–7947. [CrossRef]
38. Environment Agency. *Guidance on the Use of Soil Screening Values in Ecological Risk Assessment*; Science Report SC070009/SR2b; Environment Agency: Bristol, UK, 2008.
39. Defra. *Development of Category 4 Screening Levels for Assessment of Land Affected by Contamination—Policy Companion Document*; Defra: London, UK, 2014.
40. Environment Agency. *Soil Guideline Values for Arsenic in Soil*; Science Report SC050021/Arsenic SGV., s.l.; Environment Agency: Bristol, UK, 2009.
41. Environment Agency. *Soil Guideline Values for Cadmium in Soil*; Science Report SC050021/Cadmium SGV., s.l.; Environment Agency: Bristol, UK, 2009.
42. Environment Agency. *Soil Guideline Values for Nickel in Soil*; Science Report SC050021/Nickel SGV., s.l.; Environment Agency: Bristol, UK, 2009.
43. Vracar, R.Z.; Natasa, V.; Kamberovic, Z. Leaching of copper (I) sulphide by sulfuric acid solution with addition of sodium nitrate. *Hydrometallurgy* **2003**, *70*, 143–151. [CrossRef]

© 2018 by the authors. Licensee MDPI, Basel, Switzerland. This article is an open access article distributed under the terms and conditions of the Creative Commons Attribution (CC BY) license (http://creativecommons.org/licenses/by/4.0/).

Article

Recovery of Alkali from Bayer Red Mud Using CaO and/or MgO

Bingxin Zhou [1,2,3,4], Shaotao Cao [2,3,*], Fangfang Chen [2,3], Fangfang Zhang [2,3] and Yi Zhang [2,3]

1. School of Chemical Engineering and Technology, Tianjin University, Tianjin 300072, China; zhoubingxin@163.com
2. Key Laboratory of Green Process and Engineering, Institute of Process Engineering, Chinese Academy of Sciences, Beijing 100190, China; ffchen@ipe.ac.cn (F.C.); ffzhang@ipe.ac.cn (F.Z.); yizh@ipe.ac.cn (Y.Z.)
3. National Engineering Laboratory for Hydrometallurgical Cleaner Production Technology, Institute of Process Engineering, Chinese Academy of Sciences, Beijing 100190, China
4. National Engineering Research Center of Distillation Technology, Tianjin 300072, China
* Correspondence: stcao@ipe.ac.cn; Tel.: +86-1082544884

Received: 29 March 2019; Accepted: 26 April 2019; Published: 30 April 2019

Abstract: Recovering alkali from Bayer red mud is crucial for storage security, resource utilization and environmental protection. In this study, the addition of MgO and/or CaO was conducted to recover alkali from red mud with a hydrothermal method for the first time. A synergistic result with a residual Na_2O/SiO_2 weight ratio of 0.03 was obtained by adding the blend of CaO and MgO at an appropriate temperature. MgO was found to be more temperature-dependent than CaO when substituting Na_2O from red mud due to their different hydration processes. The alkali recovery was controlled by a reaction at a temperature of <200 °C and by internal diffusion at a higher temperature for MgO, but controlled by internal diffusion for CaO in the whole temperature range studied. The formation of hydrotalcite-like compounds with a loose structure was verified with the help of XRD, FTIR, and SEM-EDS. It was proved that both the reaction kinetics and the characteristics of solid products have a significant influence on the recovery of alkali.

Keywords: recovering alkali; Bayer red mud; reaction kinetics; magnesium oxide; hydrotalcite-like compounds

---

1. Introduction

Red mud is the solid waste emitted from the aluminum industry. China produced 69.02 million tons of alumina in 2017, with more than 100 million tons of red mud generated [1]. The pH of red mud ranges from 9.2 to 13.0 with an average value of 11.3 ± 1.0, and the high alkalinity of red mud is the primary reason for its classification as a hazardous material [2–5]. Although most alumina producers dispose of this residue in tailing dams [6], this is in no way a long-term solution considering the associated security, environmental and utilization problems [7–10].

Several processes have been proposed to neutralize the alkali in red mud, such as acids and acid gases neutralization, seawater neutralization, microbial neutralization and revegetation [4,8,11–15], among which seawater neutralization has been highly valued for the abundance and low cost of seawater. In the seawater neutralization process, the Ca- and Mg-rich brines play an important role in the reduction of pH and in the precipitation of hydroxide, carbonate or hydroxycarbonate minerals [13,16,17]. However, this method is only suitable for coastal areas.

In addition, leaching and recovering alkali from red mud has also been widely investigated, such as the soda-lime roasting process, solvent extraction process, calcification-carbonation method, and water leaching and hydro-chemical process [8,18–20]. The addition of lime has been widely studied because of its excellent performance in replacing Na_2O, as well as its abundance and cheapness. In fact, the insoluble sodium mainly lies in cancrinite, and the sodium can be leached out from it only if the structure is broken down, or the included Na$^+$ can be removed through the opening of the cancrinite structure. Therefore, the leaching process is based on the ion exchange between Ca^{2+} and Na$^+$ [4]. There are many studies about leaching alkali from red mud with lime, including the industrial use of lime to remove alkali, such as the "Lime Bayer" process and the Sagin process [4]. Yang et al. and Luo et al. investigated the recovery ratio of alkali with the addition of lime under different conditions, and obtained the best recovery ratio of 80% [21,22]. Li et al. conducted a thermodynamic analysis of the calcification reaction and studied the calcification mechanism under different conditions focusing on the calcification-carbonation processes [23].

To summarize, CaO or MgO plays an essential part in alkali recovery and pH reduction of red mud because of its high efficiency, low cost and absence of secondary pollution, unlike Ca- and Mg-rich brines. However, few examples from the literature have systematically reported the differences in using CaO and/or MgO as additions to recover alkali from red mud. Venancio et al. added CaO/MgO before the carbonation, but only reported the change of pH [24].

In this study, the hydrothermal reaction using a different molar ratio of CaO/MgO blend to recover alkali was conducted at different temperatures to determine the alkali recovery capacity, and the best recovery efficiency reached 89.3% with a residual Na$_2$O/SiO$_2$ weight ratio (N/S) of 0.03. The reaction kinetics was regressed to ascertain the rate controlling step. Moreover, the detection of phase composition, infrared spectroscopy analysis and micro-zone analysis were used to study the differences in CaO/MgO dealkalization. This study can provide reference for the reaction mechanism investigations of alkali recovery from red mud based on ion exchange.

2. Materials and Methods

2.1. Materials

Calcium oxide (CaO, AR) and magnesium oxide (MgO, AR) were purchased from Beijing Chemical Industry Co. Ltd. and were used as received. Deionized water was used in all experiments. Red mud was supplied by Coalmine Alumina Co., Henan Province, China, dried overnight at 105 °C on the ground, and finally put through a 200-mesh sieve. The standard reference red mud was from Zhengzhou Light Metals Research Institute of CHINALCO, Henan Province, China. The flux (Li$_2$B$_4$O$_7$:LiBO$_2$ = 12:22) was purchased from Luoyang Tenai Laboratory Equipment Co., Ltd., Henan Province, China.

2.2. Experimental Methods

The experiments were carried out in 150 mL steel bomb reactors which were placed in a homogeneous bath furnace. The reactors were made of nickel metal, and the furnace was filled with glycerin when the temperature was ≤160 °C, while filled with a molten salt of 53 wt. % of KNO$_3$, 40 wt. % of NaNO$_2$, and 7 wt. % of NaNO$_3$ when the temperature was higher than 160 °C. The agitation of the reaction was driven by the rotation of the steel bombs at a speed of 30 rpm and strengthened by the addition of nickel beads. The temperature of the furnace was controlled within ±0.5 °C.

For each experiment, 15 g red mud and a blend with a specific molar ratio of CaO to MgO (100%:0%, 90%:10%, 70%:30%, 50%:50%, 30%:70%, 10%:90% and 0%:100%) was added in each steel bomb reactor. The liquid to solid (L/S) weight ratio was fixed at 4:1, which was the same as in the latest literature focusing on alkali recovery of red mud [23]. The total amount of CaO and MgO in each bomb reactor was 0.056 mol and remained constant, with the CaO and/or MgO:Na$_2$O molar ratio remaining at 4:1 [23,25]. The steel bomb reactors were then put in the furnace. The furnace was heated to 20 °C below the scheduled temperature in advance, and the start of the reaction was recorded upon reaching the experimental temperature. The steel bombs were removed after reaction for 3 h, and then quickly cooled in water. The solid cake and the filtrate were obtained by filtration. The obtained solids were washed by deionized water and dried in a laboratory oven at 105 °C for subsequent analysis.

2.3. Analytical Methods

The content of water-soluble sodium in the red mud was obtained by the titration method of the Pingguo alumina factory, Guangxi, CHALCO. The analysis process is shown as follows:

(a) An amount of 5 g red mud and 200 mL sodium chloride solution (0.1000 mol/L) were added in a 500 mL conical flask, and then the suspension was agitated for 15 min by a magnetic stirrer at an agitation speed of 300 rpm, before being filtrated.

(b) An amount of 100 mL liquid filtrate obtained by filtration was moved into a 250 mL conical flask, and the liquid was concentrated to about 50 mL, and heated in an electric furnace.

(c) Between six and eight drops of bromothymol blue were added into the concentrated liquid and then the liquid was rapidly titrated with hydrochloric acid (0.1000 mol/L). We stopped adding hydrochloric acid when the color changed from blue to yellow and recorded the volume (V, mL) of the hydrochloric acid used.

The content of water-soluble sodium in the red mud could be calculated through Equation (1):

$$Na_2O\ wt.\% = 0.1 \times \frac{V}{1000} \times \frac{200}{100} \times \frac{62}{2} \times \frac{100}{5} = 0.124 \times V \tag{1}$$

The chemical composition of the solids was determined by X-ray Fluorescence (XRF, ARL9800XP, Thermofisher, Waltham, MA, USA). The loss of sodium from the red mud could be obtained by the XRF measurement and calculation. The samples were pressed into disks before measurement: (1) the solid samples were finely ground and dried at 110 °C for 1 h; (2) 0.7000 g sample and 5.6000 g flux were mixed and melted into a disk with an automatic melting furnace (FLUXY10, Classise, Mississauga, Ontario, Canada). The disks were subsequently analyzed using the XRF instrument, operated at 30 kV and 80 mA. The elemental concentrations (Al_2O_3, Fe_2O_3, CaO, TiO_2, SiO_2, MgO, Na_2O and K_2O) were quantified based on the measurements of disks using the MVR software. Prior to the analysis of the samples, the XRF instrument was calibrated with standard reference red mud and the intensity–concentration working curve was made.

During the hydrothermal process, the silicon oxide would not be leached into the liquor, and the recovery ratio of alkali from the red mud could be calculated through Equation (2):

$$\alpha = \left(1 - \frac{N}{N_0}\right) \times 100\% \tag{2}$$

where α was the recovery ratio of alkali, N was the N/S in the solid sample, and N_0 was the N/S of the initial red mud.

X-ray diffraction patterns of the solids were identified from powder diffraction patterns (XRD, X'pert MPD Pro, PanAnalytical, Almelo, the Netherlands) recorded with Cu K_α (λ = 0.15408 nm) radiation. The structural investigation of solids was carried out by Fourier transform infrared spectroscopy (FTIR, T27-Hyperion-Vector22, Bruker, Bremen, Germany). The surface morphology of samples was detected by scanning electron microscopy (SEM, JSM-7001F, JEOL, Tokyo, Japan) with energy dispersive X-ray spectroscopy (EDS, Inca X-MAX, Oxford, UK).

3. Results and Discussion

3.1. Characterization of Bayer Red Mud

From the X-ray diffraction patterns in Figure 1, Bayer red mud mainly consists of cancrinite ($Na_8(Si_6Al_6O_{24})(H_{0.88}(CO_3)_{1.44})(H_2O)_2$), katoite ($Ca_{2.93}Al_{1.97}Si_{0.64}O_{2.56}(OH)_{9.44}$), hematite ($Fe_2O_3$), perovskite ($CaTiO_3$) and calcite ($CaCO_3$). The main chemical composition of red mud is shown in Table 1 and Na_2O accounts for 5.97 wt. % in the red mud. The content of water-soluble sodium in the red mud was obtained with the titration method as shown in Section 2.3, and the result showed that V equaled 0.95 mL, so the water-soluble sodium and water-insoluble sodium were calculated for

2.0 wt. % and 98.0 wt. % of the total sodium in the red mud, respectively. Besides, the X-ray diffraction pattern in Figure 1 shows no detectable changes after the removal of water-soluble sodium. It is thus inferred that most water-insoluble alkaline lies in the cancrinite. In Figure 2, it can be seen that the particles of the red mud show irregular appearance and the distribution of sodium atoms focuses on some specific particles, which is consistent with the previous assertion that most of the alkali lies in cancrinite.

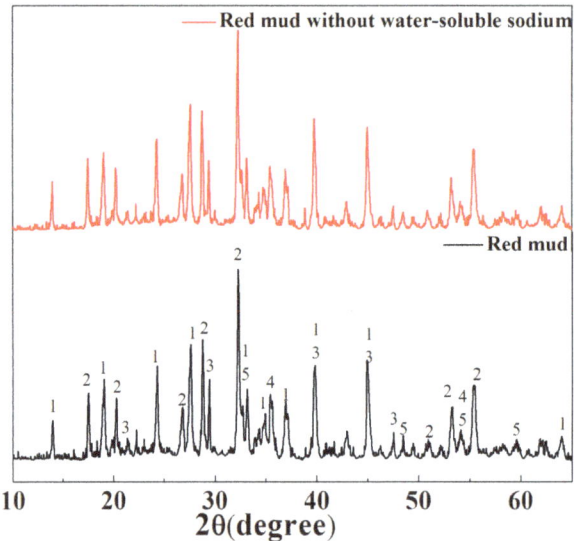

Figure 1. X-ray diffraction analysis of Bayer red mud (1—$Na_8Si_6Al_6O_{24}(H_{0.88}(CO_3)_{1.44})(H_2O)_2$; 2—$Ca_{2.93}Al_{1.97}Si_{0.64}O_{2.56}(OH)_{9.44}$; 3—$CaCO_3$; 4—$Fe_2O_3$; 5—$CaTiO_3$).

Figure 2. Morphology of the red mud (the green parts represent Na).

Table 1. Main chemical composition of Bayer red mud measured by X-ray fluorescence.

Chemicals	Al_2O_3	SiO_2	CaO	Fe_2O_3	Na_2O	TiO_2	K_2O	MgO	N/S
Content (wt. %)	24.42	21.12	18.47	8.14	5.97	4.56	1.53	0.75	0.28

3.2. Recovery of Na$_2$O Using CaO/MgO

The influence of temperature on alkali recovery efficiency was investigated. The results are illustrated in Figure 3 with pure CaO or MgO as an addition, and it is found that the recovery ratio of Na$_2$O positively correlates with temperature. The recovery ratio increased smoothly from 72.5% to 80.9% in the temperature range of 90–210 °C with CaO as an addition, but more sharply from 13.4% to 78.9% when MgO was added from 90 °C to 200 °C, and then only improved to 79.9% when the temperature further increased to 210 °C. Although the recovery efficiencies of Na$_2$O with CaO as an addition were always higher than those with MgO, the relative efficiency (MgO:CaO) increased from 18.5% to 98.8% when temperature increased from 90 °C to 210 °C. Moreover, the recovery efficiency was similar for both CaO and MgO at temperatures ≥200 °C.

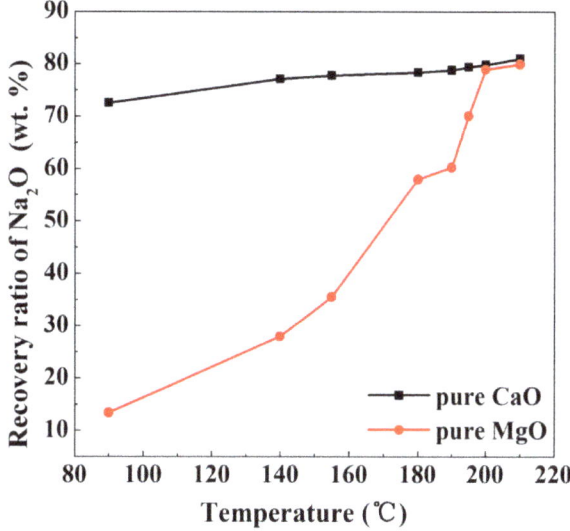

Figure 3. Influence of temperature on the recovery ratio of Na$_2$O using pure CaO/MgO (t = 3 h).

Nevertheless, the recovery of sodium oxide shows a totally different rule with the blend of CaO and MgO as an addition, as shown in Figure 4. For instance, the recovery efficiency decreased with the increasing proportion of MgO when the temperature was lower than 200 °C. However, the recovery ratios of Na$_2$O with 10% to 50% (molar ratio) MgO addition were higher than those of the single use of CaO or MgO when the temperature was higher than 200 °C, and the maximum was obtained at 50% MgO addition. The best recovery of Na$_2$O was 89.3% in the experimental range, with N/S of 0.03 and only 0.64 wt. % Na$_2$O left in the residue.

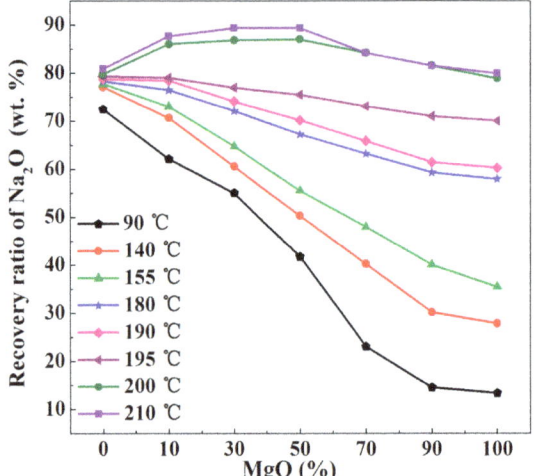

Figure 4. Influence of blending MgO and CaO on the recovery ratio of Na$_2$O (t = 3 h, % means the mole fraction of MgO addition).

3.3. Phase Differences of CaO/MgO Substituting Na$_2$O

As shown in Figure 5a, the dealkalization product bearing Ca is attributed to katoite with CaO as an addition [23]. However, as shown in Figure 5b,c, most residues bearing Mg are MgO and Mg(OH)$_2$ with MgO as an addition at a temperature lower than 200 °C, indicating the incomplete hydration of MgO and the difficulty in leaching Na$_2$O. In fact, temperature has been proven to play an important role in the hydration process of MgO [26,27]. As the alkali recovery was based on the ion exchange between Ca^{2+}/Mg^{2+} and Na$^+$ [23], the hydration of MgO and CaO to dissociate the ion was of great importance to the alkali recovery. The reaction kinetics will be discussed in Section 3.4. The XRD patterns in Figure 5 and Table 2 also show the appearance of (Mg$_{4.5}$Al$_{1.5}$)(Si$_{2.5}$Al$_{1.5}$)O$_{10}$(OH)$_8$ which is a kind of hydrotalcite-like compound and contributed to the sodium removal. The related reactions can be simplified as follows:

$$Na_8(Si_6Al_6O_{24})(H_{0.88}(CO_3)_{1.44})(H_2O)_2 + CaO + H_2O \rightarrow Ca_{2.93}Al_{1.97}Si_{0.64}O_{2.56}(OH)_{9.44} + CaCO_3 + Na^+ + OH^- \tag{3}$$

$$Na_8(Si_6Al_6O_{24})(H_{0.88}(CO_3)_{1.44})(H_2O)_2 + MgO + H_2O \rightarrow (Mg_{4.5}Al_{1.5})(Si_{2.5}Al_{1.5})O_{10}(OH)_8 + CO_3^{2-} + Na^+ + OH^- \tag{4}$$

Table 2. XRD patterns of solid products at temperatures <200 °C.

MgO%	Temperature/°C	MgO	Mg(OH)$_2$	(Mg$_{4.5}$Al$_{1.5}$)(Si$_{2.5}$Al$_{1.5}$)O$_{10}$(OH)$_8$
0	90	x	x	x
	140	x	x	x
	190	x	x	x
50	90	√	√	x
	140	√	√	x
	190	√	√	√
100	90	√	√	x
	140	√	√	√
	190	√	√	√

% means the mole fraction of MgO addition; only new phases in each sample are listed out; √: existence; x: nonexistence.

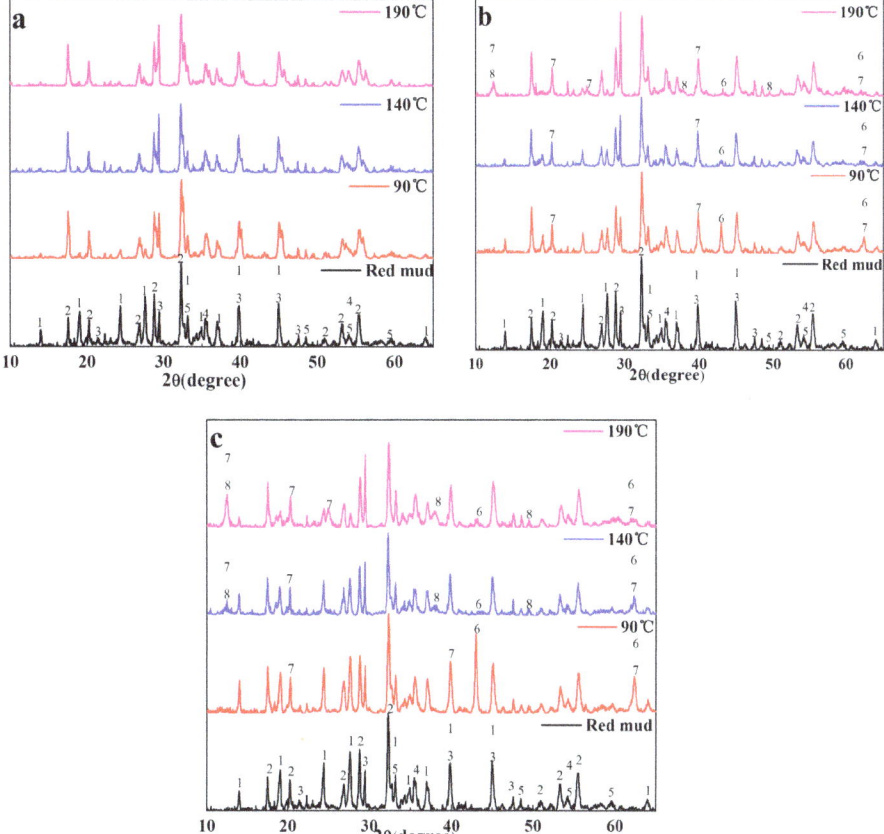

Figure 5. XRD patterns of solid products at temperatures <200 °C: (**a**) 0% MgO addition; (**b**) 50% MgO addition; (**c**) 100% MgO addition (1—$Na_8Si_6Al_6O_{24}(H_{0.88}(CO_3)_{1.44})(H_2O)_2$; 2—$Ca_{2.93}Al_{1.97}Si_{0.64}O_{2.56}(OH)_{9.44}$; 3—$CaCO_3$; 4—$Fe_2O_3$; 5—$CaTiO_3$; 6—$MgO$; 7—$Mg(OH)_2$; 8—$(Mg_{4.5}Al_{1.5})(Si_{2.5}Al_{1.5})O_{10}(OH)_8$); % means the mole fraction of MgO addition; only new phases are marked out while the same XRD pattern for the red mud is presented without labels in each figure for clarity).

When the temperature increases to 200 °C, the changes in XRD patterns are shown in Figure 6 and Table 3. New hydrotalcite-like compounds of $Ca_{10}Mg_2Al_4(SiO_4)_5(Si_2O_7)_2(OH)_4$ and $Mg_3Si_2O_5(OH)_4$ appear when MgO is added. The relating reactions can be simplified as follows:

$$Na_8(Si_6Al_6O_{24})(H_{0.88}(CO_3)_{1.44})(H_2O)_2 + MgO + H_2O \rightarrow Mg_3Si_2O_5(OH)_4 + Al(OH)_4^+ + CO_3^{2-} + Na^+ + OH^- \quad (5)$$

$$Na_8(Si_6Al_6O_{24})(H_{0.88}(CO_3)_{1.44})(H_2O)_2 + MgO + CaO + H_2O \rightarrow Ca_{10}Mg_2Al_4(SiO_4)_5(Si_2O_7)_2(OH)_4 + Na^+ + CO_3^{2-} + OH^- \quad (6)$$

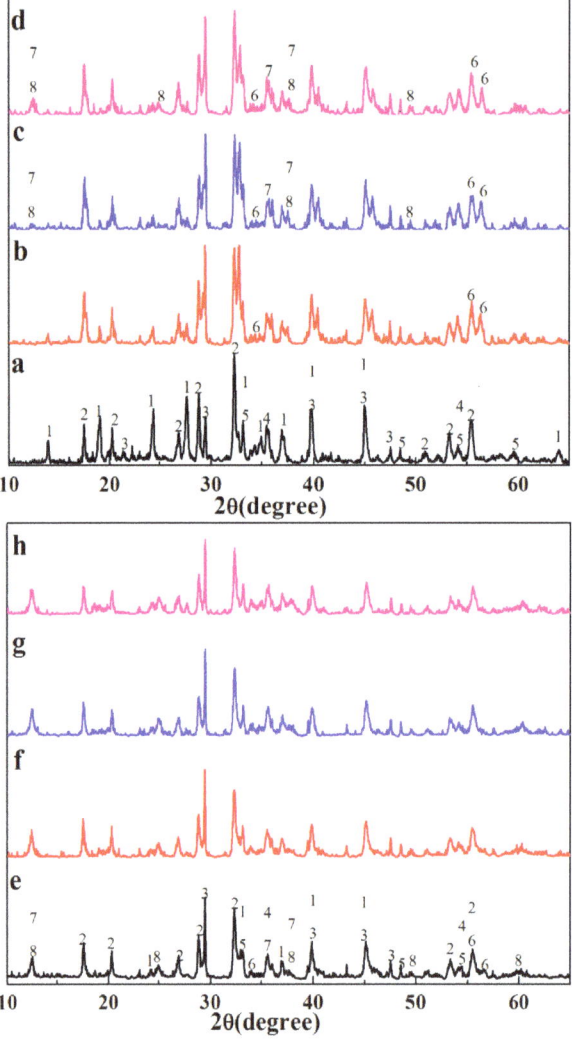

Figure 6. XRD patterns of the solid products at 200 °C: (**a**) Bayer red mud; (**b**) 0% MgO addition; (**c**) 10% MgO addition; (**d**) 30% MgO addition; (**e**) 50% MgO addition; (**f**) 70% MgO addition; (**g**) 90% MgO addition and (**h**) 100% MgO addition (1—$Na_8Si_6Al_6O_{24}(H_{0.88}(CO_3)_{1.44})(H_2O)_2$; 2—$Ca_{2.93}Al_{1.97}Si_{0.64}O_{2.56}(OH)_{9.44}$; 3—$CaCO_3$; 4—$Fe_2O_3$; 5—$CaTiO_3$; 6—$Ca_{10}Mg_2Al_4(SiO_4)_5(Si_2O_7)_2(OH)_4$; 7—$Mg_3Si_2O_5(OH)_4$; 8—$(Mg_{4.5}Al_{1.5})(Si_{2.5}Al_{1.5})O_{10}(OH)_8$); % means the mole fraction of MgO addition; only new phases are marked out while the same XRD pattern for the red mud is presented without labels in each figure for clarity).

Table 3. XRD patterns of the solid products at 200 °C.

MgO%	$Ca_{10}Mg_2Al_4(SiO_4)_5(Si_2O_7)_2(OH)_4$	$Mg_3Si_2O_5(OH)_4$	$(Mg_{4.5}Al_{1.5})(Si_{2.5}Al_{1.5})O_{10}(OH)_8$
0	x	x	x
10	√	x	x
30	√	√	√
50	√	√	√
70	√	√	√
90	√	√	√
100	√	√	√

% means the mole fraction of MgO addition; only new phases in each sample are listed out; √: existence; x: nonexistence.

3.4. Reaction Kinetics of Alkali Recovery

The reaction kinetics has been implemented to express the alkali recovery process with CaO or MgO as an addition. According to the shrinking core model (SCM) [28,29], the kinetic equation of the heterogeneous reaction can be expressed as Equation (7) if the process is controlled by internal diffusion or as Equation (8) if controlled by chemical reaction:

$$k_a t = 1 - \frac{2}{3}\alpha - (1-\alpha)^{2/3} \qquad (7)$$

$$k_b t = 1 - (1-\alpha)^{1/3} \qquad (8)$$

where t represents the reaction time (min), α is the recovery ratio of Na_2O (%), k_a is the rate constant for the process of internal diffusion, and k_b is the rate constant for the chemical reaction. The external diffusion was not considered because the reaction system was fully stirred.

As shown in Figure 7, a period of time of 30 min was used because the reaction was relatively fast. The alkali recovery process is controlled by internal diffusion with CaO as an addition and the activation energy is calculated to be 12.67 kJ/mol. In contrast, the recovery process with MgO as an addition is controlled by chemical reaction with an apparent activation energy of 40.99 kJ/mol when the temperature is lower than 200 °C, while this process is controlled by internal diffusion with an apparent activation energy of 18.04 kJ/mol when the temperature is higher than 200 °C. The supporting information showed the recovery ratio of alkali versus time (Supplementary Materials Tables S1 and S2) and the calculation of the CaO/MgO leaching reaction with the other SCM model (Figure S1) to show the differences.

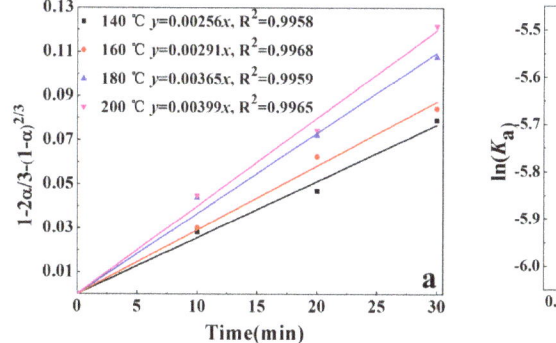
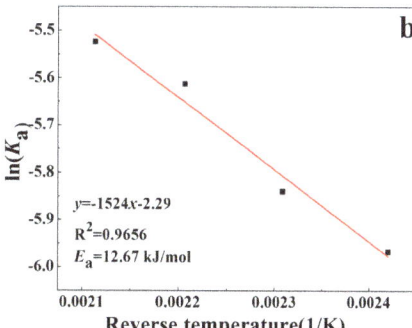

Figure 7. Cont.

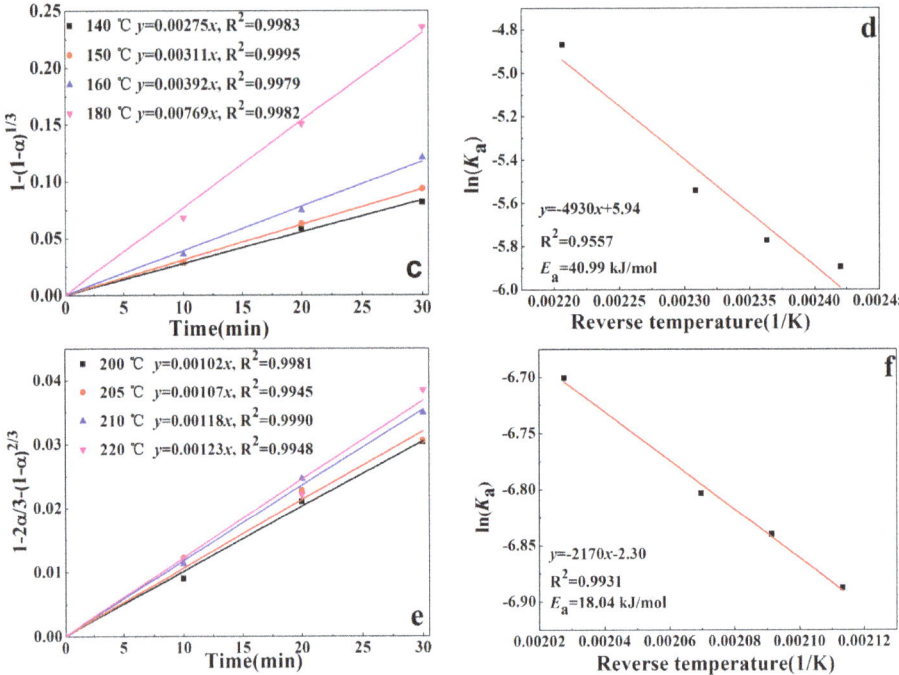

Figure 7. Kinetic fitting of (**a**) the plots of $[1 - \frac{2}{3}\alpha - (1-\alpha)^{2/3}]$ versus time for CaO, (**b**) $\ln(K_a)$ versus temperature for CaO, (**c**) the plots of $[1 - (1-\alpha)^{1/3}]$ versus time for MgO lower than 200 °C, (**d**) $\ln(K_a)$ versus temperature for MgO lower than 200 °C, (**e**) the plots of $[1 - \frac{2}{3}\alpha - (1-\alpha)^{2/3}]$ versus time for MgO higher than 200 °C and (**f**) $\ln(Ka)$ versus temperature for MgO higher than 200 °C.

The alkali recovery process can be principally divided into the following five steps: (1) the hydration of CaO/MgO; (2) the dissociation of $Ca(OH)_2$/$Mg(OH)_2$; (3) the diffusion of Ca^{2+}/Mg^{2+} from the bulk solution to the Na^+ detached site of cancrinite; (4) the ion exchange between Ca^{2+}/Mg^{2+} and Na^+; (5) the diffusion of Na^+ to the bulk solution [28]. It can be speculated that the alkali recovery process with CaO as an addition may be controlled by the third step or the fifth step in this study based on a consideration of the hydration process, the obtained rate equation and the activation energy. In contrast, the process with MgO as an addition may be mostly controlled by the first step at low temperatures, and the rate equation together with the activation energy is consistent with the results of MgO hydration under similar conditions [27]. Therefore, as the MgO hydration rate increases with increasing temperature, the recovery ratio of Na_2O with pure MgO as an addition also increases significantly.

As shown above, the alkali recovery process with MgO as an addition was controlled by internal diffusion, rather than by reaction at higher temperatures. Besides, for 30 minutes the extraction of alkali using MgO (Figure 7e) was much lower than when using CaO (as compared to Figure 7a), and it was because the hydration of MgO was relatively slow compared with CaO, so there was an induction period when using MgO. Moreover, as the solubility of calcium hydroxide is higher than that of magnesium hydroxide, the hydrolysis rate of magnesium hydroxide may be slower than that of calcium hydroxide. Theoretically, the cancrinite is hexagonal with the space group of $P6_3$ [23], and Mg^{2+} was easier to obtain through the openings as the radius of Mg^{2+} is smaller than that of Ca^{2+}. However, the experimental results show that the recovery ratio does not increase monotonously with the increasing MgO proportion, as shown in Figure 4. It can be explained by the fact that the cancrinite transformed to hydrotalcite-like compounds bearing Mg, of which the lamellar structure as well as the loosened surface was conducive to the further substitution of Na^+ [13,30]. However, the hydration of MgO was speculated to be much faster than the substitution reaction, together with the lesser usage of Me to exchange Na at high temperatures, leading to the increasing existence of $Mg(OH)_2$ (or $Ca(OH)_2$) on the surface of solid products upon further addition. This may in turn have prevented the diffusion of Na^+ from the cancrinite to the bulk solution. As a result, the recovery efficiency decreased rather than increased when the addition proportion of MgO was higher than 50%.

3.5. FTIR Spectra Analysis

To explore the frameworks and the valence-bond structures, the FTIR spectra of the red mud and all leached solid samples were collected in the range of 4000–400 cm^{-1} at room temperature. The results are shown in Figure 8.

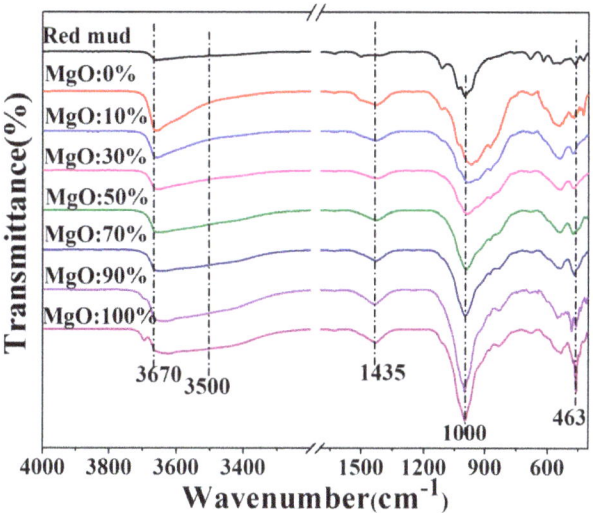

Figure 8. FTIR spectra of solid samples at 200 °C.

The peaks around 3700–3500 cm^{-1} can be attributed to the vibration of CaO–H and AlO–H bonds in the solids [31], and the peak intensity shows a sudden increase with the addition of CaO, which is consistent with the phase changes. The broad absorption bands around 3500–3400 cm^{-1} can be attributed to the -OH stretching vibration of hydroxyl groups in layer structures [32]. Owing to the overlapping of Mg–OH and Al–OH bonds related with newly formed hydrotalcite-like structures, the bands in this region became stronger and broader with the increasing proportion of MgO. The peaks between 1500 and 1400 cm^{-1} are attributed to the overtone of Mg–O and Al–O lattice vibrations [33]. The absorption bands around 1000 cm^{-1} are generally assigned to the metal–oxygen skeletal stretching and bending vibration of hydrotalcite, and the peaks around 463 cm^{-1} can be attributed to Mg–O vibrations in the mixed Mg–Al oxide. The increasing intensity of these two peaks shows strong evidence of the production of hydrotalcite-like structures [34,35]. To sum up, the formation of hydrotalcite-like structures is proved by the significant changes in FTIR spectra.

3.6. Solid Morphology

There are two types of appearance of solids after dealkalization at 200 °C, as shown in Figure 9. The polyhedral particles with a dense structure are verified to be $Ca_{2.93}Al_{1.97}Si_{0.64}O_{2.56}(OH)_{9.44}$ by EDS analysis in Figure 9a', and the dense structure further supports the speculation that the reaction is controlled by internal diffusion with CaO as an addition. Moreover, a small amount of Na was still detected. However, the solid products containing Mg show a lamellar appearance with a loose structure. For example, 30% MgO addition caused sporadical slakes distributed on the dense particles of calcium garnet, and more MgO addition led to the generation of more slakes and their aggregates. The lamellar and loose structures are consistent with the morphology of hydrotalcite-like structures which have been reported by other researchers [35].

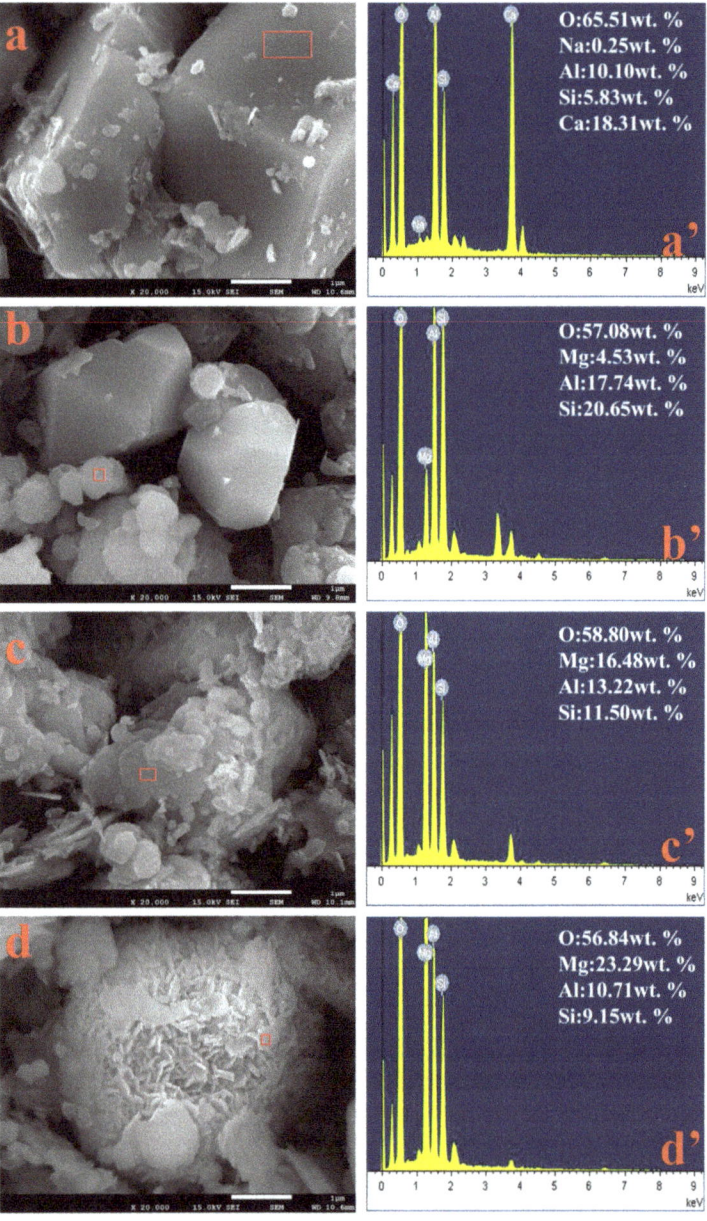

Figure 9. Morphologies of the solid products at 200 °C: (**a-a′**) 0% MgO addition; (**b-b′**) 30% MgO addition; (**c-c′**) 70% MgO addition and (**d-d′**) 100% MgO addition.

4. Conclusions

Recovery of alkali from Bayer red mud was investigated systematically with CaO and/or MgO as additions for the first time. It was found that the blend of CaO and MgO had a better capacity to substitute alkali from the red mud when the temperature was no less than 200 °C. The highest Na_2O recovery of 89.3% with N/S of 0.03 was obtained at 50% MgO addition at 200 °C. Moreover, MgO substitution to Na_2O was found to be more temperature-dependent than that of CaO through different reaction mechanisms. The Na_2O recovery process with MgO as an addition was controlled by reaction when the temperature was lower than 200 °C, but it was controlled by internal diffusion at a higher temperature. However, the process was always controlled by internal diffusion in the whole temperature range studied when CaO was used as an addition.

XRD analysis, FTIR analysis and SEM-EDS confirmed the hydrotalcite-like structure of solid products bearing Mg. To sum up, the recovery efficiency was the result of the reaction kinetics and the characteristics of solid products. This study can provide reference for the reaction mechanism investigation of alkali recovery from red mud based on ion exchange. It also provides a new idea for the treatment of red mud.

Supplementary Materials: The following are available online at http://www.mdpi.com/2075-163X/9/5/269/s1, Figure S1: Kinetic fitting of (a) the plots of $[1 - (1 - \alpha)^{1/3}]$ versus time for CaO, (b) $\ln(K_a)$ versus temperature for CaO, (c) the plots of $[1 - \frac{2}{3}\alpha - (1 - \alpha)^{2/3}]$ versus time for MgO lower than 200 °C, (d) $\ln(K_a)$ versus temperature for MgO lower than 200 °C, (e) the plots of $[1 - (1 - \alpha)^{1/3}]$ versus time for MgO higher than 200 °C and (f) $\ln(K_a)$ versus temperature for MgO higher than 200 °C, Table S1: Recovery ratio (wt. %) of alkali extraction on time with pure CaO as addition under different temperatures, Table S2: Recovery ratio (wt. %) of alkali extraction on time with pure MgO as addition under different temperatures.

Author Contributions: Investigation, B.Z. and S.C.; Supervision, F.C.; Writing—original draft, B.Z.; Writing—review & editing, B.Z., S.C., F.C., F.Z. and Y.Z.

Funding: This research was funded by the National Natural Science Foundation of China (No. 51674233); the Henan Transfer Project of CAS Scientific and Technological Achievements (No. 2019103); the Guangxi Natural Science Foundation (No. 2016GXNSFEA380002); and the Shanxi Province Coal Based Low-carbon Technology Major Projects (No. MC2016-05).

Conflicts of Interest: The authors declare no conflict of interest.

References

1. Chen, H.; Huo, Y.B.; Yuan, Y.; Shen, L.Y.; Xia, Y.Z.; Chai, Y.; Zhang, Q.; Wang, N.N. Regional statistics of alumina production in China from January to December 2017. *China Alum. Mon.* **2018**, *230*, 130–131.
2. Roosen, J.; Van Roosendael, S.; Borra, C.R.; Van Gerven, T.; Mullens, S.; Binnemans, K. Recovery of scandium from leachates of Greek bauxite residue by adsorption on functionalized chitosan-silica hybrid materials. *Green Chem.* **2016**, *18*, 2005–2013. [CrossRef]
3. Lehoux, A.P.; Lockwood, C.L.; Mayes, W.M.; Stewart, D.I.; Mortimer, R.J.G.; Gruiz, K.; Burke, I.T. Gypsum addition to soils contaminated by red mud: Implications for aluminium, arsenic, molybdenum and vanadium solubility. *Environ. Geochem. Health* **2013**, *35*, 643–656. [CrossRef] [PubMed]
4. Smith, P. The processing of high silica bauxites-review of existing and potential processes. *Hydrometallurgy* **2009**, *98*, 162–176. [CrossRef]
5. Shoppert, A.A.; Loginova, I.V.; Rogozhnikov, D.A.; Karimov, K.A.; Chaikin, L.I. Increased as adsorption on maghemite-containing red mud prepared by the alkali fusion-leaching method. *Minerals* **2019**, *9*, 60. [CrossRef]
6. Power, G.; Grafe, M.; Klauber, C. Bauxite residue issues: I. Current management, disposal and storage practices. *Hydrometallurgy* **2011**, *108*, 33–45. [CrossRef]
7. Klauber, C.; Gräfe, M.; Power, G. Bauxite residue issues: II. Options for residue utilization. *Hydrometallurgy* **2011**, *108*, 11–32. [CrossRef]
8. Liu, Z.B.; Li, H.X. Metallurgical process for valuable elements recovery from red mud—A review. *Hydrometallurgy* **2015**, *155*, 29–43. [CrossRef]

9. Enserink, M. Envrionment after red mud flood, scientists try to halt wave of fear and rumors. *Science* **2010**, *330*, 432–433. [CrossRef] [PubMed]
10. Ning, G.; Zhang, B.; Liu, C.; Li, S.; Ye, Y.; Jiang, M. Large-scale consumption and zero-waste recycling method of red mud in steel making process. *Minerals* **2018**, *8*, 102. [CrossRef]
11. Brunori, C.; Cremisini, C.; Massanisso, P.; Pinto, V.; Torricelli, L. Reuse of a treated red mud bauxite waste: Studies on environmental compatibility. *J. Hazard. Mater.* **2005**, *117*, 55–63. [CrossRef] [PubMed]
12. Grafe, M.; Power, G.; Klauber, C. Bauxite residue issues: III. Alkalinity and associated chemistry. *Hydrometallurgy* **2011**, *108*, 60–79. [CrossRef]
13. Hanahan, C.; McConchie, D.; Pohl, J.; Creelman, R.; Clark, M.; Stocksiek, C. Chemistry of seawater neutralization of bauxite refinery residues (red mud). *Environ. Eng. Sci.* **2004**, *21*, 125–138. [CrossRef]
14. Rai, S.; Wasewar, K.L.; Agnihotri, A. Treatment of alumina refinery waste (red mud) through neutralization techniques: A review. *Waste Manage. Res.* **2017**, *35*, 563–580. [CrossRef]
15. Qu, Y.; Li, H.; Wang, X.; Tian, W.; Shi, B.; Yao, M.; Zhang, Y. Bioleaching of major, rare earth, and radioactive elements from red mud by using indigenous chemoheterotrophic bacterium acetobacter sp. *Minerals* **2019**, *9*, 67. [CrossRef]
16. Johnston, M.; Clark, M.W.; McMahon, P.; Ward, N. Alkalinity conversion of bauxite refinery residues by neutralization. *J. Hazard. Mater.* **2010**, *182*, 710–715. [CrossRef] [PubMed]
17. Rai, S.; Wasewar, K.L.; Lataye, D.H.; Mukhopadhyay, J.; Yoo, C.K. Feasibility of red mud neutralization with seawater using Taguchi's methodology. *Int. J. Environ. Sci. Technol.* **2013**, *10*, 305–314. [CrossRef]
18. Li, R.B.; Zhang, T.G.; Liu, Y.; Lv, G.Z.; Xie, L.Q. Calcification-carbonation method for red mud processing. *J. Hazard. Mater.* **2016**, *316*, 94–101. [CrossRef] [PubMed]
19. Zhang, Y.; Zheng, S.L.; Wang, X.H.; Du, H.; Fang, Z.Z.; Zhang, Y. Reactive extraction of sodium hydroxide from alkali solutions for the separation of sodium and aluminum - part I. effect of extractant. *Hydrometallurgy* **2015**, *154*, 47–55. [CrossRef]
20. Zhong, L.; Zhang, Y.F.; Zhang, Y. Extraction of alumina and sodium oxide from red mud by a mild hydro-chemical process. *J. Hazard. Mater.* **2009**, *172*, 1629–1634. [CrossRef]
21. Luo, Z.T.; Xiao, Y.L.; Zhang, L.; Wang, X.; Zheng, Y.R.; Yang, J.J. Multi-stage cycle dealkalization and alkali recovery process of red mud slurry. *J. Univ. Jinan Sci. Technol.* **2013**, *27*, 369–372.
22. Yang, J.J.; Li, J.W.; Xiao, Y.L.; Luo, Z.T.; Han, Y.F. Research on dealkalization of sintering process red mud by lime process at normal atmosphere and mechanism thereof. *Inorg. Chem. Ind.* **2012**, *44*, 40–42.
23. Li, R.B.; Li, X.L.; Wang, D.X.; Liu, Y.; Zhang, T.A. Calcification reaction of red mud slurry with lime. *Powder Technol.* **2018**, *333*, 277–285. [CrossRef]
24. Venancio, L.C.A.; Souza, J.A.S.; Macedo, E.N.; Botelho, F.A.; de Oliveira, A.M.; Fonseca, R.S. Bauxite residue amendment through the addition of Ca and or Mg followed by carbonation. In *Light Metals*; Ratvik, A.P., Ed.; Springer International Publishing Ag: Phoenix, AZ, USA, 2017; pp. 53–59.
25. Zheng, X.F.; Hu, J.; Jiang, M.; Xue, Z.X. Study on optimization of dealkalization process on adding lime to red mud produced by low temperature Bayer process. *Light Metals* **2010**, *4*, 21–23.
26. Amaral, L.F.; Oliveira, I.R.; Salomao, R.; Frollini, E.; Pandolfelli, V.C. Temperature and common-ion effect on magnesium oxide (MgO) hydration. *Ceram. Int.* **2010**, *36*, 1047–1054. [CrossRef]
27. Maryska, M.; Blaha, J. Hydration kinetics of magnesium oxide—Part 3. Hydration rate of MgO in terms of temperature and time of its firing. *Ceram. Silikaty* **1997**, *41*, 121–123.
28. Li, X.F.; Ye, Y.Z.; Xue, S.G.; Jiang, J.; Wu, C.; Kong, X.F.; Hartley, W.; Li, Y.W. Leaching optimization and dissolution behavior of alkaline anions in bauxite residue. *Trans. Nonferrous Met. Soc. China* **2018**, *28*, 1248–1255. [CrossRef]
29. Safari, V.; Arzpeyma, G.; Rashchi, F.; Mostoufi, N. A shrinking particle-shrinking core model for leaching of a zinc ore containing silica. *Int. J. Miner. Process.* **2009**, *93*, 79–83. [CrossRef]
30. Couperthwaite, S.J.; Johnstone, D.W.; Millar, G.J.; Frost, R.L. Neutralization of acid sulfate solutions using bauxite refinery residues and its derivatives. *Ind. Eng. Chem. Res.* **2013**, *52*, 1388–1395. [CrossRef]
31. Qu, J.; Zhong, L.H.; Li, Z.; Chen, M.; Zhang, Q.W.; Liu, X.Z. Effect of anion addition on the syntheses of Ca-Al layered double hydroxide via a two-step mechanochemical process. *Appl. Clay Sci.* **2016**, *124*, 267–270. [CrossRef]
32. Heraldy, E.; Nugrahaningtyas, K.D.; Sanjaya, F.B.; Darojat, A.A.; Handayani, D.S.; Hidayat, Y. In effect of reaction time and (Ca plus Mg)/Al molar ratios on crystallinity of Ca-Mg-Al layered double hydroxide. In Proceedings of the 10th Joint Conference on Chemistry, Solo, Indonesia, 8–9 September 2015.

33. Del Arco, M.; Martin, C.; Martin, I.; Rives, V.; Trujillano, R. A FTIR spectroscopic study of surface acidity and basicity of mixed Mg, Al-oxides obtained by thermal decomposition of hydrotalcite. *Spectrochim. Acta Part A Mol. Spectrosc.* **1993**, *49*, 1575–1582. [CrossRef]
34. Kuśtrowski, P.; Sułkowska, D.; Chmielarz, L.; Rafalska-Łasocha, A.; Dudek, B.; Dziembaj, R. Influence of thermal treatment conditions on the activity of hydrotalcite-derived Mg–Al oxides in the aldol condensation of acetone. *Micropor. Mesopor. Mat.* **2005**, *78*, 11–22. [CrossRef]
35. Olszówka, J.E.; Karcz, R.; Bielańska, E.; Kryściak-Czerwenka, J.; Napruszewska, B.D.; Sulikowski, B.; Socha, R.P.; Gaweł, A.; Bahranowski, K.; Olejniczak, Z.; et al. New insight into the preferred valency of interlayer anions in hydrotalcite-like compounds: The effect of Mg/Al ratio. *Appl. Clay Sci.* **2018**, *155*, 84–94. [CrossRef]

© 2019 by the authors. Licensee MDPI, Basel, Switzerland. This article is an open access article distributed under the terms and conditions of the Creative Commons Attribution (CC BY) license (http://creativecommons.org/licenses/by/4.0/).

Technical Note

Recovering Iron from Iron Ore Tailings and Preparing Concrete Composite Admixtures

Chang Tang [1,2,3], Keqing Li [1,2,*], Wen Ni [1,2,*] and Duncheng Fan [1,2]

1. Key Laboratory of the Ministry of Education of China for High-Efficient Mining and Safety of Metal Mines, University of Science and Technology Beijing, Beijing 100083, China; b20160058@xs.ustb.edu.cn (C.T.); b20170035@xs.ustb.edu.cn (D.F.)
2. School of Civil and Resource Engineering, University of Science and Technology Beijing, Beijing 100083, China
3. Hebei Technology Research Center for Application of High Performance Concrete with Ultra-low Environment Load, Shahe 054100, China
* Correspondence: lkqing2003@163.com (K.L.); niwen@ces.ustb.edu.cn (W.N.)

Received: 15 March 2019; Accepted: 9 April 2019; Published: 15 April 2019

Abstract: Iron ore tailings (IOTs) are a form of solid waste produced during the beneficiation process of iron ore concentrate. In this paper, iron recovery from IOTs was studied at different points during a process involving pre-concentration followed by direct reduction and magnetic separation. Then, slag-tailing concrete composite admixtures were prepared from high-silica residues. Based on the analyses of the chemical composition and crystalline phases, a pre-concentration test was developed, and a pre-concentrated concentrate (PC) with an iron grade of 36.58 wt % and a total iron recovery of 83.86 wt % was obtained from a feed iron grade of 12.61 wt %. Furthermore, the influences of various parameters on iron recovery from PC through direct reduction and magnetic separation were investigated. The optimal parameters were found to be as follows: A roasting temperature of 1250 °C, a roasting time of 50 min, and a 17.5:7.5:12.5:100 ratio of bitumite/sodium carbonate/lime/PC. Under these conditions, the iron grade of the reduced iron powder was 92.30 wt %, and the iron recovery rate was 93.96 wt %. With respect to the original IOTs, the iron recovery was 78.79 wt %. Then, highly active slag-tailing concrete composite admixtures were prepared using the high-silica residues and S75 blast furnace slag powder. When the amount of high-silica residues replacing slag was 20%, the strength of cement mortar blocks at 7 days and 28 days was 33.11 MPa and 50 MPa, respectively, whereas the activity indices were 89 and 108, respectively. Meanwhile, the fluidity rate was appropriately 109. When the content of high-silica residues replacing slag was not more than 30%, the quality of mineral admixtures was not reduced. Last but not least, reusing the high-silica residues during iron recovery enabled the complete utilization of the IOTs.

Keywords: iron ore tailings; iron recovery; concrete composite admixtures; reuse

---

1. Introduction

Iron ore tailings (IOTs) are a form of solid waste produced during the beneficiation process of iron ore concentrate. Among all kinds of mining solid waste, IOTs are one of the most common solid wastes in the world due to their high output and low utilization ratio. In China, the generation of IOTs has increased rapidly due to recent growth in the iron and steel industries. According to some recent statistics, the output of iron tailings was around 500 million tons in 2015 [1]. Although there has already been a recent trend of gradual reduction in the amount of IOTs produced, the total accumulation of IOTs has exceeded 7.5 billion tons in China [2,3]. The enormous amount of IOTs deposited as waste incurs a high economic cost for waste management and also creates serious environmental problems and security risks. However, some studies have found that waste IOTs have an iron grade of approximately 8 wt %–12 wt % on average, and occasionally as high as 27 wt % [4,5]. If these materials contain 10 wt % iron on average, 750 million tons of metallic iron is lost during the disposal of IOTs.

With the passage of time, the earth's iron ore resources continue to decline. Due to this reason, IOTs might become valuable resources in the future. Currently, the processes for utilizing IOTs can be mainly divided into two categories, namely iron recovery from IOTs and the use of IOTs as raw materials. Recovery is a well-known process, and many recovery methods have been developed, including gravimetric separation, magnetic separation, and flotation separation [5–8]. However, these processes create new waste streams after the recovery of iron and are inefficient when using traditional mineral processing techniques. IOTs can also be used to make other products, such as ceramic tiles [9,10], engineered cementitious composites [11], IOT-filled polypropylene cement, and epoxy composites [12,13], and underground backfill mining materials [3]. The conversion of solid industrial waste into other valuable products has received considerable attention in recent years. However, in these processes, iron within the utilized IOTs is not recovered, thus wasting large amounts of iron resources.

Recently, a method involving direct reduction followed by magnetic separation has been used to recover iron from red mud, vanadium tailings, pyrite slag, copper slag, cyanide tailings, and IOTs [14–19]. This method has the advantage of high iron recovery, and is especially suitable for ore with a low iron grade. However, there is no suitable method for dealing with the secondary tailings after iron selection.

In this study, Qidashan IOTs were comprehensively utilized to generate zero-emission IOT waste. The IOTs were owned by the Anshan Iron Steel Group (AISG) in northeastern China and were discharged after the processing of iron minerals from Anshan magnetite quartzite, because iron recovery or reuse of these IOTs as raw materials for value-added products was not economically viable. Meanwhile, due to the low iron content of IOTs, the cost of recovering iron through direct reduction–magnetic separation is high. Consequently, an innovative technique called "pre-concentration followed by direct reduction and magnetic separation technology" was developed [20]. According to this method, pre-concentration is initially undertaken to obtain a pre-concentrated concentrate (PC) and high-silicon residue. The PC is used as a raw material for direct reduction after pre-concentration, and the resulting high-silicon residue from this process is used to prepare slag-tailing concrete composite admixtures that consist of high-silicon residues and blast furnace slag (BFS) discharged by the Ansteel Group Corporation during iron smelting. The reduced iron powder can be conveniently used for electric arc furnace steelmaking by the nearby Ansteel Group Corporation facility as a high-grade alternative raw material to steel scrap. The admixtures can be used to partially or fully replace cement to produce various construction concretes or filling materials for backfilling the mined-out area. Therefore, this technology can truly achieve full iron recovery from IOTs and approaches zero-tailing mining, thus resulting in substantial economic and social benefits. Notably, the amount of silicate residues remaining after direct reduction followed by magnetic separation can reach up to 20 wt % of the total IOTs, and therefore they are used as cement raw meal. In this paper, iron was recovered from IOTs using pre-concentration followed by direct reduction and magnetic separation, whereas slag-tailing concrete composite admixtures were subsequently prepared using the high-silicon residues separated from IOTs.

2. Materials and Methods

2.1. Raw Materials

2.1.1. Iron Ore Tailings

IOTs were sampled from a storage dam at the Qidashan iron ore dressing plant in Anshan, which is in the Liaoning province of China. The chemical composition of this material is presented in Table 1: The total iron content was approximately 12.61 wt %. The crystalline phases in the received IOTs are shown in Figure 1, and included quartz, hematite, and magnetite. After analyzing the iron phases, approximately 58.92 wt % of the total iron was in the form of magnetite or hematite (Table 2), whereas 37.31 wt % of the total iron was in the form of silicate minerals, such as chlorite and hornblende. It is worth noticing that the latter materials cannot be separated effectively using traditional magnetic separation methods.

Table 1. Chemical composition of the iron ore tailings (IOTs).

Components	TFe	SiO$_2$	CaO	MgO	Al$_2$O$_3$	Na$_2$O	K$_2$O	MnO	TiO$_2$	S	P
Content/wt %	12.61	75.46	1.70	1.75	1.65	0.32	0.34	0.13	0.06	0.10	0.03

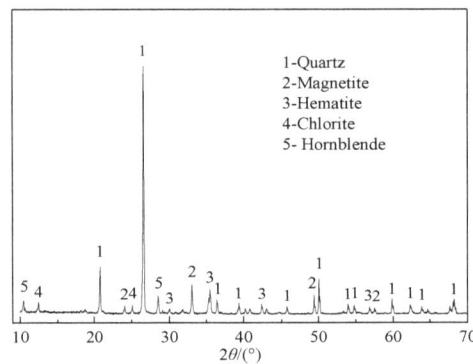

Figure 1. XRD pattern of the IOTs.

Table 2. Analysis of the iron phases in the IOTs.

Phases	Magnetic Iron	Hematite and Limonite	Iron Carbonate	Iron Silicate	Iron Sulfide	TFe
Iron content/wt %	2.29	5.16	0.39	4.72	0.08	12.64
Fraction/wt %	18.12	40.82	3.09	37.34	0.63	100.00

2.1.2. Reducing Coal

The reducing coal used in this study was bitumite. Its composition is listed in Table 3. Carbon (45.16 wt %) was the major active reagent during the direct reduction process.

Table 3. General analysis of the bitumite.

Ingredients	Moisture	Ash	Volatiles	Fixed Carbon
Content/%	9.88	16.20	28.76	45.16

2.1.3. Additives

Analytical-grade lime (CaO) and sodium carbonate (Na$_2$CO$_3$) were used as additives in this study.

2.1.4. S75, Portland Cement PI42.5, and Standard Sand

S75 ground-granulated blast furnace slag powder was produced by the Anshan Iron Steel Group in northeastern China, and the specific surface area was 353 m^2/kg. The cement used was Portland cement with a strength grade of 42.5 (complying with the Chinese National Standard GB 175-1999) and was produced by the Tangshan Jidong Cement Co., Ltd., in China. Standard sand was manufactured and packed by the Xiamen ISO Standard Sand Co., Ltd., in China.

2.2. Experimental Methods

The experimental workflow is shown in Figure 2. The workflow includes pre-concentration, direct reduction–magnetic separation, and final products. Among them, iron concentrate I and iron concentrate II represent the magnetic products after two magnetic separations (0.11 T and 0.8 T), respectively.

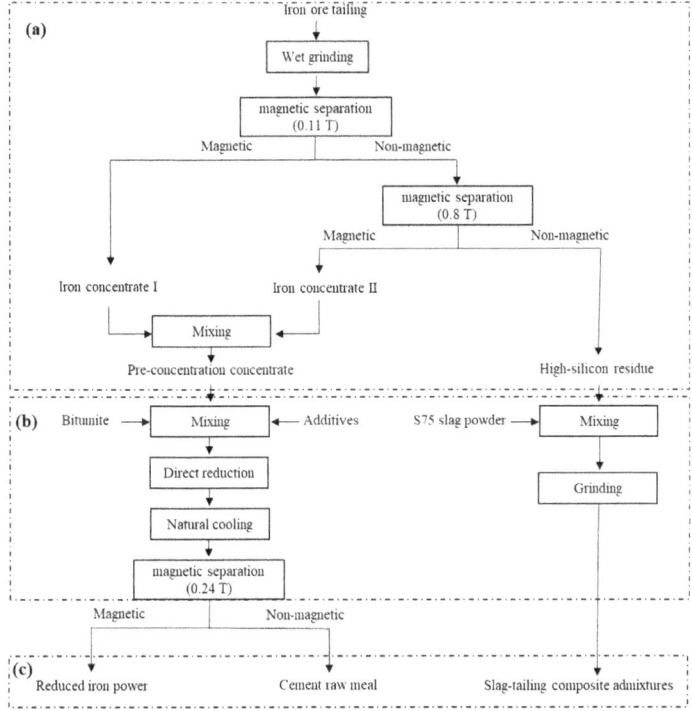

Figure 2. Experimental flowsheet: (**a**) The process of pre-concentration; (**b**) the process of iron recovery; (**c**) final products.

For pre-concentration, the IOTs were first milled to a size such that 50 wt % of particles were smaller than 38 µm. They were then separated in a low-intensity magnetic separator (model: RK/CXG; Φ = 50 mm; magnetic intensity: 0.11 T, Hualian Mining Co., Ltd., Tianjin, China) followed by a high-intensity magnetic separator (model: RK/CSQ; Φ = 50 mm × 70 mm; magnetic intensity: 0.8 T, Bureau of geology and minerals 102, Xichang, China). The magnetic product was the pre-concentrated concentrate (PC), whereas the nonmagnetic product was the high-silicon residues that would be reused as concrete admixture.

For direct reduction–magnetic separation, a direct reduction was conducted in a muffle furnace (model: CD-1400, Oriental Electrical Instrument, Yuyao, China). Six sample groups were designed to investigate the effects of various parameters on iron recovery from IOTs, including roasting temperature, roasting time, the carbon-to-IOT ratio, as well as the ratio of the additives and the grinding fineness of the roasted product. The PC was mixed with bitumite and additives at different ratios. Afterwards, the mixtures were roasted at various temperatures and times, and the roasted products were air-cooled and wet-milled in a rod mill before being separated in a low-intensity magnetic separator with a magnetic field intensity of 0.24 T. The iron grade of the reduced iron powder was chemically analyzed, and the recovery of iron was calculated based on the mass balance during the magnetic separation process. The nonmagnetic reducing slag (the chemical composition is presented in Table 4) in the roasted products was used as cement raw meal to directly burn a cement clinker, considering that the chemical composition was similar to that of cement raw materials.

Table 4. Chemical composition of the reducing slag/wt %.

Components	SiO$_2$	Fe$_2$O$_3$	FeO	Al$_2$O$_3$	CaO	MgO
Content	59.57	1.36	1.93	3.10	24.81	2.65
Components	Na$_2$O	K$_2$O	MnO	S	P	Loss
Content	0.97	0.52	0.10	0.10	0.02	4.27

The process of preparing the concrete mineral admixture using high-silicon slag was as follows: S75 was mixed with high-silica residue in different proportions, and the mixtures were milled in the same mill for 30 min, thus yielding slag-tailing composite admixtures for concrete with an ultrafine particle size, which was determined using a laser diffraction analyzer (model: LMS-30, Seishin Enterprise CO., Ltd., Osaka, Japan). Afterwards, the slag-tailing concrete composite admixtures (225 g), Portland cement (225 g), and water (225 g) were mixed together and stirred slowly for 30 s in a mixer (model: NJ-160A, Jianyi Instrument and Machinery Co., Ltd., Wuxi, China), thus forming a cement paste. Then, sand (1350 g) was added to the mixer. The mixture was stirred rapidly for 30 s and poured into prismoidal molds (40 mm × 40 mm × 160 mm) on a vibration table running at a moderate vibration rate. All the samples were demolded after curing at 24 h under 90% humidity at 20 ± 3 °C in a standard curing box (model: SHBY-40B, Luda Machinery Instrument, Shaoxing China). The samples were dipped in saturated lime water for curing under the same conditions. The compressive and flexural strengths of the cured specimens were tested after aging for 3, 7, and 28 days.

In this article, iron grade means the total iron (TFe) content, as detected by the chemical analysis (EDTA titration using the ethylenediamine tetra acetic acid as complexing agent to determine the chemical composition). The iron phases were determined using an atomic absorption spectrophotometer (model: AA-6800, Shimadzu, Kyoto, Japan) and titration. The mineral phases of the IOTs and roasted products were investigated using X-ray diffraction (Ultima IV diffractometer with a copper target, 40 kV, 40 mA, scanning speed of 20°·min^{-1}, scanning angle of 10°–100°, model: MXP21VAHF, MAC Science Co., Ltd., Osaka, Japan).

3. Results and Discussion

3.1. Pre-Concentration

The pre-concentration process was investigated using a previously described procedure [21,22]. The optimized conditions for pre-concentration were obtained from a previous work [22], and they were as follows: A particle size such that 50 wt % of the particles were smaller than 38 μm, a low magnetic field intensity of 0.11 T, and a high magnetic induction of 0.8 T. Under these conditions, the productivities of PC, the iron grade, and total iron recovery were 28.82 wt %, 36.58 wt %, and 83.86 wt %, respectively. The pre-concentration products are presented in Table 5. The chemical composition of the products is given in Table 6.

Table 5. Pre-concentration products/wt %. PC: Pre-concentrated concentrate.

Products	Productivity Rate	Iron Grade	Iron Recovery Rate
PC	28.82	36.58	83.86
High-silicon residue	71.18	2.85	16.14
IOTs	100.00	12.57	100.00

Table 6. Chemical composition of the pre-concentration products/wt %.

Products	SiO$_2$	Fe$_2$O$_3$	FeO	CaO	MgO	Al$_2$O$_3$	Na$_2$O	K$_2$O	MnO	TiO$_2$	S	P
PC *	41.71	42.83	8.18	1.82	1.69	1.82	0.41	0.27	0.14	0.072	0.1	0.02
High-silicon residue **	91.29	2.96	0.72	1.3	1.25	1.19	0.42	0.17	0.074	0.024	0.066	0.023

* The PC was composed of iron concentrate I and Iron concentrate II; ** high-silicon residue is the nonmagnetic component after the pre-concentration process.

3.2. Iron Recovery through Direct Reduction Followed by Magnetic Separation

In order to improve iron recovery, sodium carbonate (Na_2CO_3) and lime (CaO) were used as additives in the reduction process. The effects of the bituminous content, roasting time, roasting temperature, sodium carbonate content, lime content, and grinding fineness of the roasted products on iron grade and iron recovery were studied.

3.2.1. Effect of Bitumite Ratio on Iron Recovery

When studying the influence of bituminous content on iron grade and iron recovery (based on previous studies [23–27]), the optimized conditions of sodium carbonate content, lime content, roasting temperature, roasting time, and fineness of roasted products were determined. The specific parameters used in the experiments were as follows: A 100:10:5 mixing ratio for PC/lime/sodium carbonate, respectively; a roasting temperature of 1150 °C; and a roasting time of 60 min. The grinding fineness of the roasted product at a −74 μm ratio was 65%. The results are shown in Figure 3a.

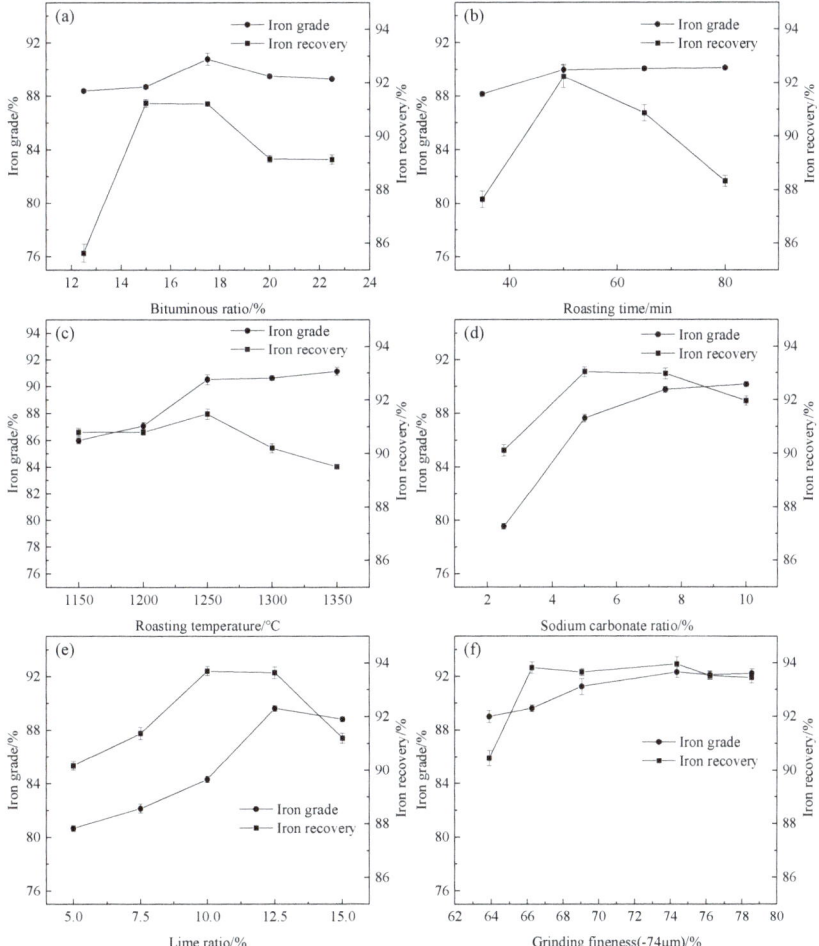

Figure 3. Effect of different influencing factors on iron recovery and iron grade: (**a**) Bituminous ratio; (**b**) roasting time; (**c**) roasting temperature; (**d**) sodium carbonate ratio; (**e**) lime ratio; (**f**) grinding fineness.

Figure 3a shows that the iron grade and recovery rate increased rapidly as the bitumite ratio increased from 12.5 wt % to 17.5 wt %, respectively. After reaching this value (17.5 wt %), both quantities decreased. The iron recovery rate increased from 85.63 wt % to 91 21 wt %, indicating that increasing the bitumite dose significantly enhanced the iron recovery rate. However, when the bitumite was in excess, the reduced metallic iron could not overtake the porous coal and other gangue minerals. The increase in size of the metallic iron particles was hindered, which reduced iron recovery [15]. During this test, the optimal bitumite content was found to be 17.5 wt %. Meanwhile, the increase in noncombustible substances in bituminous coal reduced the iron grade.

3.2.2. Effect of Roasting Time on Iron Recovery

The effect of roasting time on iron recovery was investigated under the following conditions: A 100:17.5:10:5 mixing ratio of PC/bitumite/lime/sodium carbonate, respectively; a roasting temperature of 1150 °C; and a grinding fineness the same as that reported in Section 3.2.1. The corresponding experimental results are shown in Figure 3b.

As shown in Figure 3b, the iron recovery rates peaked after roasting for approximately 50 min. As the roasting time increased from 35 to 50 min, the iron recovery of the reduced iron powder increased from 87.65 wt % to 92.23 wt % before increasing slowly from 50 to 80 min. Meanwhile, the iron grade maintained a stable trend. Consequently, the optimum roasting time was found to be 50 min for the following conditions: A 100:17.5:10:5 mixing ratio of PC/bitumite/lime/sodium carbonate, respectively; and a roasting temperature of 1150 °C.

3.2.3. Effect of Roasting Temperature on Iron Recovery

The mixtures consisting of PC, bitumite, lime, and sodium carbonate in the mixing ratio of 100:17.5:10:5, respectively, were roasted for 50 min at various temperatures. The grinding fineness was the same as that reported in Section 3.2.1. Figure 3c shows that increasing the roasting temperature up to 1250 °C induced a rapid increase in the iron grade from 90.8 wt % to 91.48 wt %. However, when the temperature exceeded 1250 °C, the rate of improvement decreased dramatically. When the roasting temperature was too low, the metallic iron particles remained small and closely associated with the gangue minerals. Therefore, the gangue mineral particles were easily mixed with the metallic iron during the magnetic separation process. It is worth noticing that further increasing the roasting temperature improved the growth of the metallic iron particles [28]. When the roasting temperature exceeded 1350 °C, however, the metallic iron particles became coarser, and the minerals softened or melted, causing difficulty during grinding [29]. Consequently, a roasting temperature of 1250 °C was recommended as the optimum temperature.

3.2.4. Effect of Sodium Carbonate Ratio on Iron Recovery

In order to optimize the sodium carbonate dosage, various sodium carbonate ratios were studied, while the other process parameters were kept constant. The conditions were maintained as follows: A 100:17.5:10 mixing ratio of PC, bitumite, and lime, and roasting at 1250 °C for 50 min. The grinding fineness was the same as that reported in Section 3.2.1. The corresponding results are presented in Figure 3d.

When the sodium carbonate ratio increased from 2.5 wt % to 7.5 wt %, the iron grade of the reduced iron powder increased from 81.55 wt % to 91.75 wt %, respectively, whereas the recovery of iron increased from 90.12 wt % to 92.97 wt %, respectively. When the sodium carbonate ratio exceeded 7.5 wt %, no obvious improvements were observed in the iron recovery. When sodium carbonate was present, it reacted with SiO_2 in the gangue minerals to form low-melting phases, such as $Na_2O·SiO_2$, $2Na_2O·SiO_2$, and $Na_2O·2SiO_2$ (melting point of 1088 °C), during the direct reduction. Furthermore, the liquid phase was composed of $Na_2O·SiO_2$, $2Na_2O·SiO_2$, and $Na_2O·2SiO_2$, which were formed locally [30]. When a liquid phase is present in roasted products, the diffusion coefficient increases, thus accelerating the diffusion and migration of iron in some cases and causing the metallic iron particles to gather and grow rapidly [28,31,32]. Based upon these results, the optimum sodium carbonate content was found to be 7.5 wt %.

3.2.5. Effect of Lime Ratio on Iron Recovery

The effect of lime content on iron recovery was also studied. The samples consisted of PC, bitumite, and sodium carbonate in a ratio of 100:17.5:7.5, respectively, and they were roasted at 1250 °C for 50 min. The grinding fineness was the same as that reported in Section 3.2.1. The corresponding results are shown in Figure 3e. It can clearly be seen that increasing the ratio of lime from 5 wt % to 12.5 wt % increased the iron grade from 80.67 wt % to 89.62 wt %, whereas the corresponding iron recovery increased from 90.17 wt % to 93.64 wt %. When the lime addition ratio exceeded 12.5 wt %, both the iron grade and recovery rate decreased. Based upon the results, the optimum sodium carbonate content was chosen to be 12.5 wt %.

Lime is a common and low-cost reaction additive that is used to improve the reduction of iron ore. Because PC contains numerous SiO_2 groups, the poor reactivity of fayalite, which forms readily during the reaction between FeO and SiO_2, is expressed using Equation (1):

$$FeO + SiO_2 = FeO \cdot SiO_2. \tag{1}$$

When the lime is added, it reacts with SiO_2 instead of FeO in the gangues to form calcium silicate ($CaO \cdot SiO_2$) during reduction roasting, and meanwhile bituminous coal participates in the reaction as a reducing agent. This reaction can be expressed using Equation (2):

$$FeO \cdot SiO_2 + CaO + C = Fe + CaO \cdot SiO_2 + CO\uparrow. \tag{2}$$

Consequently, more FeO can be reduced to metallic iron, enhancing the direct reduction process of iron oxide.

3.2.6. Effect of the Grinding Fineness of the Roasted Product on Iron Recovery

Complete mineral liberation is essential when magnetically separating metallic iron particles [15]. Consequently, the roasted product must be ground before magnetic separation. The effect of the grinding fineness of the roasted product was also studied under the following conditions: A roasting temperature of 1250 °C, a roasting time of 50 min, and a mixing ratio of 100:17.5:12.5:7.5 of PC, bitumite, lime, and sodium carbonate, respectively. The effect of the grinding fineness of the roasted product on iron recovery is shown in Figure 3f. The relationship between the grinding fineness and grinding time of the roasted product is presented in Table 7.

Table 7. Relationship between grinding fineness and grinding time of the roasted product/wt %.

Grinding Time/min	10	15	20	25	30	35
Grinding fineness (−74 μm)/wt %	63.9	66.25	69.05	74.35	76.21	78.54

Figure 3f shows that as the grinding time was prolonged and the grinding fineness increased, the iron grade of the reduced iron powder and the iron recovery initially increased and then stabilized. When 74.35 wt % of the particles were smaller than 74 μm, the metallic iron particles were almost completely liberated, generating an iron grade of 92.30 wt % and an iron recovery of 93.96 wt %. Generally, the finer the grinding fineness, the more difficult the recovery of magnetic fraction is. When the grinding fineness was −74 μm, the iron recovery was not improved relative to 66 μm, though the iron grade improved. Due to this reason, −74 μm was chosen as the optimum grinding fineness.

3.2.7. XRD Analysis of the Roasted Product

The roasted products were analyzed using XRD to explain the experimental results discussed above. The XRD pattern of the roasted product is shown in Figure 4.

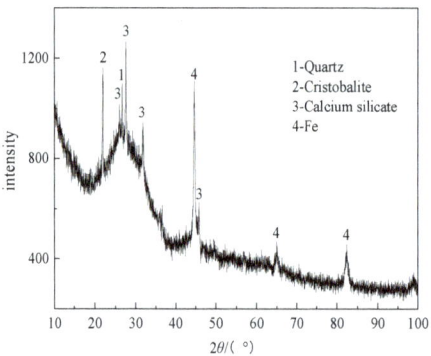

Figure 4. XRD pattern of the roasted product.

Figure 4 shows that the peaks for metallic iron appeared without evidence of any other iron compounds in the roasted product, indicating that the iron-bearing minerals were almost completely reduced to metallic iron. The impurities included calcium silicate, cristobalite, and some residual quartz, as well as amorphous forms of the roasted product. Under high temperature conditions, a liquid phase formed in large amounts during the reduction. This liquid phase accelerated the growth of metallic iron particles.

This study did not investigate the combined effects of sodium carbonate, lime, and other factors on iron recovery. The optimum parameters determined by the above single factor test were found to be a roasting temperature of 1250 °C; a roasting time of 50 min; a mixing ratio of 100:17.5:12.5:7.5 of PC, bitumite, lime, and sodium carbonate, respectively; and the proportion of −74 μm being more than 74.35 wt %. Under such conditions, the iron grade and iron recovery rate were found to be 92.30 wt % and 93.96 wt %, respectively. Specifically, 78.79 wt % of the iron could be recovered from Qidashan IOTs through pre-concentration followed by direct reduction and magnetic separation. The XRD pattern showed that the main phase in the roasted product was metallic iron.

3.3. Preparation of Slag-Tailing Concrete Composite Admixtures Using the High-Silica Residues

Since the main component of high-silica residues is SiO_2, which has no mineral activity, it is necessary to prepare concrete mineral admixtures using mechanical excitation. Therefore, a grinding test was carried out. The method for determining the optimized grinding time and method was based on the approach of Huang [33] and was as follows: First, S75 and high-silica residue were mixed according to the ratios given in Table 8, and the mixture was ground for 30 min. Then, a cement mortar test block was prepared using ground composite admixtures of cement, ISO sand, and water. The mixing proportions of the cement mortar are presented in Table 9. The specific surface area of each group mixture, fluidity, and the compressive strength of each group mortar were investigated. Group C-0 was made of pure cement and ISO sand as a blank group for a calculation of the activity index of the composite admixtures.

Table 8. Composite admixture proportions/wt %.

Number	S 75	Tailing
S-0	100	0
S-10	90	10
S-20	80	20
S-30	70	30
S-40	60	40
S-50	50	50

Table 9. Mixing proportion of cement mortar/g.

Number	Cement	Composite Admixture	ISO Sand	Water
C-0	450	0	1350	225
CS-0~CS50*	225	225		

CS-0~CS50* represents cement mortar when the composite admixtures were numbered S-0, S-10, S-20, S-30, S-40, or S-50.

3.3.1. Chemical Composition and Mineral Phases of the High-Silica Residues

According to the data shown in Table 6, the SiO_2 content of the high-silica residues was 91.29 wt %. The XRD pattern of the high-silica residues is shown in Figure 5. The main crystalline phase was quartz, with small amounts of hematite and hornblende.

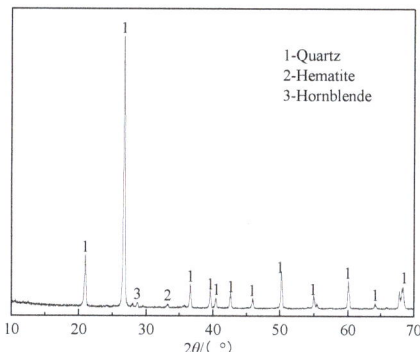

Figure 5. XRD pattern of the high-silica residues.

3.3.2. Specific Surface Area of the Slag-Tailing Composite Admixture and the Working Performance of the Mortar

Figure 6a shows the specific surface area of each group slag-tailing composite admixture after grinding for 30 min. The specific surface area of each group from S-0 to S-50 increased sequentially, indicating that the grindability of the mixture improved with an increase in the content of high-silica residue. Meanwhile, the energy consumption of the grinding could be reduced in industrial production.

The fluidity and fluidity ratio of each group mortar are shown in Figure 6b. The results show that both the fluidity and fluidity ratio increased with an increase in the content of high-silica residue. This was because the content of the fine particles increased: Due to this, the gap between the mortars could be filled, and the amount of interstitial water was reduced. The calculation of the fluidity ratio revealed that the fluidity values of each group were comparable to the reference group (Group C-0). Within a certain range, the greater the fluidity ratio, the better the performance of the mortar was. However, the compressive strength and fluidity of the mortar, which are the main properties of mineral admixtures, should be considered simultaneously for determining the optimum proportion.

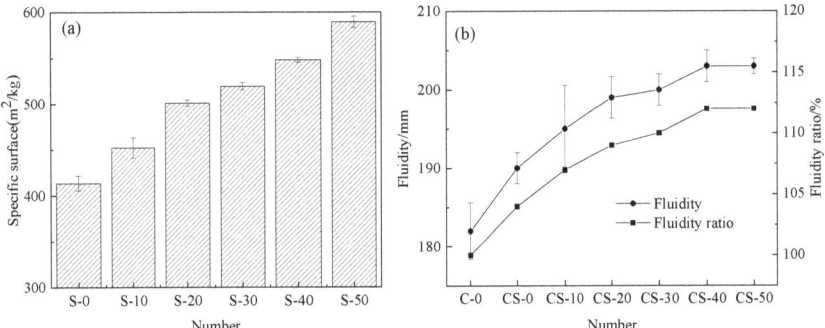

Figure 6. Specific surface area and workability of the slag-tailing: (**a**) Specific surface area when ground for 30 min; (**b**) fluidity and fluidity ratio of mortars.

3.3.3. Compressive Strength and Activity Index of Mortar

The activity index of concrete mineral admixtures should be considered when high-silica residue is used to prepare slag-tailing composite admixtures. Figure 7 shows the compressive strength of mortars and the activity index of the slag-tailing admixture.

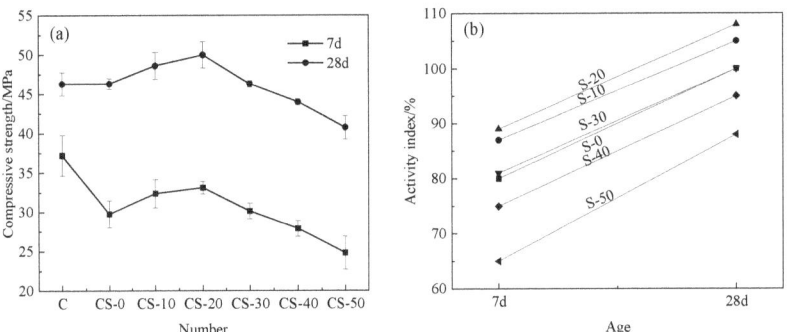

Figure 7. Compressive strength of mortars and activity index of slag-tailing admixture: (**a**) Compressive strength of mortars; (**b**) activity index of slag-tailing admixture.

From Figure 7a, it can be seen that with an increase in the content of high-silica residue, the compressive strength of each group of mortar first increased and then decreased. This was due to the reason that high-silica residues are inert materials comparable to blast furnace slag, whereas the strength of mortar can be increased by physical filling with a low content of high-silica residues. The compressive strength of 7 days and 28 days in the CS-20 group was the highest, reaching values of 33.11 MPa and 50 MPa, respectively.

The activity index is one of the main indicators for evaluating the quality of mineral admixtures, and their calculation method was based upon the compressive strength of each group compared to the blank group. Figure 7b shows the activity index of different ages of each group. It is clear that the activity index of mineral materials did not decrease when the amount of blast furnace slag replaced by tailings did not exceed 30%. Furthermore, the 7-day and 28-day activity indices of the S-20 group reached values of 89 and 108, respectively, which exceeded the activity index of the S-0 group without high-silica residues. The high activity of the S-20 group was primarily attributed to high-silica residue, which contained large quantities of ultrafine particles with diameters of less than 5 μm (the particle size distribution curve for the slag-tailing admixtures of S-20 is shown in Figure 8). These particles not only served as filling materials, thus increasing the compactness of the matrix, but also as nucleation catalysts, which accelerated the cement hydration process [34,35]. When only the quality of the slag-tailing admixture was considered, the S-20 group came out as being the best among all. However, the S-30 group may be more suitable if the maximum use of high-silica residue is considered.

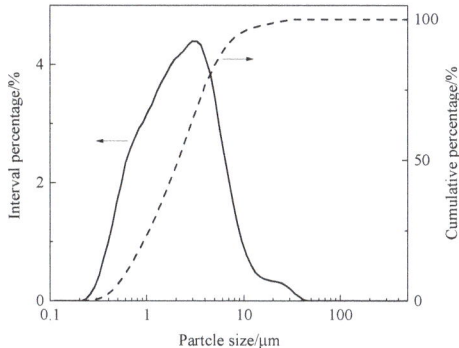

Figure 8. Particle size distribution curve for the slag-tailing admixtures of CS-20.

4. Conclusions

This paper demonstrated an innovative technical method called "pre-concentration followed by direct reduction and magnetic separation technology", which was developed to completely utilize IOTs. This method recovers a large amount of iron while solving the problem of secondary storage of reducing slag and high-silica residues. Based on the obtained results, the following conclusions can be drawn:

1. Primary grinding was followed by low-intensity magnetic separation and subsequent high-intensity magnetic separation processes to concentrate the iron in IOTs. The PC was characterized by an iron grade of 36.58 wt % and a total iron recovery of 83.86 wt % when 50 wt % of the particles were smaller than 0.038 mm, and the magnetic separation was conducted with a low magnetic field intensity of 0.11 T and a high magnetic induction of 0.8 T. Furthermore, high-silicon residues containing 91.29 wt % SiO_2 were obtained;
2. The bitumite ratio, roasting temperature, roasting time, lime ratio, sodium carbonate ratio, and particle size of the roasted product were the six major factors that influenced iron recovery. Optimal reduction conditions were obtained as follows: A roasting temperature of 1250 °C, a roasting time of 50 min, and a mixing ratio of 17.5:7.5:12.5:100 ratio of bitumite/sodium carbonate/lime/PC, respectively. Under these conditions, the iron grade of the reduced iron powder was found to be 92.30 wt %, and the iron recovery rate was 93.96 wt %. Relative to the original IOTs, the iron recovery was 78.79 wt %;
3. When the amount of high-silica residues replacing blast furnace slag was 20%, the strength of the cement mortar was 33.11 MPa and 50 MPa at 7 days and 28 days, respectively, whereas the corresponding activity index was 89 and 108, respectively, Furthermore, the fluidity rate of the mortar was approximately 109. When the content of high-silica residues replacing blast furnace slag was not more than 30%, the quality of slag-tailing concrete composite admixtures was not reduced;
4. Iron was recovered from IOTs through pre-concentration followed by direct reduction and magnetic separation processes. Moreover, the high-silica residues from the pre-concentration process were reused in slag-tailing concrete composite admixtures. Combining these recycling processes can ensure complete utilization of IOTs;

5. Incorporating an extra pre-concentration step before the direct reduction circumvented the high costs of recovering small amounts of iron through direct reduction. Consequently, this innovative technology can be extended to other IOTs with low iron content, exhibiting high potential for many other applications.

Author Contributions: C.T. performed the experiment, wrote the manuscript, and analyzed the data with the help of D.F.; K.L. revised the manuscript; W.N. designed the experiment.

Funding: This research was funded by the Ministry of Science and Technology International Science and Technology Cooperation Program, grant number 2016YFE0130700.

Acknowledgments: The authors would also like to thank the Analytical and Testing Centre of the University of Science and Technology Beijing, China University of Geoscience, and the Xi'an Tianzhou Mining Science and Technology Development Co., Ltd., of China, which supplied the facilities necessary for data collection.

Conflicts of Interest: The authors declare no conflicts of interest.

References

1. Jiang, J.H.; Ye, G.H.; Hu, S.M. The technology status and research progress of iron tailings re-beneficiation. *Min. Metall.* **2018**, *27*, 1–4.
2. Zhizhen, L.; Xinjian, G.; Resource, S.O. Current Situations of Tailings Utilization and Sustainable Development of Mines. *Miner. Eng. Res.* **2018**, *116*, 37–41.
3. Chen, H.A.; Shen, W.G.; Shan, L.; Xiong, C.B.; Su, Y.Q.; Liu, B.; Rao, J.L. Situation of discharge and comprehensive utilization of iron tailings domestic and abroad. *Concrete* **2012**, *2*, 88–91.
4. Jiabin, C.; Wenlong, J.; Lianghui, Y. Survey and Evaluation of the Iron Tailings Resources in China. *Min. Res. Dev.* **2010**, *3*, 60–62.
5. Ajaka, E.O. Recovering fine iron minerals from Itakpe iron ore process tailing. *ARPN J. Eng. Appl. Sci.* **2009**, *4*, 17–28.
6. Rao, K.H.; Narasimhan, K.S. Selective flocculation applied to Barsuan iron ore tailings. *Int. J. Miner. Process.* **1985**, *14*, 67–75.
7. Han, J.; Zhu, S.; Xun, Z.; Zhao, G. Experimental Research on Mineral Separation of a Magnetic-flotation Tailings Mixture. *Min. Res. Dev.* **2012**, *2*, 37–43.
8. Sakthivel, R.; Vasumathi, N.; Sahu, D.; Mishra, B.K. Synthesis of magnetite powder from iron ore tailings. *Powder Technol.* **2010**, *201*, 187–190. [CrossRef]
9. Das, S.K.; Kumar, S.; Ramachandrarao, P. Exploitation of iron ore tailing for the development of ceramic tiles. *Waste Manag.* **2000**, *20*, 725–729. [CrossRef]
10. Das, S.K.; Ghosh, J.; Mandal, A.K.; Singh, N.; Gupta, S. Iron ore tailing: A waste material used in ceramic tile compositions as alternative source of raw materials. *Trans. Indian Ceram. Soc.* **2012**, *71*, 21–24. [CrossRef]
11. Huang, X.; Ranade, R.; Li, V.C. Feasibility study of developing green ECC using iron ore tailings powder as cement replacement. *J. Mater. Civ. Eng.* **2013**, *25*, 923–931. [CrossRef]
12. Adedayo, S.M.; Onitiri, M.A. Mechanical properties of iron ore tailings filled-polypropylene composites. *J. Miner. Mater. Charact. Eng.* **2012**, *11*, 671. [CrossRef]
13. Adedayo, S.M.; Onitiri, M.A. Tensile properties of iron ore tailings filled epoxy composites. *West. Indian J. Eng.* **2012**, *35*, 51–59.
14. Chun, T.J.; Zhu, D.Q.; Pan, J.; He, Z. Preparation of metallic iron powder from red mud by sodium salt roasting and magnetic separation. *Can. Metall. Q.* **2014**, *53*, 183–189. [CrossRef]
15. Yang, H.; Jing, L.; Zhang, B. Recovery of iron from vanadium tailings with coal-based direct reduction followed by magnetic separation. *J. Hazard. Mater.* **2011**, *185*, 1405–1411. [CrossRef]
16. Zhang, G.F.; Yang, Q.R.; Yang, Y.D.; Wu, P.; McLean, A. Recovery of iron from waste slag of pyrite processing using reduction roasting magnetic separation method. *Can. Metall. Q.* **2013**, *52*, 153–159. [CrossRef]
17. Maweja, K.; Mukongo, T.; Mutombo, I. Cleaning of a copper matte smelting slag from a water-jacket furnace by direct reduction of heavy metals. *J. Hazard. Mater.* **2009**, *164*, 856–862. [CrossRef]
18. Zhang, Y.; Li, H.; Yu, X. Recovery of iron from cyanide tailings with reduction roasting—Water leaching followed by magnetic separation. *Adv. Mater. Res.* **2012**, *396*, 486–489. [CrossRef]

19. Li, C.; Sun, H.; Bai, J.; Li, L. Innovative methodology for comprehensive utilization of iron ore tailings: Part 1. The recovery of iron from iron ore tailings using magnetic separation after magnetizing roasting. *J. Hazard. Mater.* **2010**, *174*, 71–77. [CrossRef]
20. Ni, W.; Fu, C.H.; Fan, D.C.; Li, J.; Qiu, X.J.; Li, Y. A Method of Extracting Iron Ore Tailings by Deep Reduction after Using Strong Magnetism to Pre-concentrate the Tailings. Chinese Patent No. 201210362689.3, 11 September 2013.
21. Li, J.; Ni, W.; Fan, D.C.; Li, Y.; Fu, C.H. Process Mineralogy Research on Iron Tailings from Qidashan. *Metal Mine* **2014**, *1*, 158–162.
22. Fan, D. *Research on Pre-Concentration and Deep Reduction of Qidashan Iron Ore Tailings and the Comprehensive Utilization of Tailings*; University of Science and Technology Beijing: Beijing, China, 2018.
23. Zhu, D.Q.; Chun, T.J.; Pan, J.; He, Z. Recovery of Iron from High-Iron Red Mud by Reduction Roasting with Adding Sodium Salt. *J. Iron Steel Res. Int.* **2012**, *19*, 1–5. [CrossRef]
24. Yu, W.; Sun, Y.; Liu, Z.; Kou, J.; Xu, C. Effects of particle sizes of iron ore and coal on the strength and reduction of high phosphorus oolitic hematite-coal composite briquettes. *ISIJ Int.* **2014**, *54*, 56–62. [CrossRef]
25. Yu, W.; Sun, T.; Kou, J.; Wei, Y.; Xu, C.; Liu, Z. The function of $Ca(OH)_2$ and Na_2CO_3 as additive on the reduction of high-phosphorus oolitic hematite-coal mixed pellets. *ISIJ Int.* **2013**, *53*, 427–433. [CrossRef]
26. Zhu, D.; Chun, T.; Pan, J.; Lu, L.; He, Z. Upgrading and dephosphorization of Western Australian iron ore using reduction roasting by adding sodium carbonate. *Int. J. Miner. Metall. Mater.* **2013**, *20*, 505–513. [CrossRef]
27. Fan, D.; Ni, W.; Wang, J.; Wang, K. Effects of CaO and Na_2CO_3 on the reduction of high silicon iron ores. *J. Wuhan Univ. Technol. -Mater. Sci. Ed.* **2017**, *32*, 508–516. [CrossRef]
28. Weissberger, S.; Zimmels, Y.; Lin, I.J. Mechanism of growth of metallic phase in direct reduction of iron bearing oolites. *Metall. Trans. B* **1986**, *17*, 433–442. [CrossRef]
29. Ni, W.; Yan, J.; Cheng, Y. Beneficiation of unwieldy oolitic hematite by deep reduction and magnetic separation process. *Chin. J. Eng.* **2010**, *32*, 287–291.
30. Hao, Z.; Wu, S.; Wang, Y.; Luo, G.; Wu, H.; Duan, X. Acting mechanism of F, K, and Na in the solid phase sintering reaction of the Baiyunebo iron ore. *Int. J. Miner. Metall. Mater.* **2010**, *17*, 137–142. [CrossRef]
31. Nadiv, S.; WeissbergerI, S.; Lin, J.; Zimmels, Y. Diffusion mechanisms and reactions during reduction of oolitic iron-oxide mineral. *J. Mater. Sci.* **1988**, *23*, 1050–1055. [CrossRef]
32. Li, G.; Zhang, S.; Rao, M.; Zhang, Y.; Jiang, T. Effects of sodium salts on reduction roasting and Fe–P separation of high-phosphorus oolitic hematite ore. *Int. J. Miner. Process.* **2013**, *124*, 26–34. [CrossRef]
33. Huang, X.Y.; Ni, W.; Zhu, L.-P.; Wang, Z.-J. Grinding characteristic of Qidashan iron tailings. *J. Univ. Sci. Technol. Beijing* **2010**, *32*, 1253–1257.
34. Kronlöf, A. Effect of very fine aggregate on concrete strength. *Mater. Struct.* **1994**, *27*, 15–25. [CrossRef]
35. Lawrence, P.; Cyr, M.; Ringot, E. Mineral admixtures in mortars: Effect of inert materials on short-term hydration. *Cem. Concr. Res.* **2003**, *33*, 1939–1947. [CrossRef]

© 2019 by the authors. Licensee MDPI, Basel, Switzerland. This article is an open access article distributed under the terms and conditions of the Creative Commons Attribution (CC BY) license (http://creativecommons.org/licenses/by/4.0/).

Article

Mobility and Attenuation Dynamics of Potentially Toxic Chemical Species at an Abandoned Copper Mine Tailings Dump

Wilson Mugera Gitari [1,2,*], Rendani Thobakgale [1,2] and Segun Ajayi Akinyemi [2,3]

1. Environmental Remediation and Water Pollution Chemistry Group, School of Environmental Sciences, University of Venda, Private Bag X5050, Thohoyandou, Limpopo 0950, South Africa; rthobakgale@gmail.com
2. Department of Ecology and Resources Management, University of Venda, Private Bag X5050, Thohoyandou, Limpopo 0950, South Africa; segun.akinyemi@eksu.edu.ng
3. Department of Geology and Applied Geophysics, Ekiti State University, Ado Ekiti P.M.B. 5363, Nigeria
* Correspondence: mugera.gitari@univen.ac.za; Tel.: +27-159-628-572

Received: 18 December 2017; Accepted: 5 February 2018; Published: 12 February 2018

Abstract: Large volumes of disposed mine tailings abound in several regions of South Africa, as a consequence of unregulated, unsustainable long years of mining activities. Tailings dumps occupy a large volume of valuable land, and present a potential risk for aquatic systems, through leaching of potentially toxic chemical species. This paper reports on the evaluation of the geochemical processes controlling the mobility of potentially toxic chemical species within the tailings profile, and their potential risk with regard to surface and groundwater systems. Combination of X-ray fluorescence (XRF), X-ray diffraction (XRD), and scanning electron microscopy-energy dispersive spectroscopy (SEM-EDS) techniques, show that the tailing profiles are uniform, weakly altered, and vary slightly with depth in both physical and geochemical properties, as well as mineralogical composition. Mineralogical analysis showed the following order of abundance: quartz > epidote > chlorite > muscovite > calcite > hematite within the tailings profiles. The neutralization of the dominant alumino-silicate minerals and the absence of sulfidic minerals, have produced medium alkaline pH conditions (7.97–8.37) at all depths and low concentrations of dissolved Cu (20.21–47.9 µg/L), Zn (0.88–1.80 µg/L), Pb (0.27–0.34 µg/L), and SO_4^{2-} (15.71–55.94 mg/L) in the tailings profile leachates. The relative percentage leach for the potentially toxic chemical species was low in the aqueous phase (Ni 0.081%, Cu 0.006%, and Zn 0.05%). This indicates that the transport load of potentially toxic chemical species from tailings to the aqueous phase is very low. The precipitation of secondary hematite has an important known ability to trap and attenuate the mobility of potentially toxic chemical species (Cu, Zn, and Pb) by adsorption on the surface area. Geochemical modelling MINTEQA2 showed that the tailings leachates were below saturation regarding oxyhydroxide minerals, but oversaturated with Cu bearing mineral (i.e., cuprite). Most of the potentially toxic chemical species occur as free ions in the tailings leachates. The precipitation of secondary hematite and cuprite, and geochemical condition such as pH of the tailings were the main solubility and mobility controls for the potentially toxic chemical species, and their potential transfer from tailings to the aqueous phase.

Keywords: copper tailings dump; leaching; pH; mobility; chemical species; attenuation mechanisms; aquatic systems

1. Introduction

South Africa has a long history of mining activities dating back for a century, and this has been pivotal for the economy of the country leading to the development of infrastructures and the establishment of other manufacturing industries [1]. However, a consequence of unsustainable and unregulated long years of mining activities, resulted in a large volume of tailings dumps abounding in several regions including poor communities of the country, and this poses a risk to the surrounding environment adjacent to the disposed tailings dumps. Moreover, tailings dump disposal has been reported to occupy large hectares of valuable land that can be used for development projects ensuring employment generation [2]. In South Africa, the majority of the tailings are gold tailings dumps within the Witwatersrand basin in Johannesburg occupying an estimated surface area of 200 km^2 [3]. Other tailings dumps include the abandoned copper tailings dump in Musina occupying an estimated surface area of 4.99 km^2, and the abandoned cassiterite tailings dump located at the eastern limb of the Bushveld complex in Mokopane covering an estimated surface area of 3.9 km^2 [4,5]. Tailings dump characteristics vary accordingly depending on the type of the ore mined, mining equipment used, chemicals implemented during mineral processing, mineralogy and geochemistry, and particle size of the tailings [6]. However, tailings dumps generally contain un-wanted and un-economic minerals bound with the mined ore, potentially toxic chemical species (Cu, Pb, Zn, Co, and Ni) bound to minerals present within the tailings dump, and inorganic chemicals such as sulfuric acid used to enhance ore separation during mineral processing [7]. Nevertheless, the disposal of an enormous volume of tailings dumps poses a serious risk to the surrounding environment through air pollution due to air-dried out tailings, erosion of the tailings with the potential of valuable land degradation, and leaching of soluble inorganic potentially toxic chemical species (Cu, Ni, Pb, Zn, Cd, and Cr) occurring in a variety of minerals present in the tailings dump [8]. The leaching or release of soluble inorganic potentially toxic chemical species from tailings dumps into the adjacent surface and groundwater systems has been reported to be a major environmental concern induced by tailings dumps [7,9]. When sulfide minerals (mainly pyrite) present within the tailings are in contact with atmospheric oxygen and infiltrating rain water, the tailings become susceptible to oxidative weathering, consequently resulting in the formation of acidic mine drainages with low pH conditions laden with high concentration of sulfates (SO_4^{2-}), and potentially toxic chemical species [10–12]. Under this geochemical condition, many of the potentially toxic chemical species become highly soluble and mobile, and consequently, this may contaminate the receiving surface and groundwater systems, thereby causing changes to local groundwater dynamics in terms of both quantity and quality of water and its flow direction, and may further cause loss for aquatic habitat and the establishment of alien invasive species [13]. Groundwater in the mining district of Johannesburg in South Africa is heavily contaminated by potentially toxic chemical species, and becomes acidic because of oxidative weathering of pyrite present within the mine tailings dumps [14]. The duration of these environmental impacts can be long-term.

However, this is mainly attributed to the absence of minerals in the tailings that can neutralize the acidity generated during sulfide oxidation [9]. According to Bortnikova et al. [15], the main minerals within the tailings, decreasing the acidity, are carbonates such as calcite, dolomite, siderite, and alumino-silicates such as chlorite, epidote, and plagioclase. According to Romero et al. [7], neutralization of the acidity in the tailings occurs through the dissolution of these carbonate and alumino-silicate minerals, which consume hydrogen ions and generate neutral to medium alkaline pH conditions within the tailings systems. Under medium alkaline pH condition, the solubility and mobility of many potentially toxic chemical species becomes low, consequently, dissolved concentrations of potentially toxic chemical species in tailings leachates and pore waters becomes relatively low, mitigating the environmental severity of potentially toxic chemical species [7,9,16]. Sulfide oxidation and neutralization processes have a great influence on the solubility and mobility of potentially toxic chemical species and their retention in primary and secondary mineral phases within the tailings [9]. Thus, contamination of the adjacent surface and groundwater ecosystems is dependent on the solubility and mobility of the potentially toxic chemical species, which is a function of pH and mineralogy of the tailings [7]. In order to assess the current environmental state and potential contamination on aquatic systems induced by tailings dump disposal, it is necessary to understand the geochemical behavior of the chemical constituents and the natural geochemical conditions of the disposed tailings. Nevertheless, in South Africa, much of the previous work done on disposed tailings focused on physicochemical and mineralogical characterization of copper and gold tailings with the view to assessing the beneficial application in the construction industry [17]. Matshusa and Makgae [18] characterized the abandoned mine tailings in Limpopo, and the rehabilitation challenges. Singo [19] studied the assessment of heavy metal pollution on soils adjacent to the old copper mine tailings in Musina. Nonetheless, the challenge is to understand the leaching behavior of the potentially toxic chemical species with tailings depth, when the disposed copper tailings are in contact with atmospheric oxygen and infiltrating rain water over the long-term. This is crucial when assessing the current state of the disposed copper tailings and potential risk on the adjacent aquatic systems. The present paper reports on evaluation of the geochemical processes controlling the solubility and mobility of potentially toxic chemical species within the established copper tailings profiles. The potential transfer and risk of potentially toxic chemical species from tailings to the adjacent surface and groundwater systems is elucidated.

2. Materials and Methods

2.1. Study Area and Sampling Techniques

The study area is located at Musina (22°20'17" S and 30°02'30" E), a small town at the northern end of Limpopo Province, South Africa. The town of Musina has a long history of mining activities and is a home to several successful mining companies, some of which are not working anymore. Nevertheless, some of the exploited mineral resources in the region include iron ore, coal, copper, graphite, magnetite, asbestos, diamonds, and semi-precious stones [17]. However, copper mining in Musina dates from 1906 when the Musina Development Mining Company started to mine copper ore, and since then over 40 million tons of ore have been mined, recovering about 700,000 tons of copper from chalcopyrite, bornite, and chalcocite as the primary minerals [20]. The mined ore was processed in an erected concentrator, and copper concentrates were recovered by flotation thickening, and the waste was rejected from the concentrator as tailings dump (Figure 1). Nonetheless, copper mining activities in Musina stopped in 1992, whereby poor ventilation and water seepage into the mine have also been speculated in forcing the mine to stop operation [21]. Thus, the abandoned copper tailings dump disposal in Musina has been exposed to weathering for a period of 25 years [18]. Moreover, geologically, the Musina area is characterized by medium-high grade metamorphic rocks of which the Sand River Gneiss is the oldest (3.7 Ma) and form the basement to a sequence of rock covers of the Beit bridge complex, which according to their dominant lithology are subdivided into the Mount Dowe, Malala Drift, and Gumbu Group [22].

To meet the aim of the study, three tailings profiles were established, drilled to depths of 5 m (tailings profile A and B) and 3 m (tailings profile C) respectively. Tailings profiles A and B are located at the southern-central sector, meanwhile tailings profile C is located at the north-eastern sector of the abandoned copper tailings dump (Figure 1). However, due to the uniform mineralogical and geochemical characteristics of the tailings, only tailings profile B and C were considered for this study. A systematic sampling grid technique was commissioned by using a square grid and collecting samples from the nodes (interactions of the gridline). At each node, tailings samples were collected at regular depth intervals with the use of a hand drill auger. The sampling sites were selected according to the heterogeneity of the tailings considering color, texture differences, and degree of concretion or contact with the underlying soil or tailings soil-interface. The collected tailings samples were preserved in tightly sealed and marked oxygen diffusion-free polyethylene plastic bags to minimize dust contamination, and transported to the laboratory for sample preparation and further laboratory analyses.

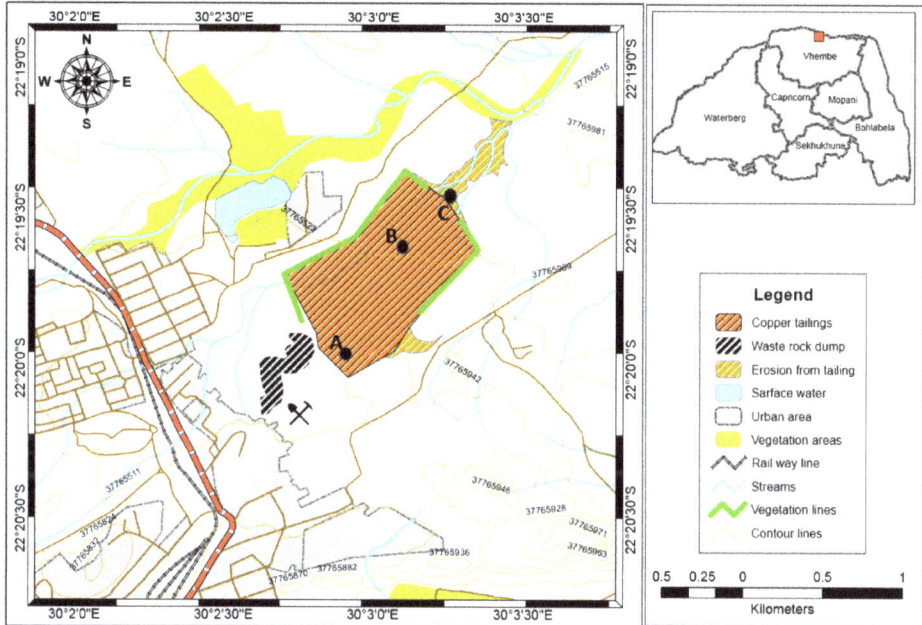

Figure 1. Location of the abandoned copper tailings dumps and the selected sampling sites in Musina, Limpopo, South Africa.

2.2. Determination of Tailings Physical Properties

The procured tailings samples were tested for bulk density, air-filled porosity, and moisture content. The bulk density was determined on air-dried samples using the gravimetric technique from which the air-filled porosity was calculated. The initial moisture content was determined as defined by the ASTM D 2216 [23].

2.3. Geochemical Analysis

2.3.1. Paste pH and Eh

Tailings paste pH and Eh values were measured in a saturated soil paste using a combined Thermo Scientific instrument (reference electrode: 8102BNUWP). A saturated soil paste was prepared by adding 20 mL of distilled water to 20 g of air dried original sample (liquid:solid ratio of 1:1). The slurry was mixed for 5 s and left to stand for 10 min.

2.3.2. Bulk Chemical Element Composition

Bulk chemical elemental composition was determined by X-ray fluorescence spectrometry (XRF) (ARL9400 XP+ Sequential XRF with WinXRF software, Thermo Fisher, Waltham, MA, USA) at the Faculty of Natural and Agricultural Sciences, Department of Geology, University of Pretoria (Pretoria, South Africa). The tailings samples were milled in a tungsten-carbide milling pot to achieve particle sizes <75 µm. The samples were dried at 100 °C and roasted at 1000 °C to determine loss on ignition (LOI) values. One gram of the samples was mixed with 6 g of lithium tetraborate flux and fused at 1050 °C to make a stable fused glass bead. For trace elements analysis the samples were mixed with PVA binder and pressed in an aluminum cup @ 10 tons. The analyzed major concentration was reported in percentage terms and trace elements in ppm. The obtained analyzed values were compared with the certified values.

2.4. Mineralogical Analysis

Qualitative and quantitative mineralogical analysis was performed by X-ray Diffraction Spectrometry (XRD) on the representative tailings samples within the respective tailings profiles at the University of Pretoria. The samples were analyzed using a X'Pert Pro powder diffractometer (PANalytical, Almelo, The Netherlands) in θ–θ configuration with an X'Celerator detector and variable divergence, and fixed receiving slits with Fe filtered Co-Kα radiation ($\lambda = 1.789$ Å). The phases were identified using X'Pert High score plus software. The relative phase amounts (wt %) were estimated using the Rietveld method (Autoquan Program). Scanning electron microscopy coupled with energy dispersive X-ray spectrometry (SEM-EDS) (Leo1450 SEM, Zeiss, Oberkochen, Germany, voltage was 10 kV, working distance 14 mm) was performed at the Department of Electron Microscope Unit, University of Cape Town (Cape Town, South Africa) to identify and understand the distribution of secondary mineral phases present within the respective tailings samples on the tailings profiles.

2.5. Batch Leaching Tests or Pore Water Chemistry

Batch extractions were performed according to the standardized leaching test European norm EN 12457-2 [24]. Ten grams of the homogenized tailings samples was suspended in a 100 mL Mili-Q + deionized water (liquid:solid ratio of 1:10) in polyethylene plastic bottles. Sample suspensions were agitated in a horizontal shaker for 24 h at a temperature of 22 ± 3 °C, at 200 rpm, until a stable reading of pH and electrical conductivity (EC) was obtained. After shaking, and pH and EC stabilization, the batches were centrifuged at 2500× g for 20 min and filtered through a 0.45 μm membrane filter. The clear aqueous extracts were divided into two subsamples. One set of samples was acidified to pH <2 with concentrated HNO_3 for cation analysis using inductively coupled plasma-mass spectroscopy (ICP-MS) technique, and the second set of samples was left un-acidified for anion analysis by ion chromatography technique (IC). The analyses were performed in duplicates to validate the accuracy of the analytical techniques. The obtained concentrations of leachates were evaluated to assess the solubility and mobility patterns of the potentially toxic chemical species, and their potential transfer from tailings to aquatic systems. The obtained concentrations of leachates were further compared with the regulatory standards for drinking water quality as defined by the World Health Organization [25].

3. Results and Discussion

3.1. Macroscopic Description and Physical Properties of Copper Tailings Samples

The procured tailings samples are uniform, varying less in color ranging from dark to light gray at shallow and deeper depths within the respective tailings profiles, and dark to reddish brownish towards the underlying soil or soil-interface. Within tailings profiles A and B, the dark to light gray color is continuous from a shallow depth of about 2.5 m to deeper zones of the tailings (≈3.5 m). At the base of the respective tailings profiles (≈4.5 m), there are spots of underlying soil ranging in color from light to darkish brown, indicating mixing with soil material below the mine tailings dump. Meanwhile, in tailings profile C the zone or contact with the underlying soil is reached at a shallow depth of ≈2.5 m (Figure 2). Although, the water table was not reached along the tailings profiles, there was a layer of noticeable increase resistance to drilling at the upper and shallow depth of the tailings profiles, which is attributed to the presence of consolidated layer commonly known as hardpan or cement layer, when considering the climatic condition (semi-arid) of the Musina area. A similar feature was observed at a shallow depth (≈0.9 m) of the tailings profile at the Chambishi site within the Zambian copperbelt, and was ascribed to the presence of cement or hardpan layer [11]. The presence of hardpan or cement layer in a vertical profile have been reported to generally result in increasing bulk density and decreasing air-filled porosity in the tailings dump [12]. The presence of cement or hardpan layers is of geochemical significance as they may incorporate large quantities of metals through adsorption and/or co-precipitation with secondary hardpan minerals, thus limiting the mobility of potentially toxic chemical species and their severity to the environment [26]. Table 1 presents some of the physical properties of the copper tailings. The relative trends for bulk density and air-filled porosity within the vertical tailings profiles are shown in Figure 2. Considering the uniform mineralogical, geochemical, and physical properties of the tailings, tailings profiles B and C were considered in this study.

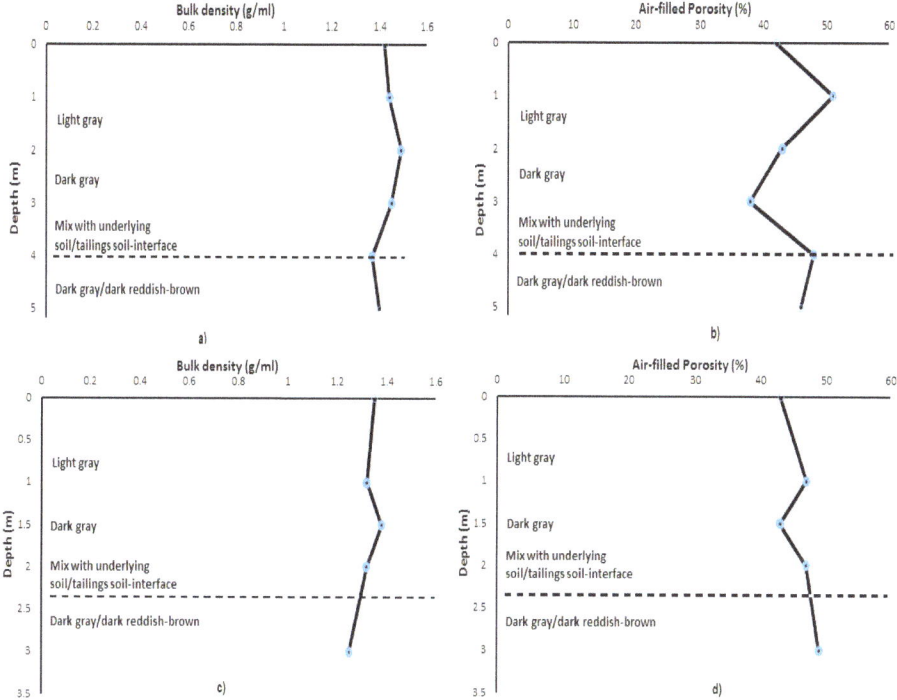

Figure 2. Simplified vertical profiles for (**a**,**b**) bulk density and air-filled porosity in tailings profile B, and (**c**,**d**) bulk density and air-filled porosity in tailings profile C.

Table 1. Physical properties of the old copper tailings within tailings profile B.

Depth (m)	Bulk Density (g/mL)	Porosity (%)	Moisture (%)
1	1.42	42	0.09
2	1.44	51	0.93
3	1.49	43	0.43
4	1.45	38	1.19
5	1.37	50	1.39
6	1.40	46	0.23

As indicated, the copper tailings are characterized by high bulk density in the upper and shallow depth of the tailings profiles, reaching about 1.5 g·mL^{-1} at a depth of ≈2 m (Table 1; Figure 2a,c). A similar bulk density trend was observed at the Chambishi site within the Zambian copperbelt as reported by Sracek et al. [11]. However, the bulk density then decreases slightly to ≈1.4 g·mL^{-1} at about ≈5 m towards the bottom of the tailings profiles, where the mine tailings material is already mixing with the underlying soil. As indicated in Table 1, the tailings material retains relatively low moisture content (<2%). This is verified by the low air-filled porosity (<50%) at shallow depth, although slightly increasing with tailings depth (Table 1; Figure 2b,d). The high bulk density and low air-filled porosity are coupled with relatively low initial moisture content at the upper and shallow depths of the respective tailings profiles. This could be attributed to the resistance to drilling or consolidated material which was previously referred to as cement or hardpan layer. The studied copper tailings have been exposed to oxidative weathering for a period of about 22 years, since the cessation of copper mining in 1995 [21]. However, the presence of cement or hardpan layer in the upper and shallow zones of the tailings profiles limited the penetration of oxidizing agents and oxidation products to un-oxidized tailings in the deeper zones of the respective tailings profiles.

3.2. Mineralogical Composition within the Respective Copper Tailings Samples

Table 2 presents the qualitative and quantitative mineralogical composition by XRD analysis. The relative mineralogical trends or patterns within the established tailings profiles are shown in Figure 3. Meanwhile, Figure 4 shows the XRD spectra along with SEM images and EDS quantification of the minerals within the tailings profiles. The mineralogical composition and patterns show low variation with depth in all the tailings profiles, and follow the order in relative abundance: quartz > epidote > chlorite > muscovite > plagioclase > calcite > hematite. Similar mineralogical patterns were observed by Gitari et al. [17] on the old copper tailings from the same site. A significant relative abundance of potential acid neutralizers such as alumino-silicate and traces of calcite minerals, and the absence of primary sulfidic minerals suggest a high neutralization capacity of the studied copper mine tailings. Consequently, the severe environmental impact of the mine tailings is moderated. However, quartz is the most predominant gangue mineral within the tailings profiles, and its composition increases with increasing tailings depth reaching about 42.4 wt % and 42.95 wt % at a depth of 5 m and 3 m in tailings profile B and C respectively, where the tailings are mixed with the underlying soil (Table 2; Figure 3). However, chlorite tends to decrease abruptly with increasing tailings depth within the tailings profiles (Figure 3). Calcite and hematite peaks are the only secondary minerals observed in the XRD spectra analysis. The mineralogical composition of calcite and hematite is relatively very low within the tailings samples, and their mineralogical trends or patterns are similar, and increase with increasing tailings depth, reaching 3.22 wt % and 2.97 wt %, respectively within the tailings profiles (Table 2; Figure 3).

Table 2. X-ray diffraction of bulk samples from tailings profile B and C.

Profile B (0–2 m)	Weight (%)	3σ Error	Profile B (2–4 m)	Weight (%)	3σ Error	Profile B (5 m)	Weight (%)	3σ Error
Actinolite	4.12	0.87	Actinolite	3.08	0.93	Actinolite	2.97	0.9
Calcite	1.74	0.33	Calcite	1.79	0.36	Calcite	2.05	0.36
Chlorite	17.34	0.84	Chlorite	18.87	0.9	Chlorite	17.38	0.93
Epidote	21.64	0.99	Epidote	20.41	1.02	Epidote	21.61	1.05
Hematite	1.58	0.33	Hematite	1.76	0.36	Hematite	1.85	0.33
Muscovite	8.41	0.72	Muscovite	8.18	0.81	Muscovite	8.2	0.84
Plagioclase	7.65	0.9	Plagioclase	3.98	0.84	Plagioclase	3.54	0.72
Quartz	37.5	0.99	Quartz	41.92	1.08	Quartz	42.4	1.08
Profile C (0–2 m)	Weight (%)	3σ Error	Profile C (2 m Top–2 m Bottom)	Weight (%)	3σ Error	Profile C (3 m)	Weight (%)	3σ Error
Actinolite	4.1	0.9	Calcite	2.24	0.39	Actinolite	2.47	0.75
Calcite	2	0.39	Chlorite	16.7	0.84	Calcite	3.22	0.36
Chlorite	16.82	0.87	Epidote	23.56	0.93	Chlorite	13.31	0.78
Epidote	22.23	0.93	Hematite	2.28	0.3	Epidote	19.94	0.87
Hematite	1.62	0.3	Muscovite	8.55	0.75	Hematite	2.79	0.3
Muscovite	7.13	0.78	Plagioclase	6.59	0.87	Muscovite	8.68	0.69
Plagioclase	3.16	0.75	Quartz	40.07	0.96	Plagioclase	6.64	0.75
Quartz	42.94	1.05	-	-	-	Quartz	42.96	1.05

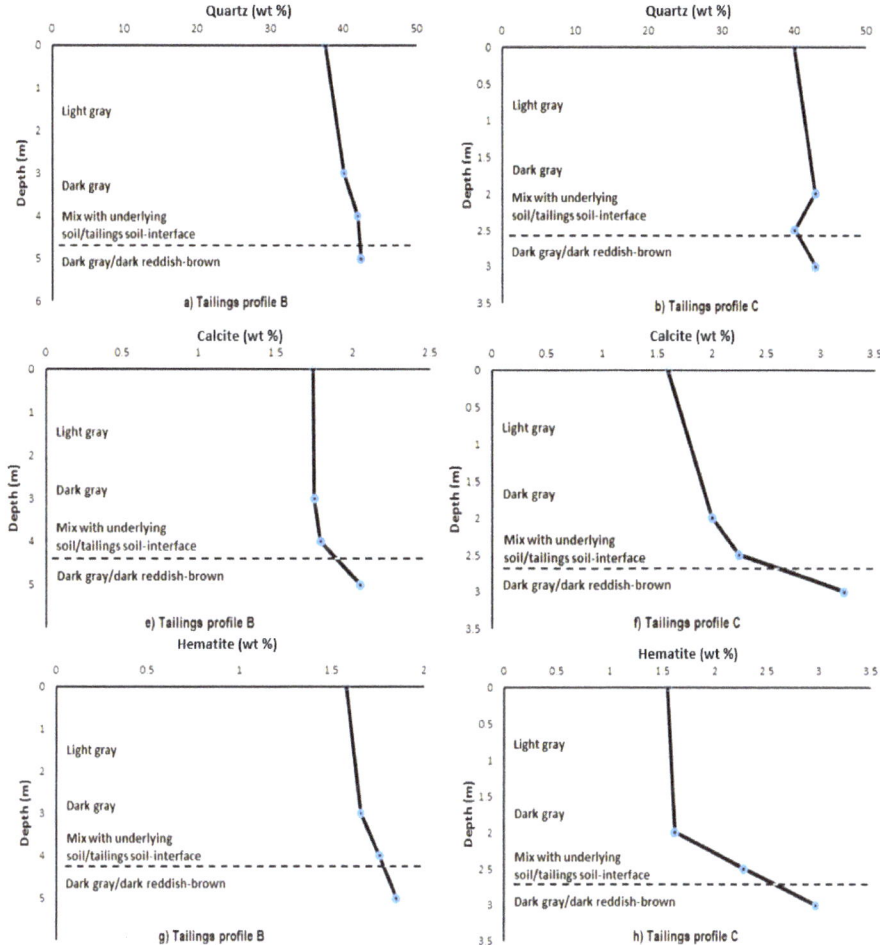

Figure 3. Mineralogical composition trends or patterns within the respective tailings profiles.

The presence of secondary minerals such as calcite and hematite within the tailings is necessary for the retention of potentially toxic chemical species by adsorption on the surface area and co-precipitation. Thus, the potential transfer of potentially toxic chemical species from tailings to the aqueous phase becomes limited [27].

To validate XRD spectra analysis, SEM-EDS analysis was performed on representative tailings samples (Figure 4). SEM microphotograph show large silver white crystals surrounded by dark minerals which are identified as quartz and Al-silicates respectively. The EDS quantification of the mineral crystals is dominated by elements of average composition, O (48.91 wt %), Si (28.32 wt %), C (13.03 wt %), Fe (11.03 wt %), Al (6.63 wt %), Na (4.92 wt %), Mg (3.39 wt %), and substantial amounts of Ca (0.98 wt %). High quantities of Si, O, Al, Na, and Mg, attest to the high quantities of quartz and alumino-silicate minerals. Simultaneously, the quantities of Fe and Ca, confirm the traces of calcite and hematite secondary minerals respectively.

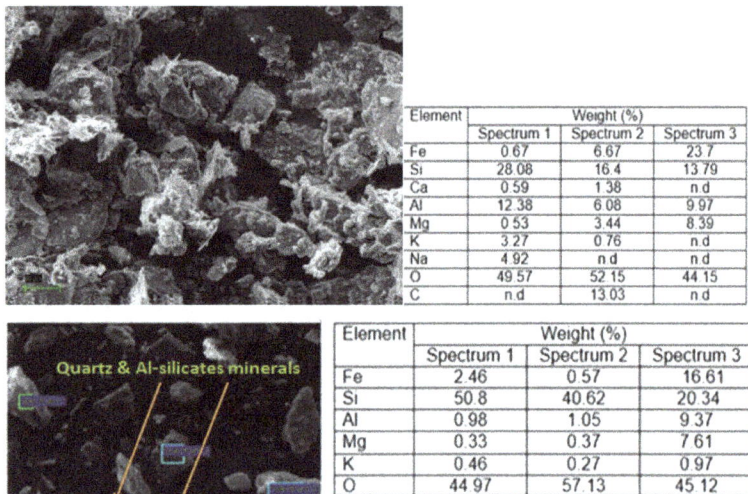
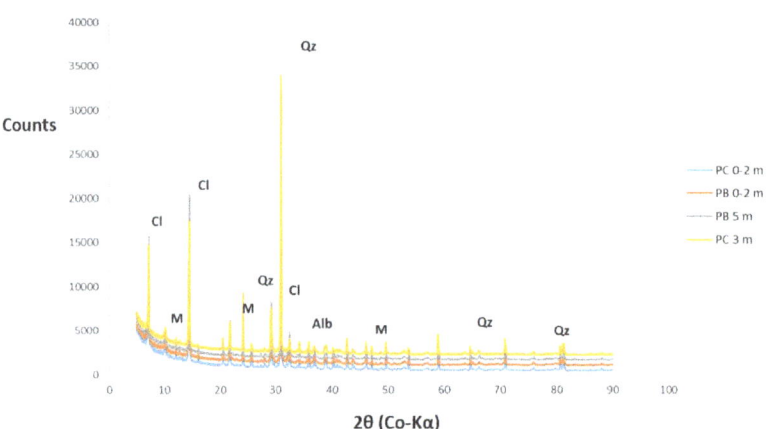

Figure 4. Representative scanning electron microscopy-energy dispersive spectroscopy (SEM-EDS) under 10 and 100 μm magnification and X-ray diffraction (XRD) patterns within tailings profile C at a depth of 2.5 m. Abbreviations: Qz—quartz, Cl—clinochlore, M—muscovite, Alb—albite, n.d—not detected.

3.3. Tailings Paste pH and Eh

Figure 5 depicts paste pH values and patterns within the tailings profiles. The tailings paste pH values are medium alkaline (i.e., 7.97–8.37). Similar paste pH values were observed in neutral, low sulfide/high carbonate tailings impoundment in eastern Slovakia [9]. The medium alkaline pH condition of the tailings could be ascribed to the dominant potential acid-neutralizers or pH buffering minerals such as alumino-silicate minerals and traces of carbonate (i.e., calcite) present within the tailings system. Nevertheless, as expected the measured paste Eh values are lower in the tailings profiles (Eh is equal from −90.8 to −61.9 mV) indicating a more reducing or anoxic geochemical environment (Table 3).

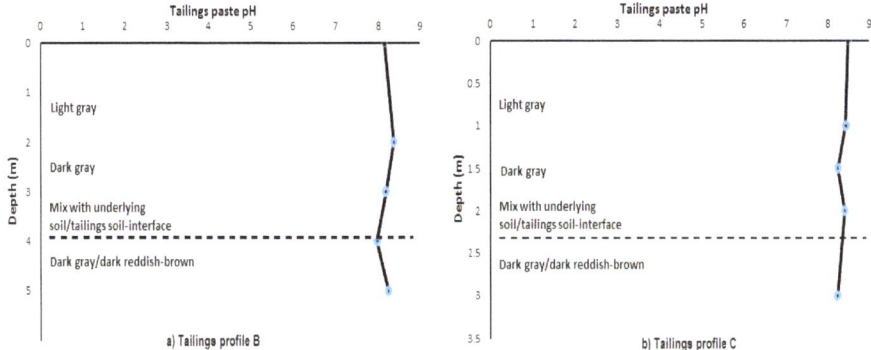

Figure 5. Tailings paste pH content within the respective tailings profiles.

3.4. Bulk Chemical Elemental Composition within the Tailings Profiles

Table 3 shows the concentration of major and trace elements within the tailings profiles. The concentration of potentially toxic chemical species was further compared with the South African soil quality guidelines for all land uses (2012). However, the bulk chemical element concentration for major and trace elements varies less in the tailings profiles. The major elements are in the following order of quantities; Si > Al > Fe > Ca > Mg > K > Na (Table 3). The major element constituents within the tailings profiles agree very well with the mineralogical composition of the tailings. This shows that quartz and alumino-silicates are the most abundant minerals within the established tailings profiles (Table 2). The concentration of silica increases with the quartz content which is the main carrier of Si in the tailings profiles (Table 2; Figure 6). The presence of other major oxides such as Fe, Mg, Ca, K, and Na correspond obviously to the moderate presence of epidote, chlorite, muscovite, calcite, and hematite as the main elemental constituent, and identified in the XRD spectra analysis (Table 2; Figure 6).

Table 3. Total concentration for major and trace elements within tailings profile B.

Paste pH	8.14	8.37	8.18	7.97	8.24	
Eh (mV)	−71.3	−83.9	−73.3	−61.9	−77.2	
Depth (m)	0–1	1–2	2–3	3–4	4–5	
Colour	Dark gray	Light/dark gray	Light/dark gray	Light/dark gray	Dark-reddish brown	
wt %						
SiO_2	58.1176	58.1431	58.2518	58.4938	58.5819	
TiO_2	0.798	0.9266	0.9151	0.9105	0.9079	
Al_2O_3	14.1711	13.9807	13.9559	13.9314	13.564	
Fe_2O_3	10.5766	11.3313	11.2364	11.0953	11.7616	
MnO	0.0502	0.0524	0.0544	0.0537	0.0556	
MgO	2.6372	2.4388	2.5485	2.6219	2.66	
CaO	7.3781	7.6648	7.5943	7.3809	7.2605	
Na_2O	0.3797	0.324	0.3671	0.361	0.4474	
K_2O	1.0167	0.8587	0.8272	0.8588	0.8066	
LOI	3.93	3.23	2.96	3.08	2.8908	
TOTAL	99.5507	99.509	99.1803	99.2741	99.4133	
ppm	0–1 m	1–2 m	2–3 m	3–4 m	4–5 m	South African Soil Quality Guidelines [28]
Cu	1063	862	602	816	515	16
Ni	36	16	20	35	14	91
Pb	-	-	-	-	-	20
Zn	25	13	15	24	10	240
Zr	151	223	187	256	123	-
Sr	236	192	197	252	133	-

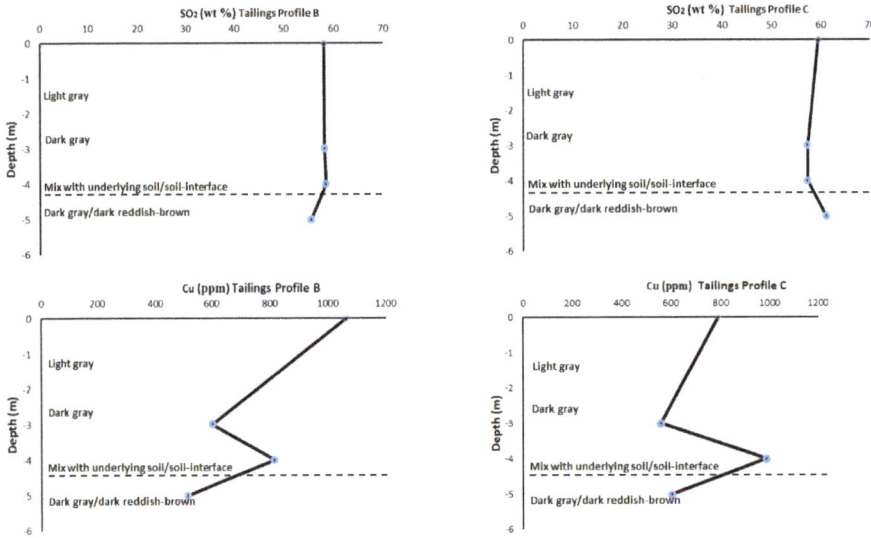

Figure 6. Relative concentration trends or patterns for selected potentially toxic elements with tailings depth.

However, the relative abundance of potentially toxic chemical species is in the following order of quantity; Cu > Sr > Zr > Ni > Zn. Copper has the highest concentration, reaching 1063 ppm at the upper zone of the tailings profile (Table 3; Figure 6). The elevated concentration of copper (Cu) could be attributed to chalcopyrite, chalcocite, and bornite which are considered as primary minerals from which copper (Cu) was recovered during past copper mining activities in Musina [21]. However, the elevated concentration of copper (Cu) decreases moderately with depth, reaching 515 ppm where the tailings are already mixed with the underlying soil or soil-interface (Figure 6). Nevertheless, when compared with the Canadian soil quality guidelines for the protection of the environment and industrial land use, copper (Cu) exceeds the guideline values by a large magnitude (Table 3). Elevated concentrations of copper (Cu) were also reported within the neutral, low-sulfide/high carbonate tailings at eastern Slovakia, and Chambishi site within the Zambian copperbelt, respectively [9,11].

When tailings are in contact with infiltrating rainwater, the elevated concentration of potentially toxic chemical species such as copper (Cu) in the tailings may leach to the adjacent surface and groundwater systems, thereby disturbing aquatic life, and further impacting the health of humans depending on groundwater sources for survival. Nevertheless, moderate concentration of Zr, along with traces of Ni and Zn were also noticeable within the tailings profiles, although these concentrations did not exceed the South African soil quality guidelines for all land uses (Table 3).

3.5. Pore-Water Chemistry, Solubility, and Mobility Patterns for Potentially Toxic Chemical Species with Tailings Depth

Water soluble elements within the tailings are considered as more useful indicators of their potential hazards to the surrounding environment than total concentrations of the elements, as water soluble elements are implicitly mobile and bioavailable [9]. The potential transfer of potentially toxic chemical species from tailings to the aqueous phase was evaluated using the standardized leaching test EN 12457-2 [24]. Romero et al. [7] successfully used a similar approach on abandoned Pb and Zn flotation tailings at "El Fraile" impoundments in Taxco, central-southern Mexico. The obtained leachates concentration of the potentially toxic chemical species and sulfates were used to evaluate the geochemical processes controlling the relative solubility and mobility of potentially toxic chemical species within the established tailings profiles, and further assess the potential transfer of potentially toxic chemical species from tailings to the aqueous phase. Moreover, the obtained leachates concentration of potentially toxic chemical species was compared with the limit guideline values for drinking water quality prescribed by the World Health Organization [25].

The physicochemical parameters, relative percentage leach, or mobility of the potentially toxic chemical species, saturation index of mineral phases, and percentages of main species within pore-water solution calculated by MINTEQA2 are shown in Tables 4–6 respectively. The results give a simulation of the chemical constituents that are likely to be present in the generated plume after a rainfall event when the tailings interact with rainfall with the potential risk to surface and groundwater contamination. The tailings leachates have a medium alkaline pH condition (8.36–8.41). The tailings leachate pH is uniform and constant with tailings depth, and corresponds very well with the tailings paste pH (7.97–8.37) (Figure 5). The medium alkaline tailings leachates pH could be ascribed to the neutralization potentials of the dominant alumino-silicate minerals, traces of carbonates as calcite, and the absence of sulfide minerals (Table 2) as indicated in the XRD spectra analysis. Electrical conductivity increased with increasing tailings depth ranging from 156.5 to 364 µS/cm. On the other hand, the total dissolved solids (TDS) is uniform and constant throughout the tailings depth, averaging 18.8 mg/L. The order of relative abundance of average dissolved concentration of water soluble elements is as follows: Al > Fe > Cu > Mn > Zn > Ni > Pb (Table 5). This observed pattern does not agree with the order of their total solid-phase concentration (Al > Fe > Cu > Ni > Zn) (Table 3). The dissolved concentrations of Al and Fe are high at the upper and shallow depth of the tailings profiles peaking at ≈235.85 and 143.52 µg/L respectively (Table 5; Figure 7). However, the dissolved concentration of Al and Fe decreases with tailings depth, reaching ≈157.86 and 100.8 µg/L respectively at the point where the tailings are in contact with the underlying soil. This could indicate that Al and Fe show similar trends or patterns of solubility and mobility within the tailings profiles which tend to decrease with increasing tailings depth. The precipitation and dissolution of alumino-silicates bearing Al and Fe such as epidote and muscovite under medium alkaline pH condition are the main solubility and mobility control for Fe and Al species in the tailings leachates. In comparison with WHO standards for drinking water quality [25], only Al concentration is above the permissible limit ($\leq$100 µg/L) in the tailings leachates (Table 5). The dissolved concentration of potentially toxic chemical species is very low in the aqueous phase of the tailings, and follows the order: Cu (20.21–47.9 µg/L) > Zn (0.88–1.8 µg/L) > Ni (0.54–0.72 µg/L) > Pb (0.27–0.34 µg/L) (Table 5). The dissolved concentration of Cu follows a similar trend or pattern as Al and Fe which is high at the upper and shallow depths of the tailings profiles peaking at approximately ≈25.65 µg/L. Cu decreases with increasing tailings depth reaching a concentration maxima (≈20.21 µg/L) at the tailings-soil interface (Figure 7). The low water-soluble concentration of dissolved potentially toxic chemical species in the aqueous phase of the tailings could be attributed to: (a) the medium alkaline pH condition of the tailings leachates (8.36–8.46); (b) adsorption and co-precipitation on secondary resistance minerals such as hematite identified during mineralogical analysis; hematite is a very stable mineral, resistant to weathering or any change in the redox condition, and has been reported to retain a large quantity of potentially toxic chemical species [7,11]; and (c) the absence of soluble sulfide minerals hosting these potentially toxic chemical species within the tailings systems. Dold [29] documented that the solubility of many potentially toxic chemical species increases with decreasing or acidic pH, and more dissolved potentially toxic chemical species become more soluble, mobile, and bioavailable to the adjacent receiving surface and groundwater systems.

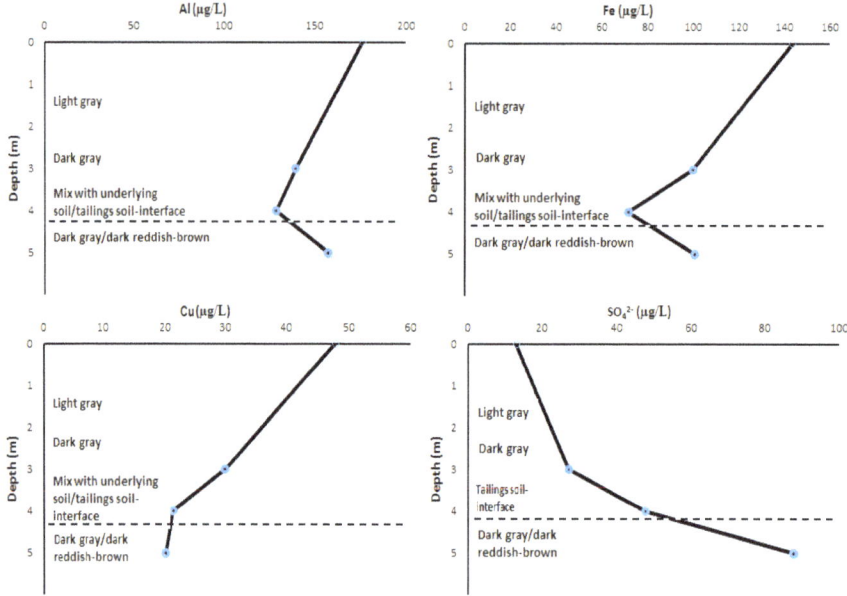

Figure 7. Representative tailings leachate concentration of selected metals as a function of depth within tailings profile B.

Table 4. Extractabilities of potentially toxic chemical species expressed as the percentage leach of the total solid phase concentration within tailings profile B.

Sample Name	Relative Mobility or % Leach of Chemical Species					
	Cu	Co	Ni	Cd	Zn	Pb
PB (0–2 m)	0.006	n.m	0.081	n.m	0.005	n.m
PB (2–4 m)	0.003	n.m	0.003	n.m	0.015	n.m
PB (4–5 m)	0.005	n.m	0.006	n.m	0.027	n.m

n.m—not measured due to low detection limit of the solid phase.

However, a better indication for the mobility of potentially toxic chemical species can be obtained by the estimation of percentages leached of their total concentration within the tailings (Table 4). Based on the calculated mobility percentages, the potentially toxic chemical species show the lowest mobility within the aqueous phase (Ni 0.081%, Cu 0.006%, and Zn 0.05%). This observed pattern confirms the low concentration of dissolved potentially toxic chemical species in the tailings leachates and pore-water solution. This indicates that the potential transfer load of potentially toxic chemical species from tailings to the adjacent receiving surface and groundwater systems is very low.

Table 6 corroborates the low dissolved concentration, and low solubility and mobility of the potentially toxic chemical species in the tailings leachates as calculated by MINTEQA2. The tailings leachates are oversaturated with cuprite (SI = 1.403–5.488), and below saturation regarding oxyhydroxide minerals (Table 6). This shows that the formation of secondary cuprite (Cu_2O) is the other important factor controlling the solubility and mobility for Cu ions in the tailings profiles. The prevailing species of Cu are free Cu^{2+} ion (100%) and sulfate complexes $CuSO_3^-$ (98.79%) in the tailings profiles suggesting that the number of sulfate containing compounds such as sulfuric acid used during past copper mining activities to enhance ore separation is the other solubility and mobility controlling phase for Cu in the tailings leachates. However, the prevailing species for other potentially toxic chemical species are free Co^{2+} (≈96.85%), Ni^{2+} (≈97.99%), and Cd^{2+}

(≈98.84%) (Table 6) indicating the absence of soluble sulfide complexes which are the main carrier for these potentially toxic chemical species. This observed pattern further verified the low dissolved concentrations and low solubility and mobility for Co, Ni, and Cd in the tailings profiles. Nonetheless, the dissolved concentration of sulfate in the tailings leachates is high compared to other anions, and increases abruptly with tailings depth within the tailings profiles peaking at an average of 87.76 µg/L at the point of contact between the tailings and underlying soil (Table 5; Figure 7). The high dissolved concentration of sulfates could be ascribed to the sulfate complexes calculated by MINTEQA2 which are connected to the sulfuric acid used to enhance ore separation during beneficiation and mineral processing of the mined copper ore.

Table 5. Physicochemical parameters of tailings pore-water for tailings profile B.

Bulk Chemical Composition	Units	PB (0–2 m)	PB (2–4 m)	PB (4–5 m)	WHO [25]
pH$_{H2O}$	-	8.46 ± 0.098	8.36 ± 0.021	8.41 ± 0.04	-
EC	µS/cm	156.5 ± 16.82	312 ± 7.07	364 ± 1.44	-
TDS	mg/L	18.8 ± 0.07	18.75 ± 0.06	18.85 ± 0.06	-
Al	µg/L	176.23 ± 59.67	129.37 ± 25	157.86 ± 0.26	≤100
Fe	µg/L	143.51 ± 95.58	71.65 ± 2.99	100.83 ± 1.47	≤1000–3000
Ti	µg/L	1.18 ± 0.15	0.78 ± 0.14	0.87 ± 0.12	-
V	µg/L	1.18 ± 0.04	0.63 ± 0.08	1.02 ± 0.03	-
Cr	µg/L	0.90 ± 0.26	0.60 ± 0.04	1.13 ± 0.04	≤50
Mn	µg/L	17.68 ± 2.40	37.10 ± 8.07	16.63 ± 1.80	≤500
Co	µg/L	0.09 ± 0.02	0.08 ± 0.01	0.09 ± 0.01	≤500
Ni	µg/L	0.54 ± 0.01	0.72 ± 0.08	0.62 ± 0.02	≤70
Cu	µg/L	47.9 ± 4.27	21.43 ± 1.74	20.21 ± 0.01	≤2000
Zn	µg/L	0.88 ± 0.21	1.54 ± 1.02	1.8 ± 0.52	≤5000
As	µg/L	0.31 ± 0.22	0.22 ± 0.01	0.29 ± 0.12	≤10
Se	µg/L	17.41 ± 1.23	20.93 ± 1.51	10.81 ± 2.88	-
Sr	µg/L	57.54 ± 1.96	123.88 ± 10.1	106.33 ± 4.22	-
Mo	µg/L	41.4 ± 2.9	59.44 ± 2.98	66.11 ± 1.22	-
Cd	µg/L	0.02 ± 0.01	0.01	0.02 ± 0.01	≤10
Ba	µg/L	13.1 ± 1.86	19.02 ± 1.98	16.17 ± 0.88	-
Hg	µg/L	0.28 ± 0.08	0.15 ± 0.07	0.10 ± 0.02	≤6
Pb	µg/L	0.27 ± 0.08	0.28 ± 0.09	0.34 ± 0.01	≤5-10
Ca	mg/L	26.71 ± 0.7	55.515 ± 3.92	30.75 ± 2.07	≤150
K	mg/L	9.09 ± 1.35	9.11 ± 1.9	8.45 ± 1.92	-
Mg	mg/L	4.88 ± 0.04	12.67 ± 1.77	8.01 ± 0.41	≤70
Na	mg/L	2.08 ± 0.13	4.4 ± 0.2	4.88 ± 0.11	≤200
S	mg/L	15.71 ± 0.03	55.94 ± 2.24	25.56 ± 1.98	-
SO_4^{2-}	ppm	13.02 ± 3.16	47.74 ± 0.93	87.76 ± 4.463	≤500
Cl^-	ppm	2.106 ± 0.35	16.93 ± 0.83	24.38 ± 0.44	≤300

Table 6. Saturation index of mineral phases (SI) and percentages of main species calculated by MINTEQA2 within the tailings pore water.

Saturation Index	Profile B	Profile C
Cd (OH)$_2$ (s)	−7.074	−3.06
CoO (s)	−6.092	−5.605
Cr (OH)$_2$ (s)	−5.33	−5.022
Cuprite	5.488	1.403
Fe(OH)$_2$ (am)	−2.78	−2.453
Ni (OH)$_2$ (s)	−4.611	−4.06
Quartz	−2.925	−2.909
Zn(OH)$_2$ (am)	−4.171	−3.535
Main Species (%)	**Profile B**	**Profile C**
Cd	-	-
Cd^{2+}	98.845	97.869
$CdOH^+$	1.145	2.013

Table 6. *Cont.*

Saturation Index	Profile B	Profile C
Co	-	-
Co^{2+}	96.854	94.082
$CoOH^+$	2.817	4.861
$Co(OH)_2$ (aq)	0.328	1.007
Cr	-	-
Cr^{2+}	0.084	0.047
$CrOH^+$	99.916	99.953
Cu	-	-
Cu^+	100	1.21
$CuSO_3^-$	N/A	98.79
Fe	-	-
Fe^{2+}	94.508	90.637
$FeOH^+$	5.485	9.343
Ni	-	-
Ni^{2+}	97.992	96.181
$NiOH^+$	1.798	3.135
$Ni(OH)_2$ (aq)	0.21	0.65

4. Summary and Conclusions

An assessment on the mobility patterns and attenuation dynamics of potentially toxic chemical species within copper tailings profiles has been carried out. The tailings were characterized by high bulk density and low air-filled porosity, along with low moisture content at the upper and shallow depths of the tailings profiles. Simultaneously, the solid phase content of potentially toxic chemical species (Cu 602–1063 ppm > Ni 20–36 ppm > Zn 15–25 ppm) was high at the upper and shallow depths of the tailings profiles, and decreased with increasing tailings depth. The physical properties of the tailings have been significant in the retention and mobility attenuation of the potentially toxic chemical species within the studied tailings profiles. The concentration of Cu exceeded the South African soil quality guidelines for all land uses, consequently this could present a potential risk for the adjacent aquatic systems during leaching, mainly when the tailings become in contact with infiltrating rain-water, and furthermore the land could become unsuitable for agricultural purposes. The relative abundance of the identified mineral peaks followed the order of quantity: quartz > epidote > chlorite > muscovite > plagioclase > calcite > hematite. The neutralization of the dominant alumino-silicate minerals, traces of carbonates as calcite, and the absence of sulfide minerals in the tailings resulted in medium alkaline pH leachates (7.97–8.37) at all depths, EC (364–156.5 µS/cm), TDS (18.8 mg/L), and the low concentrations of dissolved Cu (20.21–47.9 µg/L), Zn (0.88–1.80 µg/L), Pb (0.27–0.34 µg/L) in the tailings profiles leachates, indicating that the transport load of potentially toxic chemical species from tailings to aqueous phase is very low. This implies that the solubility and mobility of potentially toxic chemical species is very low within the aqueous phase as corroborated by their relative percentage leach (Ni 0.081%, Cu 0.006%, and Zn 0.05%). MINTEQA2 speciation calculation showed that many of the potentially toxic chemical species occur as free ions in the tailings leachates, indicating the absence of soluble complexes or ligands such as sulfides or organic matter which are known to be the main solubility and mobility control for Co, Ni, Zn, and Cd in the tailings. However, the dissolved concentration of sulfates was high (SO_4^{2-} 15.71–55.94 mg/L) in the tailings leachates, and increased with increasing tailings depth, and this could lead to high salt content in the adjacent surface and groundwater systems. The increase in sulfates content in the tailings leachates is attributed to the possible application of sulfuric acid to enhance ore separation during past copper mining activities in Musina. The water-soluble concentrations of base metals followed the order: Al > Fe > Mn > Ca > K > Na within the tailings profiles, which was consistent with the relative abundance of alumino-silicate minerals. Al and Fe were the metal species with the highest water-soluble concentrations (129–175 µg/L and 71.6–143.5 µg/L, respectively). This indicates the presence of soluble alumino-silicates and absence of sulfidic minerals in the tailings profiles. Mineralogical analysis identified hematite peak as the secondary mineral, although MINTEQA2 speciation calculation showed that the tailings leachates were below saturation regarding oxyhydroxides and were oversaturated with cuprite (Cu_2O). The precipitation of secondary hematite and cuprite under medium alkaline pH condition plays an important known ability to trap and attenuate the mobility of potentially toxic chemical species (Cu, Zn, and Pb) through adsorption on the surface area. Therefore, neutralization, dissolution, and precipitation of the dominant mineralogical constituents as a function of the tailings pH were identified to be the geochemical controls for the solubility and mobility of potentially toxic chemical species, and their potential transfer from tailings to the aqueous phase.

Acknowledgments: The authors express their respective gratitude to the Department of Ecology and Resource Management, and the Department of Mining and Environmental Geology at the University of Venda for their respective interest and access to logistic support. The authors will also want to acknowledge the support from the Environmental Remediation and Water Chemistry Group. Furthermore, we sincerely thank the National Research Foundation (NRF), Sasol Inzalo, ESKOM their financial support.

Author Contributions: Rendani Thobakgale and Segun Ajayi Akinyemi, conceived and designed the experiments; Rendani Thobakgale performed the experiments; Rendani Thobakgale and Wilson Mugera Gitari analyzed the data; Wilson Mugera Gitari contributed reagents/materials/analysis tools; Wilson Mugera Gitari wrote the paper.

Conflicts of Interest: The authors declare no conflict of interest.

References

1. Department of Mineral Resources (DMR). *The National for the Management of Derelict and Ownerless Mines in South Africa*; Department of Mineral Resources: Pretoria, South Africa, 2009.
2. Fosso-Kankeu, E. *Ni^{2+} Extraction from Low Grade Leachate of Tailings Dumps Material Using Cloned Indigenous Bacterial Species*; University of Johannesburg: Johannesburg, South Africa, 2011.
3. Godfrey, L.; Oelofse, S.; Phiri, A.; Nahman, A.; Hall, J. *Mineral Waste: The Required Governance Environment to Enable Reuse*; Final Report, CSIR/NRE/ PW/IR/2007/0080/C; Council of Scientific and Industrial Research: New Delhi, India, 2007.
4. Thobakgale, R. Evaluation of the Geochemical and Mineralogical Transformation at an Old Copper Mine Tailings Dump, Limpopo Province, South Africa. Master's Thesis, University of Venda, Thohoyandou, South Africa, 2017.
5. AngloGold Ashanti. Stakeholder Involvement in the Closure Planning Process at Ergo. In *Report to Society for 2004*; AngloGold Ashanti: Johannesburg, South Africa, 2004; pp. E45–E48. Available online: https://thevault.exchange/?get_group_doc=143/1502780448-Reporttosociety2004.pdf (accessed on 7 February 2018).
6. Lottermoser, B.G. *Mine Wastes: Characterization, Treatment and Environmental Impacts*, 3rd ed.; Springer: Berlin, Germany, 2010; pp. 204–240.
7. Romero, F.M.; Armienta, M.A. Gonzales-Hernandez, G. Solid-phase control on the mobility of potentially toxic elements in an abandoned lead/zinc tailings impoundment, Taxco, Mexico. *Appl. Geochem.* **2007**, *22*, 109–127. [CrossRef]
8. Arenas-Lago, D.; Andrade, M.L.; Lago-Vila, M.; Rodríguez-Seijo, A.; Vega, F.A. Sequential extraction of heavy metals in soils from a copper mine: Distribution in geochemical fractions. *Geoderma* **2014**, *10*, 108–118. [CrossRef]
9. Hiller, E.; Petrák, M.; Tóth, R.; Voleková, B.L.; Jurkovic, L.; Kučerová, G.; Radková, A.; Šottník, P.; Vozár, J. Geochemical and mineralogical characterization of neutral, low-sulfide/high-carbonate tailings impoundments, Markušovce, eastern Slovakia. *Environ. Sci. Pollut. Res.* **2013**, *20*, 7627–7642. [CrossRef] [PubMed]
10. Rice, K.C.; Herman, J.S. Acidification of Earth: An assessment across mechanisms and scales. *Appl. Geochem.* **2012**, *27*, 1–14. [CrossRef]
11. Sracek, O.; Mihaljevic, M.B.; Krˇibek, B.; Majer, V.; Veselovsky, F. Geochemistry and mineralogy of Cu and Co in mine tailings at the Copper belt, Zambia. *J. Afr. Earth Sci.* **2010**, *57*, 14–30. [CrossRef]

12. McGregor, R.G.; Blowes, D.W. The physical, chemical and mineralogical Properties of three cemented layers within sulfide-bearing mine tailings. *J. Geochem. Explor. Can.* **2002**, *76*, 195–207. [CrossRef]
13. Ashton, P.; Love, D.; Mahachi, H.; Dirks, P. *An Overview of the Impact of Mining and Mineral Processing Operations on Water Resources and Water Quality in the Zambezi, Limpopo and Olifants Catchments in Southern Africa*; Project by CSIR Environmentek and Geology Department, University of Zimbabwe, Harare, Zimbabwe; Report No. ENV-P-C 2001-042; Mining, Minerals and Sustainable Development: Pretoria, South Africa, 2001.
14. Naicker, K.; Cukrowska, E.; McCarthy, T.S. Acid mine drainage arising from gold mining activity in Johannesburg, South Africa. *Environ. Pollut.* **2003**, *122*, 29–40. [CrossRef]
15. Bortnikova, S.; Bessonova, E.; Gaskova, O. Geochemistry of arsenic and metals in stored tailings of a Co-Ni arsenide-ore. Khovu-Aksy area, Russia. *Appl. Geochem.* **2012**, *27*, 2238–2250. [CrossRef]
16. Conesa, H.M.; Robinson, B.H.; Schulin, R.; Nowack, B. Metal extractability in acidic and neutral mine tailings from the Cartagena-La Unión mining district (SE Spain). *Appl. Geochem.* **2008**, *23*, 1232–1240. [CrossRef]
17. Gitari, M.W.; Akinyemi, S.A.; Thobakgale, R.; Ngoejana, P.C.; Ramugondo, L.; Matidza, M.; Mhlongo, S.E.; Dacosta, F.A.; Nemapate, N. Physicochemical and Mineralogical Characterization of Musina Mine Copper and New Union Gold Mine Tailings: Implications for Fabrication of Beneficial Geopolymeric Construction Materials. *J. Afr. Earth Sci.* **2018**, *137*, 218–228. [CrossRef]
18. Matshusa, K.; Makgae, M. Overview of Abandoned Mines in the Limpopo Province, South Africa: Rehabilitation Challenges. *J. Environ. Sci. Eng.* **2014**, *15*, 156–161.
19. Singo, N.K. An Assessment of Heavy Metal Pollution near an Old Copper Mine Tailings Dump in Musina, South Africa. Master's Thesis, University of South Africa, Johannesburg, South Africa, 2013.
20. Beale, C.O. Copper in South Africa-Part II. *J. S. Afr. Inst. Min. Metall.* **1985**, *85*, 109–124.
21. Cairncross, B.; Dixon, R. *Minerals of South Africa*; Geological Society of South Africa: Johannesburg, South Africa, 1995; ISBN 0-620-19324-7.
22. Brandl, G. *The Geology of the Musina Area*; Government Printer: Pretoria, South Africa, 1981.
23. ASTM D2216. *Standard Test Methods for Laboratory Determination of Water (Moisture) Content of Soil and Rock by Mass*; ASTM D2216-98; ASTM International: West Conshohocken, PA, USA, 1998.
24. The European Committee for Standardization. *Characterization of Waste. Leaching. Compliance Test for Leaching of Granular Waste Materials and Sludge. One Stage Batch Test at a Liquid to Solid Ratio of 10 L/kg for Materials with Particle Size below 4 mm (without or with Size Reduction)*; BS EN 12457-2:2002; The European Committee for Standardization (CEN): Brussels, Belgium, 2002.
25. World Health Organization (WHO). *Guidelines for Drinking Water Quality*, 4th ed.; World Health Organization (WHO): Geneva, Switzerland, 2011.
26. Gilbert, S.E.; Cooke, D.R.; Hollings, P. The effect of hardpan layers on the water chemistry from the leaching pyrrhotite-rich tailings material. *Environ. Geol.* **2003**, *44*, 687–697. [CrossRef]
27. Lottermoser, B.G.; Ashley, P.M. Mobility and retention of trace elements in hardpan-cemented cassiterite tailings, North Queensland, Australia. *Environ. Geol.* **2006**, *50*, 835–846. [CrossRef]

28. Government Gazette of South Africa. *South African Soil Quality Guidelines for All Land Uses*; Report No. 35160 (561); Government Gazette: Pretoria, South Africa, 2012.
29. Dold, B. Basic concepts of environmental geochemistry of sulfide mine-waste. In *Mineralogia, Geoquimica y Geomicrobiologia para el Manejo Ambiental de Desechos Mineros, Proceedings of the XXIV Curso Latinoamericano de Metalogenia UNESCO-SEG, Lima, Peru, 22 August–2 September 2005*; Society of Economic Geologists (SEG): Littleton, CO, USA, 2005.

© 2018 by the authors. Licensee MDPI, Basel, Switzerland. This article is an open access article distributed under the terms and conditions of the Creative Commons Attribution (CC BY) license (http://creativecommons.org/licenses/by/4.0/).

Article

In Situ Effectiveness of Alkaline and Cementitious Amendments to Stabilize Oxidized Acid-Generating Tailings

Abdellatif Elghali *, Mostafa Benzaazoua *, Bruno Bussière and Thomas Genty

Research Institute on Mines and Environment (RIME), Université du Québec en Abitibi Témiscamingue, 445 Boul. Université, Rouyn-Noranda, QC J9X 5E4, Canada; Bruno.bussiere@uqat.ca (B.B.); thomas.genty@uqat.ca (T.G.)
* Correspondence: abdellatif.elghali@uqat.ca (A.E.); Mostafa.benzaazoua@uqat.ca (M.B.); Tel.: +1-819-762-0971#4203 (A.E.); +1-819-762-0971#2404 (M.B.)

Received: 27 March 2019; Accepted: 18 May 2019; Published: 22 May 2019

Abstract: This study investigates the effectiveness of alkaline and cementitious additives in the in situ stabilization of localized acid-generating tailings from a closed gold mine in Abitibi–Témiscamingue, Québec (Eagle/Telbel mine site). Five field cells (including one control) were constructed and equipped with mechanisms for collecting vertical water infiltration and surface runoff. The five cells included: (C1) Control cell; (C2) 5 wt % limestone amendment; (C3) 10 wt % limestone amendment; (C4) 5 wt % half ordinary Portland cement and half fly ash amendment; and (C5) 5 wt % ordinary Portland cement amendment. The control cell showed an acidic behavior (pH < 4.5) with variable concentrations of Fe, Al, Zn, and Cu. The amendments were used to neutralize the acidic leachates and decrease dissolved metal concentrations. Leachates from surface runoff samples of amended cells were less loaded with metals compared to samples of vertical infiltration. All amendment formulations increased the pH of the leachates from approximately 4 to circumneutral values. Furthermore, metal and metalloid concentrations were greatly limited, except for Cr and As for the carbonate-based amendments. Metal(-oid) stabilization was successfully achieved using the different amendment formulations, with the exception of C2, which still released As.

Keywords: acid mine drainage; alkaline amendments; cementitious amendments; kinetic testing; Joutel mine

1. Introduction

In Canada, environmental regulations require restoration and stabilization of tailings and effluents before the final closure of mine sites. The stabilization of acid-generating tailings and contaminants released from oxidized tailings can be achieved using stabilization/solidification (S/S) techniques. Such techniques consist of adding amendments, such as alkaline materials, cementitious materials, and/or industrial sub-products, to mine wastes [1–14]. Mining amendments can be divided into two categories, depending on the objective of the amendments: (i) Alkaline amendments, which are used for increasing the neutralization potential of acid-generating tailings [15–17], and (ii) cementitious amendments, which are used for the impermeabilization/solidification of acidic tailings, but can also increase the neutralization potential [18–23].

Alkaline amendments were successfully used to control and neutralize acid mine drainage under laboratory conditions during kinetic tests [1,8,11,24–27]. Limestone is one of the most commonly used materials in alkaline amendments. The dissolution of alkaline amendments under acidic conditions increases the alkalinity of leachates and reduces the mobility of contaminants through mechanisms including: (i) Precipitation of low-solubility iron oxyhydroxides; (ii) co-precipitation in and adsorption on oxyhydroxides [11,28–31]; (iii) sulfide surface passivation, which inhibits oxygen diffusion to and from reactive sulfide cores [32–35]; and (iv) reduction of bacterial activity under circumneutral conditions [36–38]. Other materials that have successfully been used as alkaline amendments include red mud bauxite and cement kiln dust [11,12,39,40].

Cementitious amendments are used to solidify hazardous wastes and contaminated soils [7,20,41,42]. Cementitious amendments can reduce the contaminant mobility by the same mechanisms as alkaline amendments. However, the application of this technique allows for the solidification of reactive grains and ensures the physical trapping of contaminants (i.e., fixation). Moreover, the cementation of tailings leads to: (i) Physical encapsulation of mobile contaminants by increasing the cohesive properties of the mixture and (ii) improvements to the cohesion and the long-term impermeability of the tailings, which reduce available reactive surface area [4,7,20,22,23,42–45]. Various materials have successfully been used as cementitious additives for these purposes, although ordinary Portland cement (OPC) is the most common. In some cases, industrial sub-products, such as fly ash (FA), slag, lime, and cement kiln dust, have been used to partially replace OPC in cemented paste backfills [1,3,5,44,46–52].

In this study, limestone, fly ash, and ordinary Portland cement were tested at the former Eagle/Telbel mine site, which is now referred to as Joutel mine. The main objective of this study was to test the in situ effectiveness of these materials to stabilize acid-generating tailings. To accomplish this, five experimental cells were constructed and equipped with two types of water collectors, one for vertical infiltration and one for surface/subsurface runoff. This instrumentation allowed for time-series measurements of the chemical compositions of drainage waters, as well as calculations of the two main components of the water balance.

2. Materials and Methods

2.1. Joutel Mine Site

Joutel is a closed gold mine site located at the north of Abitibi–Témiscamingue (Québec, Canada). Operations at the mine took place between 1974 and 1994. The gold was associated with a sulfidic deposit that contained mainly pyrite, as well as traces of pyrrhotite, chalcopyrite, sphalerite, and galena. Gold was extracted using sulfide bulk flotation, followed by cyanidation. Ore treatment produced finely ground tailings that were deposited over 120 ha in a tailings storage facility (TSF) [53]. The TSF is divided into two zones: The northern zone, which is older and relatively elevated, and the southern zone, which is younger [54].

2.2. Amendments Formulation

Both alkaline and cementitious amendments were tested in in situ experimental cells. The alkaline amendments consisted of limestone. Two formulations were tested based on preliminary laboratory tests, i.e., 5 wt % and 10 wt % limestone. The amendment was mixed in situ with oxidized tailings and then deposited in the field cells. The limestone used in this study had a maximum particle size of 6.25 mm to ensure a mixture of fine and coarse particle sizes that would provide both short- and long-term reactivity. Calculations of the amount of limestone needed to neutralize the oxidized tailings were based on the neutralization potential (NP) and acid-generating potential (AP) of the limestone and the reactive tailings, as expressed in Equation (1):

$$\%R = 100 \times \frac{\left(NP_{tailings} - \left(f \times AP_{tailings}\right)\right)}{\left(f \times \sum_{i=1}^{n} X_i \times APi\right) - \left(\sum_{i=1}^{n} X_i \times NP_i\right)} \quad (1)$$

where $NP_{tailings}$ is the neutralization potential of the mine tailings; $AP_{tailings}$ is the acidification potential of the mine tailings; f is the target NP/AP ratio; X_i is the proportion of each amendment material used (equals 1 if only one amendment is used); AP_i is the acidification potential of the amendment material; and NP_i is the neutralization potential of the amendment material. For the studied tailings, the calculated ratio was ~8% for an NP/AP ratio of 1.

Cementitious amendments in this study consisted of OPC and FA. The two formulations used were 1/2 OPC + 1/2 FA 5 wt % and OPC 5 wt % at a dosage relative to the total dry weight of the tailings. The FA used in this study was taken from the Boralex thermal energy station (Senneterre, QC, Canada), which generates wood-residue thermal energy. The amendment formulations for the five cells are indicated in Table 1.

Table 1. Amendment formulations used in the field cells.

Cell	Formulation
C1	Reference (oxidized tailings)
C2	5 wt % limestone
C3	10 wt % limestone
C4	5 wt % (1/2 OPC + 1/2 FA)
C5	5 wt % OPC

2.3. Field Cell Construction

Field cells were constructed in an acidic area of the Joutel TSF. The cells, which were shaped as inverted truncated pyramids, were 4 m wide, 4 m long, and 0.3 m deep. Only oxidized tailings were used in the cells. The tailings were homogenized in situ using a mechanical loader to avoid, as much as possible, heterogeneities in the chemical and mineralogical properties of the materials. Amendments were mixed into the tailings using the bucket of a mechanical loader (Figure S1E–H). The cells were then excavated in the TSF and linear low-density polyethylene geomembranes were installed at the base and on the sides of each cell to control exfiltration [55] (Figure S1A,B,D,F). Two systems of water collection were installed at each cell. One system collected vertical infiltration, using 5-cm PVC pipes, and the other collected surface and subsurface runoff, using a combination of 5-cm PVC pipes and a gutter (Figure S1C). Each collection system was connected to an external reservoir. The cells using binders were covered with 20 cm of sand as a protection layer (Figure S1I). The drains for the collection systems were installed at a 2% slope. A schematic representation of the cells is shown in Figure S2. Each cell was also equipped with an EC5 volumetric water content (VWC) probe, which obtains water content by measuring the dielectric constant of the media through capacitance/frequency domain technology. Each probe was calibrated in the laboratory, using the same material present in the field. Calibration curves are presented in the supplementary materials (Figure S3). Sampling was not possible during winter; average daily temperatures are presented in the supplementary materials (Figure S4).

2.4. Physical, Chemical, and Mineralogical Analyses

Grain-size distributions of the studied samples were evaluated using a laser analyzer (Malvern Mastersizer, Malvern Panalytical, Canada). Bulk chemical compositions of the samples were determined by ICP-AES (Perkin Elmer Optima 3100 RL, USA), following an $HNO_3/Br_2/HF/HCl$ digestion. Major minerals in the initial oxidized tailings were analyzed by X-ray diffraction (XRD, Bruker Ltd, Canada; Brucker D8 Advance, with a detection limit and precision of approximately 1–5%, operating with a copper cathode, Kα radiation). Results were interpreted using the DIFFRACT.EVA software (Version 3.1, Bruker, Milton, ON, Canada) and quantified using TOPAS (Version 4.2, Bruker, Milton, ON, Canada).

Leachates collected from the cells were analyzed for pH, Eh, and electrical conductivity (EC), using pH/Eh/EC meters, and their chemical compositions were analyzed by ICP-AES, following an addition of 2% HNO_3. Iron–sulfur pH–Eh diagrams were calculated for the control cell and for the four cells with alkaline and cementitious amendments at 21 °C, using the Geochemist's workbench (GWB) database (student edition, Version 12.0.1, Aqueous Solutions LLC, Richmond, VA, USA). Iron and sulfur activities were calculated using Visual Minteq (Version 3.1, KTH, Stockholm, Sweden) based on average Fe and S concentrations. Then, the pH and Eh data from the different cells were projected onto the Fe–S pH–Eh diagrams.

After two years, a composite solid sample was collected from each cell. The chemical composition of these samples was analyzed using ICP-AES. The mineralogy of these samples was investigated using Quantitative Evaluation of Minerals by Scanning Electron Microscopy (QEMSCAN, FEI, Hillsboro, OR, USA). QEMSCAN is an automated mineralogy system that produces particle maps through the collection of rapidly acquired X-rays. The maps and corresponding data quantify modal mineralogy, texture, grain sizes, mineral liberation, and elemental deportment. In this study, carbonate and sulfide liberation were evaluated in order to quantify the coating of each phase and, ultimately, to determine the long-term geochemical behavior of the amended tailings (especially for the limestone amendments). Polished sections were prepared for each sample and analyzed using particle mineralogy analysis mode (PMA). Measurement resolutions varied from 2.5 µm to 6 µm, depending on the particle-size distribution of the sample. The number of analyzed particles was approximately 47391, 78089, 43077, 41851, and 42699 for samples from C1, C2, C3, C4, and C5, respectively.

3. Results

3.1. Chemical, Mineralogical, Static Test, and Physical Characterizations of Solid Samples

Table 2 shows the results of the chemical, mineralogical, and static test, and physical characterizations of the mine tailings, limestone, OPC, and fly ash. The grain-size distribution (GSD) of the mine tailings was finer compared to the limestone, OPC, and FA. The D_{90}, which corresponds to 90 wt % passing on the cumulative GSD curve, was approximately 30.5 µm for the mine tailings sample, 4500 µm for limestone, 46.7 µm for the OPC, and 1500 µm for the FA. Initial water content was about 15 wt % for the mine tailings sample, 6.1 wt % for the limestone sample, 0 wt % for the OPC, and 3.5 wt % for the fly ash. Mineralogical analyses showed that the mine tailings were highly oxidized. They were mainly composed of secondary minerals, such as goethite (22 wt %). Carbonates and sulfides constituted about 18 and 12 wt % of the tailings, respectively. The limestone sample was mostly composed of calcite (76 wt %) and dolomite (22 wt %). The fly ash sample contained about 15.5 wt % calcite, 3.5 wt % siderite, 5.3 wt % quartz, and 75 wt % of various silicate minerals. The reconciliation of the mineralogical composition of fly ash was based only on wt % of the crystalline phases. However, fly ash is known to contain amorphous alumina–silicate glass. Consequently, the sum of wt % of phases was less than 100 wt %. Chemical characterizations showed that the mine tailings sample was high in Fe (27%), S (4.40%), and Ca (3.7%), confirming the mineralogical composition. The limestone sample was composed mainly of Ca (34%) and Mg (2.50%). The OPC sample contained mostly Ca (49%), Al (2.75%), and Fe (2.25%). The fly ash sample contained mainly Ca (7.50%), Al (4.75%), and Fe (2.25%). Other elements were present as trace concentrations within the four samples and the complete chemical composition of these samples is presented in Table 2. NP values, which were determined using the Sobek method as modified by Bouzahzah et al. (2015) [56], were ~183 kg $CaCO_3$/t and 880 kg $CaCO_3$/t for the tailings and limestone, respectively. The AP of the tailings was ~195 kg $CaCO_3$/t.

Table 2. Physical, chemical, and static test, and mineralogical characterizations of the mine tailings, limestone, ordinary Portland cement, and fly ash samples. DL = detection limit.

Characterization	Parameter	Units	MT	Limestone	OPC	FA
Physical characteristics	D10	μm	1.8	50	4.2	82
	D30		4.7	250	11.3	180
	D90		30.4	4500	46.7	1500
	initial water content	wt %	15	6.1	dry	3.52
Chemical composition	Al	%	2.02	0.295	2.75	4.69
	Ca		3.72	33.82	49.07	7.58
	Mg		0.15	2.350	1.18	1.05
	Mn		0.42	0.031	0.06	0.42
	Na		1.07	0.125	0.16	1.96
	K		0.26	0.245	0.43	1.85
	Fe		27.24	0.483	2.23	2.24
	Si*		17.76		7.76	14.82
	Li		≤0.0005	≤0.0005	≤0.0005	0.002
	Pb		0.013	≤0.0005	≤0.0005	≤0.0005
	As		0.091	≤0.0005	0.005	≤0.0005
	Cr		0.0024	0.004	0.007	0.006
	Cu		≤0.001	≤0.0001	0.007	≤0.0001
	Zn		0.01	≤0.0055	0.05	0.07
	S (total)		6.80	0.93	1.74	0.43
	S (sulfates)		0.27	-	-	-
	C		2.2	-	-	-
Static tests	NP	kg CaCO$_3$/t	183	880	-	-
	AP	kg CaCO$_3$/t	195	-	-	-
Mineralogical composition	Quartz	wt %	27.84	2.07	Not applicable	5.32
	Calcite		7.82	75.76		15.51
	Dolomite		0.60	22.17		
	Muscovite		1.03			
	Siderite		11.41			3.65
	Orthoclase		0.99			8.34
	Biotite		0.23			7.52
	Albite		12.78			20.51
	Gypsum		1.45			
	Goethite		22.29			
	Pyrite		12.22			
	Labradorite		0.30			22.88
	Chlorite		0.74			
	Corundum		0.41			
	Anhydrite					1.83

Si* was analyzed using XRF-whole rock.

3.2. Field Cell Monitoring

3.2.1. Vertical Infiltration

Vertical infiltration leachates from the different cells showed different behaviors with respect to pH and EC (Figure 1A,B). Leachates from the reference cell were acidic, with pH values between 1.7 and 4.3. However, the leachates from the amended tailings were circumneutral, with pH values between 6.7 and 8.4 for C2, 7.1 and 8.1 for C3, 6.6 and 8.6 for C4, and 8.8 and 10.3 for C5 (Figure 1A). All amendment formulations successfully buffered the acid produced by the oxidized tailings. The acidic leachates react with the limestone and binders, which dissolve and neutralize acidity [7,11,17,26,31,57–59]. Eh values showed oxidizing conditions (Eh > 100 mV) in all five cells (Figure 1C). Electrical conductivity values (Figure 1B), which illustrate the chemical quality of the leachates, showed that leachates from C1 were more loaded in terms of dissolved ions. EC values for C1 ranged between 8 and 33 mS/cm. In contrast, EC values from C2, C3, C4, and C5 were between 5 and 20 mS/cm, 5 and 12 mS/cm, 1 and 7 mS/cm, and 2 and 5 mS/cm, respectively. The cementitious amendments showed the lowest EC values compared to the control cell and limestone-amended cells; hardening of tailings, due to cementation processes, reduces the available surface area, which reduces the leaching rates of species from mineral surfaces [4,7,10,20,43,52,60,61].

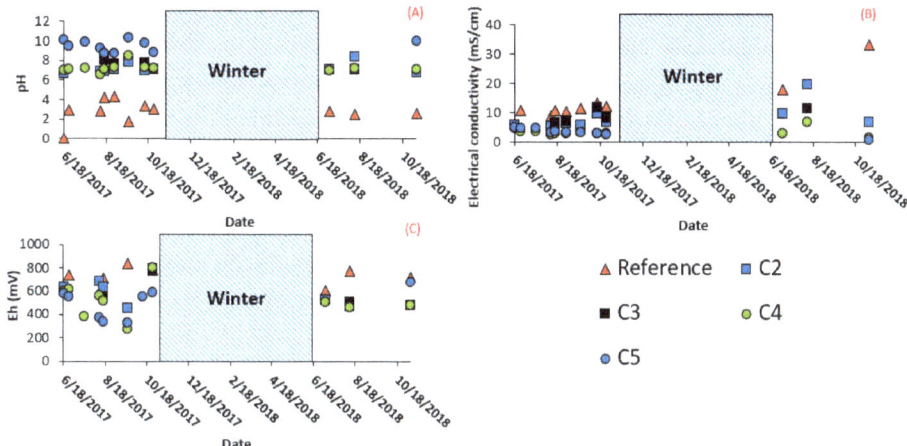

Figure 1. Evolution of (**A**) pH, (**B**) electrical conductivity, and (**C**) Eh for leachates from vertical infiltration.

Vertical infiltration leachates from the five cells were analyzed for concentrations of Ca, Mg, Mn, Al, Fe, S, Zn, Cu, As, and Cr to evaluate carbonate dissolution, sulfide oxidation, and metal(-oid) release rates (Figure 2). The end of the test was characterized by increased leaching of chemical species, including Al, S, Cr, and Zn. However, this could potentially be explained by the lower volumes of water collected at the end of the tests (October). Calcium concentrations differed slightly among the cells, with average concentrations of about 379 mg/L, 440 mg/L, 431 mg/L, 470 mg/L, and 502 mg/L for C1, C2, C3, C4, and C5, respectively (Figure 2A). Average Mg concentrations were about 3756 mg/L, 1790 mg/L, 2151 mg/L, 39 mg/L, and 3.5 mg/L for C1, C2, C3, C4, and C5, respectively (Figure 2B). Average Mn concentrations were about 1421 mg/L, 45 mg/L, 44 mg/L, 1.5 mg/L, and 0.7 mg/L for C1, C2, C3, C4, and C5, respectively (Figure 2C). The average Al concentration was about 16 mg/L for C1 and did not exceed 0.45 mg/L for the amended cells (Figure 2D). The same trend was observed for Fe, which had an average concentration of 1034 mg/L in C1 and ≤0.2 mg/L for the amended tailings (Figure 2E). Average S concentrations were about 7523 mg/L, 2838 mg/L, 3289 mg/L, 707 mg/L, and 890 mg/L for C1, C2, C3, C4, and C5, respectively (Figure 2F). Zinc concentrations were increased in leachates from C1, with an average concentration of about 17.6 mg/L. Concentrations in the amended tailings did not exceed 1.4 mg/L (Figure 2G). Copper concentrations were higher for C1, with an average of 0.93 mg/L, while concentrations in the amended tailings were negligible (Figure 2H). Arsenic was only detected in leachates from C1, C2, and C3, with average concentrations of 0.28 mg/L, 0.29 mg/L, and 0.20 mg/L, respectively (Figure 2I). Finally, average Cr concentrations were about 0.42 mg/L, 0.028 mg/L, 0.029 mg/L, 0.001 mg/L, and 0.015 for C1, C2, C3, C4, and C5, respectively (Figure 2J).

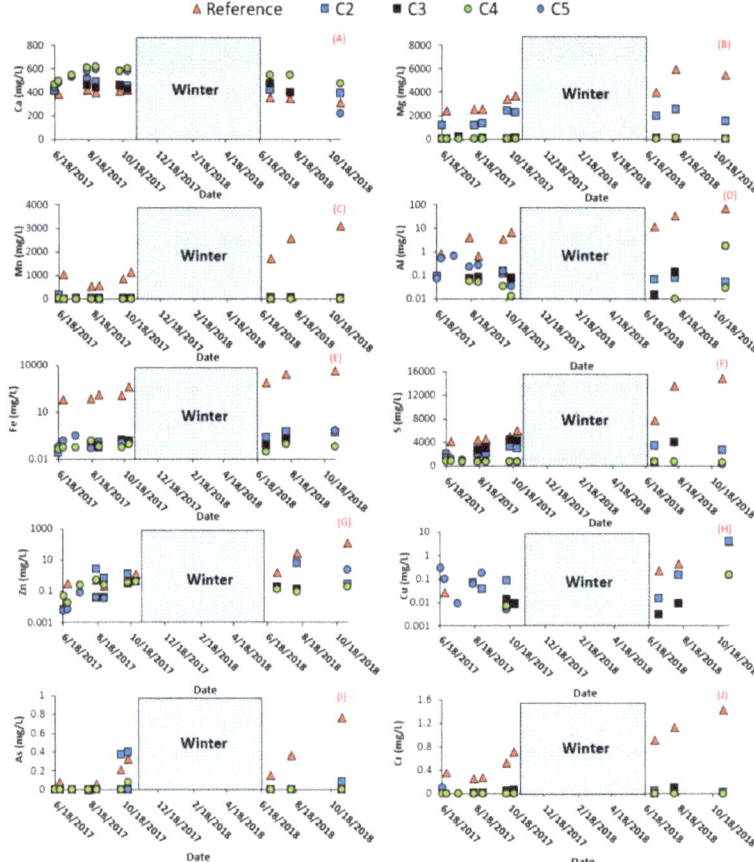

Figure 2. Evolution of concentrations of (**A**) Ca, (**B**) Mg, (**C**) Mn, (**D**) Al, (**E**) Fe, (**F**) S, (**G**) Zn, (**H**) Cu, (**I**) As, and (**J**) Cr in vertical infiltration waters.

3.2.2. Surface and Subsurface Runoff

Leachates collected from surface and subsurface runoff were analyzed for the same parameters as the leachates from vertical water infiltration. During the monitoring period, no leachates were collected from C4. Surface runoff from the control cell was acidic, with pH values less than 4 (Figure 3A), while the amended cells had pH values less than 6. Thus, the different amendment formulations successfully neutralized the acidity produced by the oxidized tailings. Average EC values for all cells were below 3.6 mS/cm (Figure 3B) and Eh values were higher than 100 mV, suggesting oxidizing conditions (Figure 3C).

Leachates from surface and subsurface runoff showed a different chemical quality compared to leachates from vertical infiltration. In general, the runoff leachates showed lower concentrations for most cations. Average Ca concentrations were about 357 mg/L, 403 mg/L, 444 mg/L, and 344 mg/L for C1, C2, C3, and C5, respectively (Figure 4A). Average Mg concentrations were about 66 mg/L, 153 mg/L, 79 mg/L, and 27 mg/L for C1, C2, C3, and C5, respectively (Figure 4B). Average Mn concentrations were about 27 mg/L for C1 and <13 mg/L for the amended tailings (Figure 4C). Average Al concentrations were <0.8 mg/L for all cells (Figure 4D). Iron was leached in relatively high concentrations from C1, with an average concentration of about 37 mg/L, but did not exceed 2 mg/L from the amended tailings

(Figure 4E). Average S concentrations were about 500 mg/L for all cells (Figure 4F). Zn, Cu, As, and Cr were detected in low concentrations (Figure 4G–J).

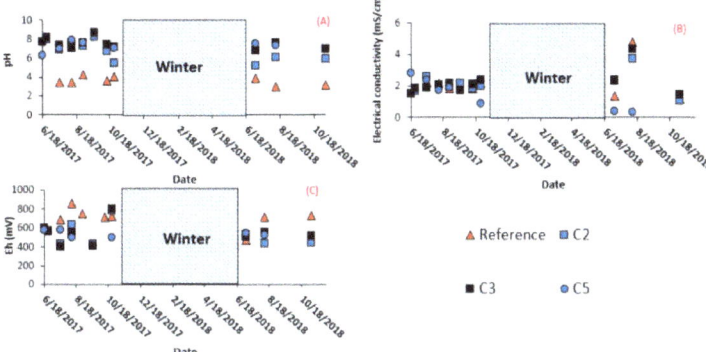

Figure 3. Evolution of (**A**) pH, (**B**) EC, and (**C**) Eh for leachates from surface runoff.

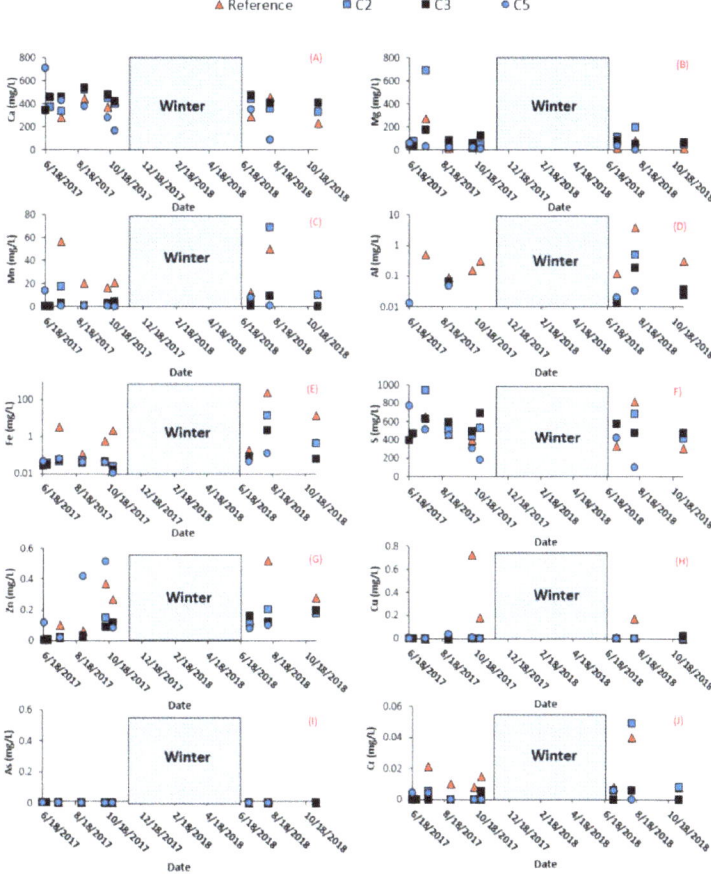

Figure 4. Evolution of concentrations of (**A**) Ca, (**B**) Mg, (**C**) Mn, (**D**) Al, (**E**) Fe, (**F**) S, (**G**) Zn, (**H**) Cu, (**I**) As, and (**J**) Cr for leachates from surface runoff.

3.3. Water Content Evolution

The results of water content monitoring for the five cells are illustrated in Figure 5. Water contents showed the same seasonal variation for all five cells. Water contents decreased during winter and increased during summer. Cell C5 (with 5% cement) showed the lowest VWC value at the end of the monitoring period. Cell C4 showed the highest VWC values over the two-year monitoring period. During the monitoring period, average VWCs were about 0.23, 0.40, 0.24, 0.43, and 0.26% for C1, C2, C3, C4, and C5, respectively. Additionally, the hydrogeological behavior of the cells observed in the field during sampling was very different. After a rain event, a water pool was observed within the cementitious cells. In addition, the volume of water collected in the barrel was always the smallest within C5 and C4, compared to that collected in the other cells.

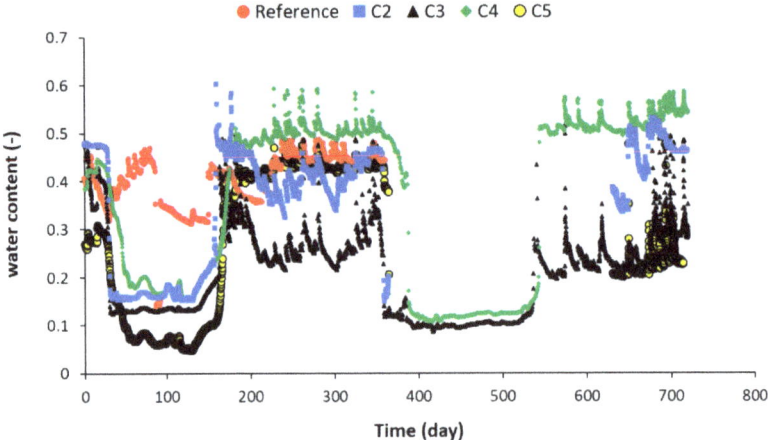

Figure 5. Evolution of volumetric water content (VWC) within the five field cells.

3.4. Field Cell Dismantlement

Samples taken from the field cells following the end of the monitoring period showed relatively similar mineralogical and chemical compositions (Table 3 and Figure 6A). The chemical compositions of the solid samples after the experiments were slightly variable among the cells. Al concentrations varied between 1.5 and 2.2 wt %, Fe concentrations were between 20 and 29 wt %, and S concentrations ranged between 4 and 7.5 wt % (other elements are presented in Table 3). The chemical differences among the samples are likely due to their different compositions (Table 1) and the initial differences due to the high volume of tailings used to fill the different cells.

Figure 6A shows a summary of the mineralogical compositions of the samples after two years of leaching tests. Sulfides consisted primarily of pyrite. Carbonates were mainly present as siderite, with moderate concentrations of calcite. Secondary minerals mainly included Fe oxides, gypsum, and Fe sulfates. The rest of the mineralogical composition comprised various silicates and other oxides.

Pyrite varied between 5 and 9 wt % and carbonate contents varied between 19 and 42 wt %. Calcite was detected in high concentrations in samples from the C3, which initially contained 10% limestone. Iron oxides were detected in high concentrations in samples from C1. This could be explained by precipitation of secondary Fe phases, due to the high Fe concentrations within the reference cell, especially as compared to the other cells (Figure 2A). Iron oxides were present in the form of coatings on the surface of carbonates, sulfides, and silicates. Gypsum contents varied between 0.5 and 6 wt %.

The mineralogical composition of the studied samples shows Ca occurring primarily in gypsum in the sample from C1 and in carbonates in samples from the amended cells (Figure 6B). More than 70 wt % of Fe occurred in siderite and Fe oxides in all samples (Figure 6C) and about 10 wt % of Fe occurred in pyrite. Sulfur was primarily associated with pyrite and gypsum (Figure 6D).

Sulfides and carbonates showed different degrees of liberation and associations (Figure 7). Mineral liberation analyses indicate the exposed areas of mineral. Pyrite was approximately 24, 13, 14, 28, and 18 wt % liberated in samples from C1, C2, C3, C4, and C5, respectively (Figure 7A). Pyrite liberation analyses for C4 and C5 only took into account pyrite locking by mineral phases and not pyrite locking by the cementation of the tailings. Therefore, the pyrite liberation within cementitious amendment was overestimated in this case. The majority of the pyrite consisted of binary associations with Ca–Mg carbonates and Fe oxides (Figure 8). Ca–Mg carbonates were about 0.62, 37, 51, 27, and 13 wt %, liberated in samples from C1, C2, C3, C4, and C5, respectively (Figure 7B). The majority of Ca–Mg carbonates consisted of binary associations with Fe oxides/siderite; they were about 24, 48, 44, 57, and 68 wt % for samples from C1, C2, C3, C4, and C5, respectively.

Table 3. Chemical composition of solid samples after field kinetic testing.

Elements (ppm)	Al	As	Ba	Be	Bi	Ca	Cd	Co	Cr	Cu	Fe	K	Li	Mg	Mn	Mo	Na	Ni	Pb	S	Ti	Zn
Detection limit (ppm)	60	5	5	5	5	60	5	5	5	10	10	1	5	15	5	5	1	5	5	200	25	55
C1	20,200	910	38	<5	108	37,230	39	<5	24	<10	272,400	2600	<5	1515	4237	8	10,700	25	128	75,350	627	99
C2	18,250	783	36	<5	108	50,910	37	<5	24	<10	256,300	2580	<5	7839	6104	7	9360	27	102	71,600	317	96
C3	15,360	654	31	<5	109	106,000	30	<5	22	<10	221,000	2530	<5	10,940	5122	9	7810	23	99	66,060	486	85
C3-duplicate	15,150	673	32	<5	105	100,900	30	<5	22	<10	208,900	2520	<5	11,100	5082	9	8090	24	98	64,600	501	83
C4	21,930	399	66	<5	104	33,660	37	<5	36	<10	291,200	2210	<5	9660	7698	6	12,900	21	105	57,920	865	115
C5	21,970	426	49	<5	100	40,720	38	<5	40	<10	281,800	2110	<5	7711	6775	8	12,600	21	136	43,100	756	107

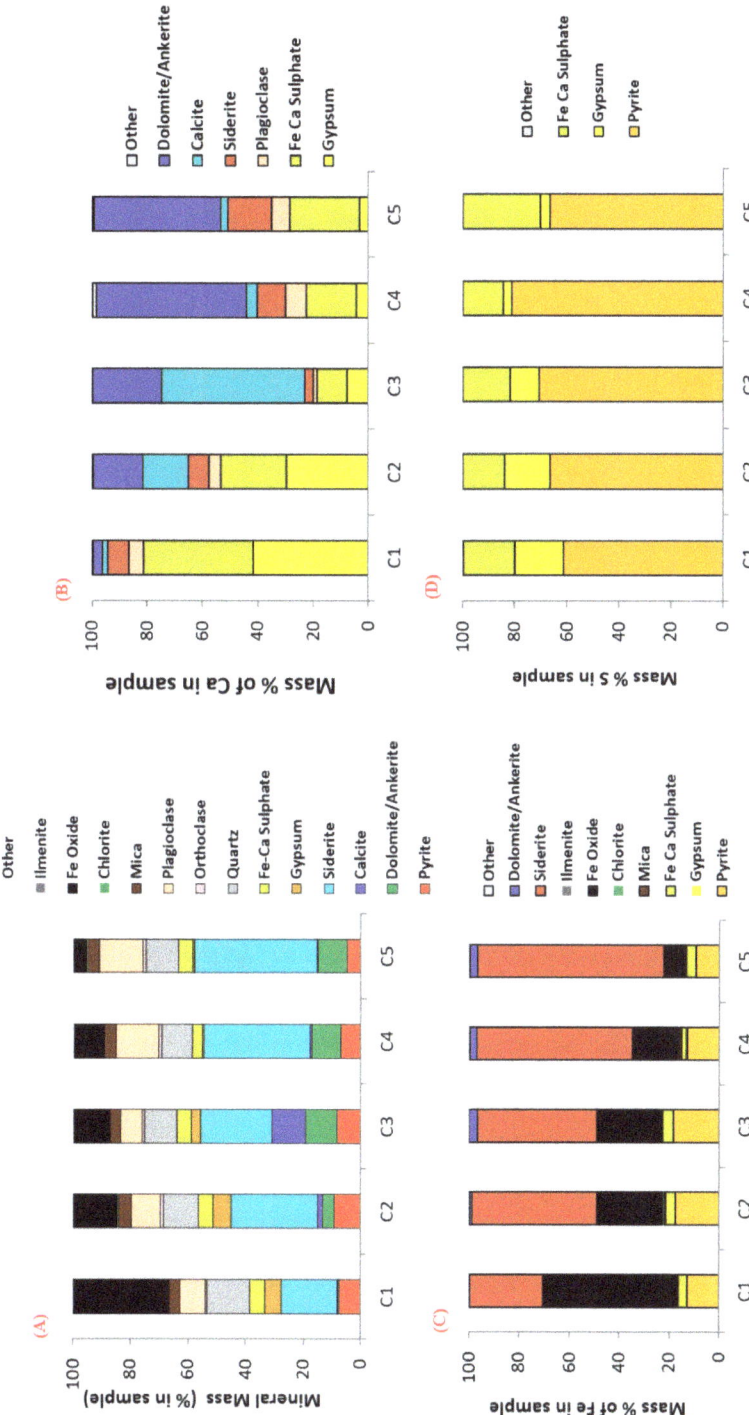

Figure 6. (A) Modal mineralogy for the dismantled samples; (B) Ca deportment; (C) Fe deportment; and (D) S deportment within the dismantled samples.

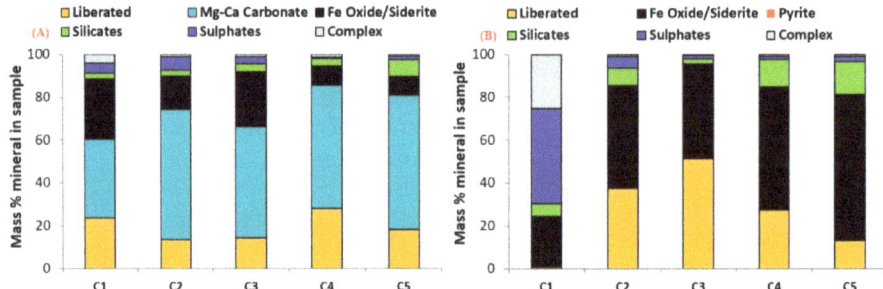

Figure 7. Degree of liberation for (**A**) carbonates and (**B**) sulfides.

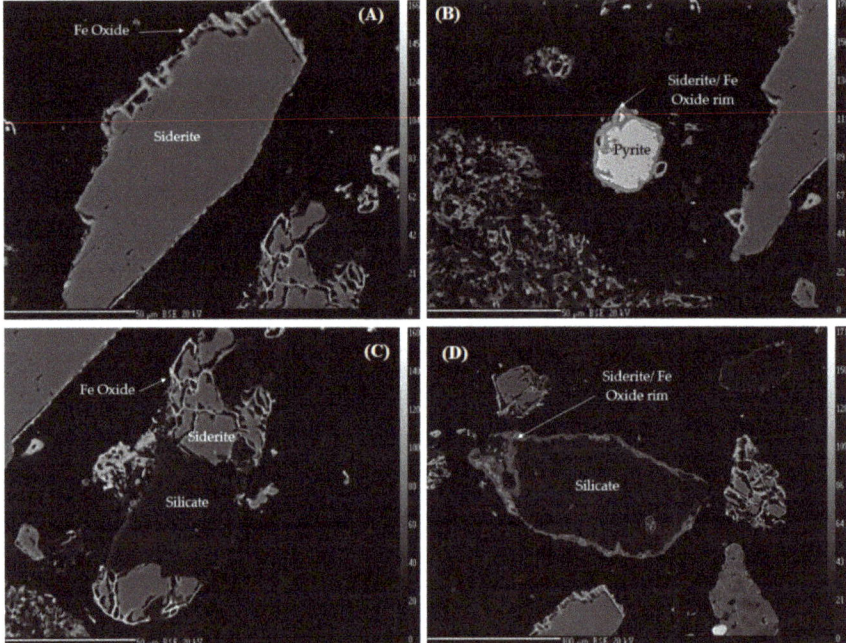

Figure 8. SEM images showing (**A**) siderite coated by Fe oxides, (**B**) pyrite coated by siderite and Fe oxides, and (**C**) and (**D**) silicates coated by Fe oxides and siderite.

4. Discussion

The oxidized tailings showed acidic pH values (<4) and, together with other monitoring data, the pH measurements suggest that surface area and contact time considerably influence final water quality. Overall, leachates from vertical infiltration samples were more acidic and more loaded in chemical species than those from runoff samples. This makes sense as the contact time between water and tailings is greater during infiltration than during surface and subsurface runoff. Vertical water infiltration rates depend upon the physical and hydrogeological properties of the tailings [62–66]. However, surface runoff rates are greatly controlled by the morphological properties of the TSF (e.g., slope gradient, slope length) and the precipitation intensity [67–69].

The geochemical behavior of amended tailings showed that alkaline and cementitious amendments are both promising techniques that could be used for stabilizing and neutralizing acid-generating tailings. The leachates from the amended cells were characterized by circumneutral pH values and were less charged with chemical species than the control cell, except in the case of Ca, which was released more from cell C5 (5 wt % OPC) (Summary of chemical analysis is presented in Table S1). In fact, the dissolution of carbonates and OPC produces alkalinity and buffers the pH of leachates [11,15,31,38,57–59,70–79]. The pH was established early, due to high dissolution rates of limestone and cement in the acidic media. The dissolution of carbonates (calcite) is explained by Equations (2) and (3):

$$CaCO_{3(s)} + 2H^+_{(aq)} \rightarrow H_2CO_{3(aq)} + Ca^{2+} \quad (2)$$

$$CaCO_{3(s)} + H^+_{(aq)} \rightarrow HCO^-_{3(aq)} + Ca^{2+} \quad (3)$$

Ca leaching was higher within C5, which contained OPC as a binder. In fact, the dissolution of tricalcium silicate (C3S), which is contained in OPC, is explained with the following reaction (Equation (4)):

$$2Ca_3SiO_5 + 6H_2O \rightarrow 6Ca^{2+} + 8OH^- + 2H_2SiO_4^{2-} \quad (4)$$

Otherwise, Ca was released by gypsum and by carbonate dissolution in the control cell.

Other chemical species (As, Fe, Al, Li, Pb, Cr, Li, and Zn) were immobilized within the amended cells. The mechanisms responsible for the attenuation of these metal(-oid)s include: Precipitation, co-precipitation, and sorption related to the formation of secondary Fe oxyhydroxides at circumneutral pH values [2,11,24]. The Fe–S Eh–pH diagram illustrated in Figure 9 shows that Fe should precipitate as Fe oxides under the conditions observed in this study. However, for the cementitious amendments (C4 and C5), in addition to the precipitation of secondary Fe phases, two other mechanisms could be responsible for the attenuation of chemical species: (i) Physical trapping and (ii) reduction of water/tailings contact surface, due to hardening of the tailings by binders (FA and OPC). The application of cementitious additives enhances the mechanical resistance of tailings and increases their long-term impermeability [20,44,60,61]. As shown in Figure 10, all amendments used in this study were capable of decreasing the mobility of the analyzed metal(-oid)s, compared to the reference cell (with the exception of As and Cr in the vertical infiltration and runoff leachates from C2, respectively). A reduction factor was calculated for each analyte using Equation (5):

$$RF = 100 \times \left(1 - \frac{C_a}{C_0}\right) \quad (5)$$

where RF is the reduction factor, C_a is the concentration of analyte in the amended cell, and C_0 is the concentration of the analyte in the reference cell.

The amendments reduced the S release from the C2, C3, C4, and C5 with respect to the control. However, more S was released from C2 and C3 than from C4 and C5. This could potentially be explained by the hardening of the tailings–cement mixture. Limestone amendments reduce the leaching of chemical species only by chemical process (e.g., precipitation of secondary phases); however, cementitious amendments reduce the leaching of chemical species by both chemical and physical processes (e.g., reducing the surface contact).

Iron was almost completely immobilized within the different amendment formulations, with RF values between 80 and 100%. Aluminium concentrations were reduced by more than 90%, with a maximum RF of 100% for C3, C4, and C5. Fe oxides precipitated in all cells, as illustrated in the mineralogical composition presented in Figure 6A. However, in C1, there were higher iron oxide concentrations. This could be explained by the pH–Eh conditions. The pH in the reference cell was acidic, which means high reactivity of sulfides would be expected under these conditions. Zinc mobility was greatly reduced for infiltrating waters, however surface and subsurface runoff waters showed RF values between 23 and 67%. Arsenic concentrations were completely reduced for the cementitious formulations with an RF of 100%, except for the formulation C3 (10% limestone) where it still released. Lead was successfully immobilized within RF values greater than 77% in all cases. The cells with cement showed the highest effectiveness regarding the As immobilization, due to: Hardening of tailings, which reduces the reactive surface, when the As is trapped within stable C–S–H phases [80]. The cementitious amendments offer more mechanisms to reduce the As leaching compared to alkaline amendments. Nickel showed RF values greater than 53%. Finally, Cr was successfully stabilized by amendments C3, C4, and C5, while C2 showed appreciable concentrations in surface and subsurface runoff samples. The overall order of effectiveness of the different amendments was C5 = C4 ≥ C3 ≥ C2. Therefore, the cementitious amendments appear to show a greater capacity to immobilize chemical species with respect to the tested alkaline amendments. The incorporation of fly ash to partially substitute the cement could reduce the costs related to the use of cementitious amendments. Considering both economic factors and overall effectiveness, the formulation with fly ash could be considered the best option. Furthermore, the cementitious amendments include the physical trapping of contaminants such as arsenic [44,47]. Indeed, the evolution of the VWC presented in Figure 5 shows that cell C5 had the lowest water content at the end of the test. This suggests that the infiltration speed is lower due to the hardening of the tailings.

Quantitative mineralogical analyses of the five cells showed that carbonates were more liberated compared to sulfides (Figure 7). Mineral liberation is recognized as a key factor in the reactivity of mine wastes [81–83]. After two years of monitoring, the reactive fraction of carbonates was higher than that of the sulfides.

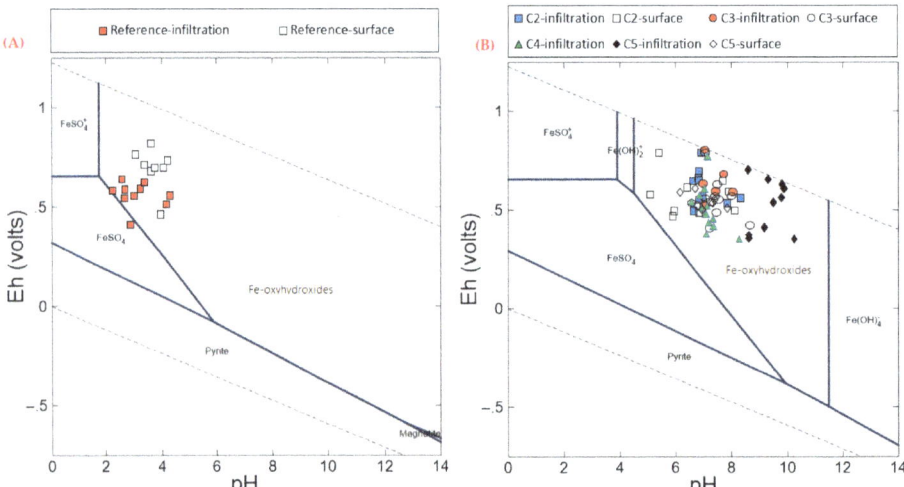

Figure 9. Fe and S pH–Eh diagrams for the leachates from (**A**) the reference cell and (**B**) amended tailings.

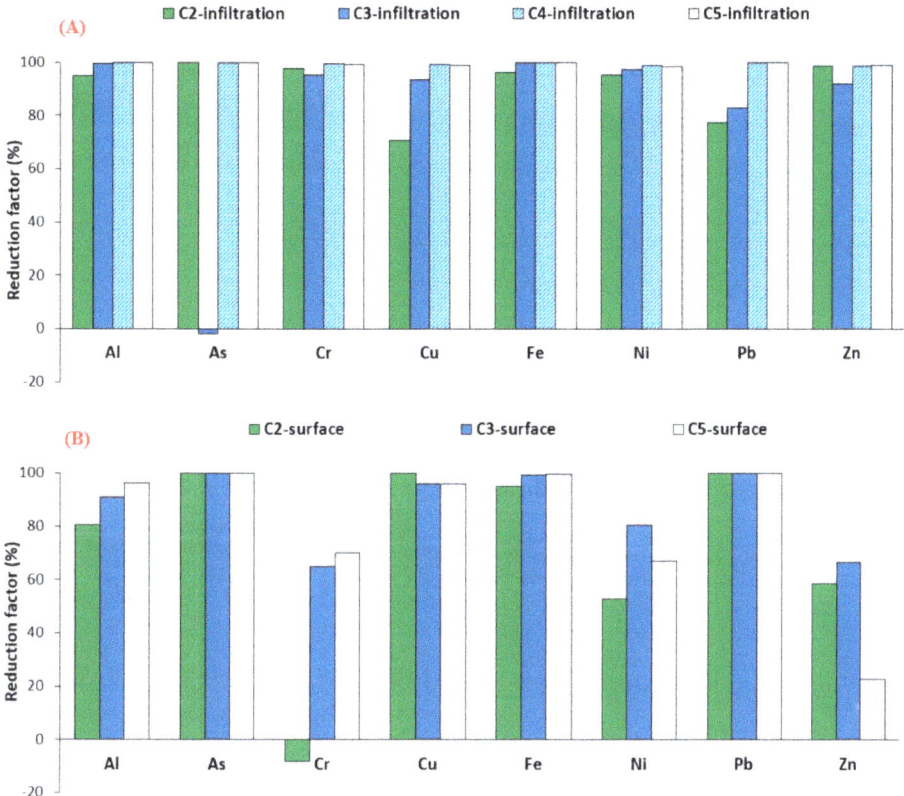

Figure 10. Reduction factors for Al, As, Cr, Cu, Fe, Ni, Pb, and Zn for the different amendment formulations. (**A**) Vertical infiltration leachates; (**B**) surface and subsurface runoff leachates.

5. Conclusions

Reducing risks, such as acid generation and contaminant release from mine tailings, is a serious challenge facing the mining industry and requires the development of cost-effective techniques and technologies. The application of limestone as an alkaline amendment or ordinary Portland cement and fly ash as cementitious amendments showed promising results to neutralize the acid produced and contaminants released from oxidized tailings, based on short-term, in situ tests. Furthermore, OPC was successfully partially substituted with fly ash, which could reduce the costs associated with cementitious amendments. In this study, cementitious amendments showed a high effectiveness compared to alkaline amendments. In fact, metal and metalloid concentrations were reduced more by both tested cementitious amendments than by the tested limestone amendments, which failed to attenuate Zn and As releases when applied at 5% and 10%, respectively. However, the applicability of mining amendments as a reclamation technique must consider two major points: i) The long-term behavior of amended tailings and for how long the contaminants will be immobilized, and ii) the cost related to applying the amendment. Economic factors will be greatly affected by the availability of the amendment materials at a close proximity to the mine site.

Supplementary Materials: The following are available online at http://www.mdpi.com/2075-163X/9/5/314/s1, Figure S1: Images showing field cells construction and amendments mixing with oxidized tailing, Figure S2: Schematic representation of the cells constructed in the TSF (image not to scale), Figure S3: Calibration curves of the volumetric water content probes, Figure S4: Average daily temperature, Table S1: Summary of chemical concentrations of the leachates from the different cells (≤LD means that the concentration was inferior to the detection limit of ICP-AES).

Author Contributions: Conceptualization, A.E., M.B., B.B., and T.G.; formal analysis and investigation, A.E. and T.G.; writing—original draft preparation, A.E. and M.B.; writing—review and editing, A.E., M.B., B.B., and T.G.; project administration, M.B. and B.B.

Funding: Funding for this study was provided by Mitacs (https://www.mitacs.ca/fr, project number IT10661) and the NSERC-UQAT Industrial Research Chair on Mine Site Reclamation and its partners.

Acknowledgments: The authors thank Mitacs (https://www.mitacs.ca/fr), Agnico-Eagle, and the NSERC UQAT Industrial Research Chair on Mine Site Reclamation and its partners for supporting this study. The authors thank URSTM staff for their support in the laboratory and the field and Gary Schudel for comments on the manuscript.

Conflicts of Interest: The authors declare no conflict of interest.

References

1. Rodriguez, L.; Gómez, R.; Sánchez, V.; Villaseñor, J.; Alonso-Azcárate, J. Performance of waste-based amendments to reduce metal release from mine tailings: One-year leaching behaviour. *J. Environ. Manag.* **2018**, *209*, 1–8. [CrossRef] [PubMed]
2. Yi, Y.; Wen, J.; Zeng, G.; Zhang, T.; Huang, F.; Qin, H.; Tian, S. A comparative study for the stabilisation of heavy metal contaminated sediment by limestone, MnO_2 and natural zeolite. *Environ. Sci. Pollut. Res.* **2017**, *24*, 795–804. [CrossRef]
3. Chen, Q.; Tyrer, M.; Hills, C.; Yang, X.; Carey, P. Immobilisation of heavy metal in cement-based solidification/stabilisation: A review. *Waste Manag.* **2009**, *29*, 390–403. [CrossRef]
4. Wang, F.; Shen, Z.; Al-Tabbaa, A. PC-based and MgO-based binders stabilised/solidified heavy metal-contaminated model soil: Strength and heavy metal speciation in early stage. *Géotechnique* **2018**, *68*, 1025–1030. [CrossRef]
5. Falciglia, P.P.; Romano, S.; Vagliasindi, F.G. Stabilisation/Solidification of soils contaminated by mining activities: Influence of barite powder and grout content on γ-radiation shielding, unconfined compressive strength and 232Th immobilisation. *J. Geochem. Explor.* **2017**, *174*, 140–147. [CrossRef]
6. Mitchell, K.; Trakal, L.; Sillerova, H.; Avelar-González, F.J.; Guerrero-Barrera, A.L.; Hough, R.; Beesley, L. Mobility of As, Cr and Cu in a contaminated grassland soil in response to diverse organic amendments; a sequential column leaching experiment. *Appl. Geochem.* **2018**, *88*, 95–102. [CrossRef]
7. Wang, F.; Wang, H.; Jin, F.; Al-Tabbaa, A. The performance of blended conventional and novel binders in the in-situ stabilisation/solidification of a contaminated site soil. *J. Hazard. Mater.* **2015**, *285*, 46–52. [CrossRef] [PubMed]
8. AlKattan, M.; Oelkers, E.H.; Dandurand, J.-L.; Schott, J. An experimental study of calcite and limestone dissolution rates as a function of pH from −1 to 3 and temperature from 25 to 80 °C. *Chem. Geol.* **1998**, *151*, 199–214. [CrossRef]
9. Fatahi, B.; Khabbaz, H. Influence of Chemical Stabilisation on Permeability of Municipal Solid Wastes. *Geotech. Geol. Eng.* **2015**, *33*, 455–466. [CrossRef]
10. Pesonen, J.; Yliniemi, J.; Illikainen, M.; Kuokkanen, T.; Lassi, U. Stabilization/solidification of fly ash from fluidized bed combustion of recovered fuel and biofuel using alkali activation and cement addition. *J. Environ. Chem. Eng.* **2016**, *4*, 1759–1768. [CrossRef]
11. Doye, I.; Duchesne, J. Neutralisation of acid mine drainage with alkaline industrial residues: Laboratory investigation using batch-leaching tests. *Appl. Geochem.* **2003**, *18*, 1197–1213. [CrossRef]
12. Paradis, M.; Duchesne, J.; Lamontagne, A.; Isabel, D. Using red mud bauxite for the neutralization of acid mine tailings: A column leaching test. *Can. Geotech. J.* **2006**, *43*, 1167–1179. [CrossRef]
13. Ahmaruzzaman, M. A review on the utilization of fly ash. *Prog. Energy Combust. Sci.* **2010**, *36*, 327–363. [CrossRef]
14. Fleri, M.A.; Whetstone, G.T. In situ stabilisation/solidification: Project lifecycle. *J. Hazard. Mater.* **2007**, *141*, 441–456. [CrossRef]
15. Cravotta III, C.A.; Trahan, M.K. Limestone drains to increase pH and remove dissolved metals from acidic mine drainage. *Appl. Geochem.* **1999**, *14*, 581–606. [CrossRef]

16. de Andrade, R.P.; Figueiredo, B.R.; de Mello, J.W.V.; Santos, J.C.Z.; Zandonadi, L.U.; Andrade, R.P.; Mello, J.W.V. Control of geochemical mobility of arsenic by liming in materials subjected to acid mine drainage. *J. Soils Sediments* **2008**, *8*, 123–129. [CrossRef]
17. Hakkou, R.; Benzaazoua, M.; Bussière, B. Laboratory Evaluation of the Use of Alkaline Phosphate Wastes for the Control of Acidic Mine Drainage. *Mine Water Environ.* **2009**, *28*, 206–218. [CrossRef]
18. Du, Y.J.; Jiang, N.J.; Liu, S.Y.; Jin, F.; Singh, D.N.; Puppala, A.J. Engineering properties and microstructural characteristics of cement-stabilized zinc-contaminated kaolin. *Can. Geotech. J.* **2013**, *51*, 289–302. [CrossRef]
19. Li, X.D.; Poon, C.S.; Sun, H.; Lo, I.M.; Kirk, D.W. Heavy metal speciation and leaching behaviors in cement based solidified/stabilized waste materials. *J. Hazard. Mater.* **2001**, *82*, 215–230. [CrossRef]
20. Nehdi, M.; Tariq, A. Stabilization of sulphidic mine tailings for prevention of metal release and acid drainage using cementitious materials: A review. *J. Environ. Eng. Sci.* **2007**, *6*, 423–436. [CrossRef]
21. Kogbara, R.B.; Yi, Y.; Al-Tabbaa, A. Process envelopes for stabilisation/solidification of contaminated soil using lime–slag blend. *Environ. Sci. Pollut. Res.* **2011**, *18*, 1286–1296. [CrossRef]
22. Yilmaz, E.; Benzaazoua, M.; Bussière, B.; Pouliot, S. Influence of disposal configurations on hydrogeological behaviour of sulphidic paste tailings: A field experimental study. *Int. J. Process.* **2014**, *131*, 12–25. [CrossRef]
23. Gilles, B.; Benzaazoua, M.; Maqsoud, A.; Bussière, B. Long term hydro-geochemical behaviour of surface paste disposal in field experimental cells. In Proceedings of the Conference Canadienne de Ge'otechnique, GeoRegina, CD-Rom, Regina, Slovakia, 28 September–2 October 2014; Volume 1, p. 19.
24. Komnitsas, K.; Bartzas, G.; Paspaliaris, I. Efficiency of limestone and red mud barriers: Laboratory column studies. *Miner. Eng.* **2004**, *17*, 183–194. [CrossRef]
25. Mylona, E.; Xenidis, A.; Paspaliaris, I. Inhibition of acid generation from sulphidic wastes by the addition of small amounts of limestone. *Miner. Eng.* **2000**, *13*, 1161–1175. [CrossRef]
26. Holmstrom, H.; Ljungberg, J.; Ohlander, B. Role of carbonates in mitigation of metal release from mining waste. Evidence from humidity cells tests. *Environ. Geol.* **1999**, *37*, 267–280. [CrossRef]
27. Duchesne, J.; Reardon, E. Determining controls on element concentrations in cement kiln dust leachate. *Waste Manag.* **1998**, *18*, 339–350. [CrossRef]
28. Acero, P.; Ayora, C.; Torrentó, C.; Nieto, J.-M. The behavior of trace elements during schwertmannite precipitation and subsequent transformation into goethite and jarosite. *Geochim. Cosmochim. Acta* **2006**, *70*, 4130–4139. [CrossRef]
29. Asta, M.P.; Ayora, C.; Román-Ross, G.; Cama, J.; Acero, P.; Gault, A.G.; Charnock, J.M.; Bardelli, F. Natural attenuation of arsenic in the Tinto Santa Rosa acid stream (Iberian Pyritic Belt, SW Spain): The role of iron precipitates. *Chem. Geol.* **2010**, *271*, 1–12. [CrossRef]
30. Benjamin, M.M.; O Leckie, J. Competitive adsorption of cd, cu, zn, and pb on amorphous iron oxyhydroxide. *J. Colloid Interface Sci.* **1981**, *83*, 410–419. [CrossRef]
31. Blowes, D.W.; Ptacek, C.J.; Jambor, J.L.; Weisener, C.G. The geochemistry of acid mine drainage. In *Treatise on Geochemistry*; Elsevier: Amsterdam, The Netherlands, 2014.
32. Belzile, N.; Maki, S.; Chen, Y.-W.; Goldsack, D. Inhibition of pyrite oxidation by surface treatment. *Sci. Total Environ.* **1997**, *196*, 177–186. [CrossRef]
33. Cai, M.-F.; Dang, Z.; Chen, Y.-W.; Belzile, N. The passivation of pyrrhotite by surface coating. *Chemosphere* **2005**, *61*, 659–667. [CrossRef] [PubMed]
34. Harrison, A.L.; Dipple, G.M.; Power, I.M.; Mayer, K.U. Influence of surface passivation and water content on mineral reactions in unsaturated porous media: Implications for brucite carbonation and CO_2 sequestration. *Geochim. Cosmochim. Acta* **2015**, *148*, 477–495. [CrossRef]
35. Kang, C.U.; Jeon, B.H.; Park, S.S.; Kang, J.S.; Kim, K.H.; Kim, D.K.; Choi, U.-K.; Kim, S.J. Inhibition of pyrite oxidation by surface coating: A long-term field study. *Environ. Geochem. Health* **2016**, *38*, 1137–1146. [CrossRef] [PubMed]
36. Evangelou, V.P.; Zhang, Y.L. A review: Pyrite oxidation mechanisms and acid mine drainage prevention. *Crit. Rev. Environ. Sci. Technol.* **1995**, *25*, 141–199. [CrossRef]
37. Nedem. Diversité Microbiologique Dans La Production De Drainage Minier Acide À La halde Sud De La Mine Doyon. *Mend/Medem* **1997**.
38. Nordstrom, D.K.; Southam, G. Geomicrobiology of sulfide mineral oxidation. *Rev. Mineral.* **1997**, *35*, 361–390.

39. Lamontagne, A. Étude De La Méthode D'empilement Des Stériles Par Entremêlement Par Couches Pour Contrôler Le Drainage Minier Acide. Ph.D. Thesis, Université Laval, Quebec City, QC, Cannada, 2001.
40. Mackie, A.L.; Walsh, M.E. Investigation into the use of cement kiln dust in high density sludge (HDS) treatment of acid mine water. *Water Res.* **2015**, *85*, 443–450. [CrossRef] [PubMed]
41. Duchesne, J.; Laforest, G. Remediation of electric arc furnace dust leachate by the use of cementitious materials: A column-leaching test. *Chin. J. Geochem.* **2006**, *25*, 99. [CrossRef]
42. Ichrak, H.; Mostafa, B.; Abdelkabir, M.; Bruno, B. Effect of cementitious amendment on the hydrogeological behavior of a surface paste tailings' disposal. *Innov. Infrastruct. Solut.* **2016**, *1*, 19. [CrossRef]
43. Tariq, A.; Yanful, E.K. A review of binders used in cemented paste tailings for underground and surface disposal practices. *J. Environ. Manag.* **2013**, *131*, 138–149. [CrossRef]
44. Benzaazoua, M.; Marion, P.; Picquet, I.; Bussière, B. The use of pastefill as a solidification and stabilization process for the control of acid mine drainage. *Miner. Eng.* **2004**, *17*, 233–243. [CrossRef]
45. Deschamps, T.; Benzaazoua, M.; Bussière, B.; Aubertin, M. Les effets d'amendements alcalins sur des résidus miniers sulfureux entreposés en surface: Cas des dépôts en pâte. *Déchets Sci. Tech.* **2009**, 19. [CrossRef]
46. Ciccu, R.; Ghiani, M.; Serci, A.; Fadda, S.; Peretti, R.; Zucca, A. Heavy metal immobilization in the mining-contaminated soils using various industrial wastes. *Miner. Eng.* **2003**, *16*, 187–192. [CrossRef]
47. Coussy, S.; Benzaazoua, M.; Blanc, D.; Moszkowicz, P.; Bussière, B. Arsenic stability in arsenopyrite-rich cemented paste backfills: A leaching test-based assessment. *J. Hazard. Mater.* **2011**, *185*, 1467–1476. [CrossRef]
48. Criado, M.; Fernández-Jiménez, A.; Palomo, A. Alkali activation of fly ash: Effect of the SiO2/Na2O ratio. *Microporous Mesoporous Mater.* **2007**, *106*, 180–191. [CrossRef]
49. Kim, J.W.; Jung, M.C. Solidification of arsenic and heavy metal containing tailings using cement and blast furnace slag. *Environ. Geochem. Health* **2011**, *33* (Suppl. 1), 151–158. [CrossRef]
50. Kumpiene, J.; Lagerkvist, A.; Maurice, C. Stabilization of As, Cr, Cu, Pb and Zn in soil using amendments—A review. *Waste Manag.* **2008**, *28*, 215–225. [CrossRef]
51. Park, C.-K. Hydration and solidification of hazardous wastes containing heavy metals using modified cementitious materials. *Cem. Concr. Res.* **2000**, *30*, 429–435. [CrossRef]
52. Peyronnard, O.; Benzaazoua, M. Alternative by-product based binders for cemented mine backfill: Recipes optimisation using Taguchi method. *Miner. Eng.* **2012**, *29*, 28–38. [CrossRef]
53. Elghali, A.; Benzaazoua, M.; Bussière, B.; Kennedy, C.; Parwani, R.; Graham, S. The role of hardpan formation on the reactivity of sulfidic mine tailings: A case study at Joutel mine (Québec). *Sci. Total Environ.* **2019**, *654*, 118–128. [CrossRef]
54. Elghali, A.; Benzaazoua, M.; Bussière, B.; Genty, T. Spatial Mapping of Acidity and Geochemical Properties of Oxidized Tailings within the Former Eagle/Telbel Mine Site. *Minerals* **2019**, *9*, 180. [CrossRef]
55. Bussière, B.; Mbonimpa, M.; Molson, J.W.; Chapuis, R.P.; Aubertin, M. Field experimental cells to evaluate the hydrogeological behaviour of oxygen barriers made of silty materials. *Can. Geotech. J.* **2007**, *44*, 245–265. [CrossRef]
56. Bouzahzah, H.; Benzaazoua, M.B.; Bussiere, P.B. A quantitative approach for the estimation of the "fizz rating" parameter in the acid-base accounting tests: A new adaptations of the Sobek test. *J. Geochem. Explor.* **2015**, *153*, 53–65. [CrossRef]
57. Blowes, D.W.; Ptacek, C.J.; Jambor, J.L.; Weisener, C.G. The geochemistry of acid mine drainage. *Treatise Geochem.* **2003**, *9*, 612.
58. Blowes, D.W.; Jambor, J.L.; Alpers, C.N. *The Environmental Geochemistry of Sulfide Mine-Wastes*; Mineralogical Association of Canada: Québec, QC, Canada, 1994; Volume 22.
59. Blowes, D.W.; Ptacek, C.J.; Jambor, J. Mineralogy of mine wastes and strategies for remediation. In *Environmental Mineralogy*; Vaughan, D.J., Wogelius, R.A., Eds.; European Mineralogical Union: Jena, Germany, 2013; Volume 13, pp. 295–338.
60. Benzaazoua, M.; Belem, T.; Bussière, B. Chemical factors that influence the performance of mine sulphidic paste backfill. *Cem. Concr. Res.* **2002**, *32*, 1133–1144. [CrossRef]
61. Benzaazoua, M.; Ouellet, J.; Servant, S.; Newman, P.; Verburg, R. Cementitious backfill with high sulfur content Physical, chemical, and mineralogical characterization. *Cem. Concr. Res.* **1999**, *29*, 719–725. [CrossRef]
62. Beven, K.; Germann, P. Macropores and water flow in soils. *Water Resour. Res.* **1982**, *18*, 1311–1325. [CrossRef]

63. Bussière, B.; Aubertin, M.; Julien, M. Couvertures avec effets de barrière capillaire pour limiter le drainage minier acide: Aspects théoriques et pratiques. *Vecteur Environ.* **2001**, *34*, 37–50.
64. Childs, E.C.; Bybordi, M. The vertical movement water in stratified porous material: 1. Infiltration. *Water Resour. Res.* **1969**, *5*, 446–459. [CrossRef]
65. Bear, J. *Hydraulics of Groundwater*; Courier Corporation: Chelmsford, MA, USA, 2012.
66. Whitaker, S. Flow in porous media I: A theoretical derivation of Darcy's law. *Transp. Porous Media* **1986**, *1*, 3–25. [CrossRef]
67. Getter, K.L.; Rowe, D.B.; Andresen, J.A. Quantifying the effect of slope on extensive green roof stormwater retention. *Ecol. Eng.* **2007**, *31*, 225–231. [CrossRef]
68. Dunne, T.; Zhang, W. Effects of Rainfall, Vegetation, and Microtopography on Infiltration and Runoff. *Water Resour. Res.* **1991**, *27*, 2271–2285. [CrossRef]
69. Vermang, J.; Norton, L.; Huang, C.; Cornelis, W.; Da Silva, A.; Gabriels, D. Characterization of Soil Surface Roughness Effects on Runoff and Soil Erosion Rates under Simulated Rainfall. *Soil Sci. Soc. J.* **2015**, *79*, 903–916. [CrossRef]
70. Jambor, J. Mineralogy of sulfide-rich tailings and their oxidation products. *Environ. Geochem. Sulfide Mine Wastes* **1994**, *22*, 59–102.
71. Jambor, J.; Dutrizac, J.; Groat, L.; Raudsepp, M. Static tests of neutralization potentials of silicate and aluminosilicate minerals. *Environ. Geol.* **2002**, *43*, 1–17.
72. Lapakko, K.A. Evaluation of neutralization potential determinations for metal mine waste and a proposed alternative. In Proceedings of the Third International Conference on the Abatement of Acidic Drainage, Pittsburgh, PA, USA, 24–29 April 1994.
73. Benzaazoua, M.; Bussière, B.; Dagenais, A. Comparison of kinetic tests for sulfide mine tailings. In Proceedings of the Tailings and Mine Waste 01, Fort Collins, CO, USA, 6–19 January 2001; Balkema: Fort Collins, CO, USA, 2001; pp. 263–272.
74. Benzaazoua, M.; Dagenais, A.-M.; Archambault, M. Kinetic tests comparison and interpretation for prediction of the Joutel tailings acid generation potential. *Environ. Geol.* **2004**, *46*, 1086–1101. [CrossRef]
75. Blowes, D.W.; Jambor, J.L.; Hanton-Fong, C.J.; Lortie, L.; Gould, W. Geochemical, mineralogical and microbiological characterization of a sulphide-bearing carbonate-rich gold-mine tailings impoundment, Joutel, Québec. *Appl. Geochem.* **1998**, *13*, 687–705. [CrossRef]
76. Bussière, B.; Benzaazoua, M.; Aubertin, M.; Mbonimpa, M. A laboratory study of covers made of low-sulphide tailings to prevent acid mine drainage. *Environ. Geol.* **2004**, *45*, 609–622. [CrossRef]
77. Caldeira, C.L.; Ciminelli, V.S.; Osseo-Asare, K. The role of carbonate ions in pyrite oxidation in aqueous systems. *Geochim. Cosmochim. Acta* **2010**, *74*, 1777–1789. [CrossRef]
78. Jamieson, H.E.; Walker, S.R.; Parsons, M.B. Mineralogical characterization of mine waste. *Appl. Geochem.* **2015**, *57*, 85–105. [CrossRef]
79. Nicholson, R.V.; Gillham, R.W.; Reardon, E.J. Pyrite oxidation in carbonate-buffered solution: 1. Experimental kinetics. *Geochim. Cosmochim. Acta* **1988**, *52*, 1077–1085. [CrossRef]
80. Coussy, S.; Benzaazoua, M.; Blanc, D.; Moszkowicz, P.; Bussière, B. Assessment of arsenic immobilization in synthetically prepared cemented paste backfill specimens. *J. Environ. Manag.* **2012**, *93*, 10–21. [CrossRef] [PubMed]
81. Elghali, A.; Benzaazoua, M.; Bouzahzah, H.; Bussière, B.; Villarraga-Gómez, H. Determination of the available acid-generating potential of waste rock, part I: Mineralogical approach. *Appl. Geochem.* **2018**, *99*, 31–41. [CrossRef]
82. Elghali, A.; Benzaazoua, M.; Bussière, B.; Bouzahzah, H. Determination of the available acid-generating potential of waste rock, part II: Waste management involvement. *Appl. Geochem.* **2019**, *100*, 316–325. [CrossRef]
83. Erguler, Z.A.; Erguler, G.K. The effect of particle size on acid mine drainage generation: Kinetic column tests. *Miner. Eng.* **2015**, *76*, 154–167. [CrossRef]

© 2019 by the authors. Licensee MDPI, Basel, Switzerland. This article is an open access article distributed under the terms and conditions of the Creative Commons Attribution (CC BY) license (http://creativecommons.org/licenses/by/4.0/).

Article

Effect of High Mixing Intensity on Rheological Properties of Cemented Paste Backfill

Liuhua Yang [1,2,*], Hongjiang Wang [1,*], Hong Li [1] and Xu Zhou [1]

1. School of Civil and Resource Engineering, University of Science and Technology Beijing, Beijing 100083, China; lihongzxl@126.com (H.L.); pedrozhxu@gmail.com (X.Z.)
2. Institut für Keramik, Glas-und Baustofftechnik der TU Bergakademie Freiberg, 09599 Freiberg, Germany
* Correspondence: yanglh2005@163.com (L.Y.); b20150043@xs.ustb.edu.cn (H.W.); Tel.: +86-010-62334680 (L.Y.); Tel.: +86-010-62333864 (H.W.)

Received: 13 March 2019; Accepted: 10 April 2019; Published: 18 April 2019

Abstract: Cemented paste backfill (CPB) consists of a mixture of fine particles, mainly consisting of tailings and cement dispersed in water. Therefore, it is necessary to introduce an intensive shearing force into the paste during mixing in order to maintain an equilibrium between agglomeration and dispersion. It is influential for the macroscopical fluidity and rheological properties when changes occur in the microstructure of CPB under shear. However, the research on how mixing affects the properties of CPB is still in its infancy. This paper puts an insight into the relation between the mixing intensity and the rheological behavior of the CPB. It can be demonstrated that two threshold mixing intensities exist in this process. After passing the first or lower threshold, the rheological parameters (yield stress and viscosity) of the paste decrease. After passing the second threshold, a continued increase is observed. The changes in rheological properties are connected with physical and chemical changes in the microstructure of the CPB. The results are discussed in light of the three concepts "structural breakdown", "thixotropic breakdown", and "thixotropic behavior" of rheological properties of CPB.

Keywords: rheology; microstructure; flow behavior; cemented paste backfill; mixing

1. Introduction

A growing interest is being paid to the rheological behavior of concentrated suspensions [1], such as cemented paste backfill (CPB), in both theoretical and applied fields. CPB is a mixture of unclassified tailings, crushed aggregates, powder, water, and chemical admixtures [2,3]. CPB technology is also considered superior to conventional slurry backfill methods in terms of both the environmental and economic benefits [4]. These components are generally blended on the surface of a mine and then transported by gravity or pumped to the underground goaf [5]. In the cemented backfill, mixing is an essential process for the acquisition and management of the properties of backfilling materials [6]. More than homogenizing the paste constituents, it plays an important role in characterizing the rheological properties of fresh CPB [7,8]. As the paste backfilling technology has already been a priority option for tailings disposal under policies emphasizing environmental conservation, improper mixing is currently one of the urgent concerns for mining engineers [9,10]. The aim of minimizing the mixing intensity and mixing time without damaging the quality and productivity is widely pursued [11], even in the cementitious materials related industries [12]. Hence, a well-designed mixing technique is crucial for CPB preparation.

As a complex of tailings, cement, and water, CPB is a kind of granular fluid and which, when in fresh state, could be regarded as a suspension of tailing particles suspended in the slurry [13,14]; however, the slurry is not even heterogeneous due to the existence of abundant cement particles and hydrates. Therefore, the rheological properties and macroscopic flow behavior of CPB are significantly influenced by the internal structure of the paste matrix [15,16]. Consistent with other concentrated suspensions, CPB also has the rheological properties of yielding, shear thickening, shear thinning, rheopexy, and thixotropy [17,18]. When under shearing, the intrinsic network structure of CPB responds to the shear-induced stresses with the interference of interparticle forces, leading to changes in the rheological behavior [19]. Thus, to understand the influence mechanism of mixing speed on the properties of CPB is a priority for the improvement of the mixing technique. Under different mixing intensities, CPB specimens tend to present different rheological properties. The phenomenon is affected by the interparticle forces, and it could be explained by the Particle Flow Interaction theory (PFI-theory), which consists of several coupled partial differential equations (PDEs) and is well explained in previous studies [20]. These equations describe the processes of coagulation, dispersion, and re-coagulation of the particles (giving a true thixotropic behavior) [21]. The microstructural changes that lead to thixotropy are called thixotropic breakdown. These concepts were proposed by Hattori and Izumi [22,23] and by Tattersall and Richie [24,25], respectively.

In the PFI-theory, the thixotropic behavior has a relation to coagulation, dispersion, and re-coagulation of the particles, where the coagulation refers to the touch between two or more particles. In order to improve the technological properties of CPB, the particles need to be separated which poses a challenge to CPB mixing. These particles are forced together by their total potential energy, as illustrated in Figure 1, where the term D_s is the distance between two solid surfaces of the particles [26]. The total potential energy interaction V_T originates from combined forces of van der Waals attraction, steric hindrance, and electrostatic repulsion [27]. Under a shear force that is larger than the V_T, the particles will be separated and the rheological properties of CPB decrease, which is called thixotropic breakdown [28].

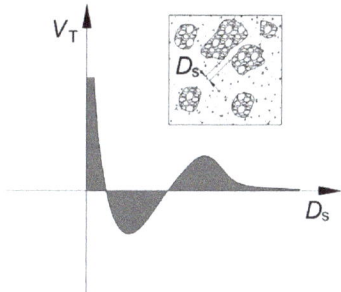

Figure 1. Total potential energy interaction V_T between particles.

The structural breakdown phenomenon seems to be rarely mentioned compared to the well-known thixotropy. Since no recovery of torque was observed, structural breakdown was considered to be different from thixotropic breakdown [29]. As mentioned in the previous paragraph, thixotropic breakdown is related to behavior generated by the total potential energy that exists between particles, while the structural breakdown is not [30]. It is attributed to the shrinking electric double layer of cement particles and/or breaking the connections between the particles formed by the hydration process [31]. That is, when the cement was added to the mixer and contacted with water, the hydration product covers the surface of the particles in the form of membrane [32], causing the particles to contact together. As the CPB is mixed, this connection between the particles is broken and the particles are separated. If the mixing intensity is high enough, the shrinking electric double layer will also occur and the chemical environment of CPB will be changed [33]. Currently, it is the complexity of the intrinsic network structure and the interparticle interactions that make the mixing theory lag behind its application [34,35].

The paste is generally produced by blending the constituents and then transforming the mixture from a wet granular state into a granular suspension microstructure. The high-intensity mixing technology of CPB (without crushed aggregates) usually includes two steps, as shown in Figure 2 [36], where the tailings are firstly mixed with binding materials to form a stiff paste that is similar to a conventional concrete mixing process, and then the high-intensity mixing is applied to make the paste become soft or fluid. This technology allows the CPB to possess a better performance while the mechanism of action is still unclear. Therefore, the impact and its mechanism of varying blending intensities on the rheological properties of fresh CPB were investigated, providing some theoretical support for further development of the high-intensity mixing technology.

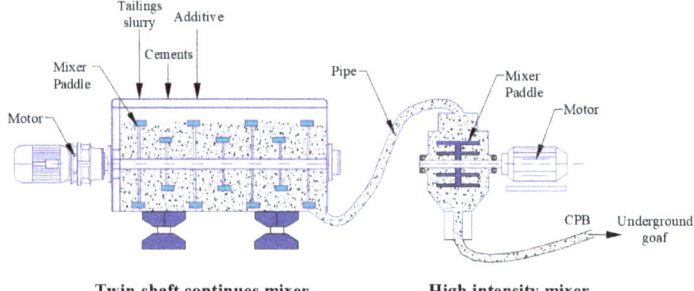

Figure 2. The high-intensity mixing technology of cemented paste backfill (CPB).

2. Materials and Methods

2.1. Materials

2.1.1. Tailings

The tailings sampled to prepare CPB were the full process tailings obtained from a copper mine, of which the specific gravity was 2.69. The chemical characteristics of the tailings were analyzed by X-ray fluorescence (XRF) (NEX CG, Rigaku International Corp., Tokyo, Japan) and are shown in Table 1.

Table 1. Chemical characteristics of the tailings.

Compound	Pb	Zn	S	As	Au	Ag	CaO	MgO	Al_2O_3	SiO_2
Content/%	0.035	0.003	0.39	0.057	0.03	1.59	9.16	1.4	6.19	64.69

The particle size distribution is shown in Figure 3 which was tested with a Topsizer 2000 Laser Particle Characterization System made by OMEC, Ltd., (Zhuhai, China). The measurement covers a range of 0.02 to 2000 μm, with 1% precision. It was observed that the fine particles (<20 μm) accounted for about 29.98 wt % and the tailings could be classified as coarse tailings [37].

2.1.2. Cement

The components of the cement used (obtained by X-ray fluorescence (XRF)) are shown in Table 2, and the cement belongs to ordinary Portland cement (OPC) CEMI 32.5R type. The main chemical components of the cement were CaO and MgO, and a small amount of K_2O. The Blaine fineness was 402 m²/kg and specific gravity 3.14, according to the cement factory report.

Table 2. Characteristics of the cement.

Items	MgO	SiO_2	Na_2O	K_2O	Al_2O_3	SO_3	Fe_2O_3	CaO
Amount (%)	1.40	20.70	0.18	0.48	4.50	2.60	3.30	65.10

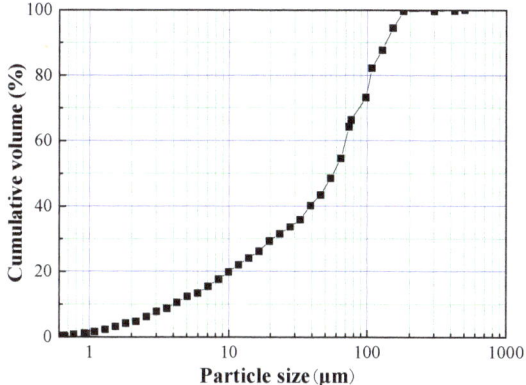

Figure 3. Particle size distribution of the unclassified tailings.

2.2. Mixture Contents

The CPB samples were produced using deionized water. The proportions are provided in Table 3 where four kinds of CPB of three different solid contents and a constant water-cement (w/c) ratio of 2.75 were tested in CPB-A, CPB-B, and CPB-C, and a relatively low water-cement (w/c) ratio of 2.25 was tested in CPB-D. The t/c refers to the tailings-to-cement ratio and is based on the mass of tailings versus the mass of cement. Composition of different CPB samples is given in Table 3.

Table 3. Mixture proportions of CPB.

Code	Water w/c (by mass)	Cement (wt %)	Tailings t/c (by mass)	Solids Content (wt %)	Solids Content (vol. %)
CPB-A	2.75	8.12	8.56	77.66	56
CPB-B	2.75	8.64	7.83	76.25	54
CPB-C	2.75	9.17	7.16	74.79	52
CPB-D	2.25	11.17	5.70	74.86	52

2.3. Preparation of Samples

In the experiment, a cement mortar mixer was used for initial mixing, and a high-shearing-type mixer as shown in Figure 4 was used for secondary mixing, where the velocity was under control of the pre-set computer programs at a room temperature of 20 °C. The equipment, as illustrated in Figure 4, includes a glass bottle containing CPB, a mixer, and a control computer. It is not trivial to measure the shear rates of CPB samples in the mixer, with a rotary propeller approximate value (Equation (1)) [38], however, an indication of the value of the shear rate applied to the mixture in the mixing vessel at varying intensity was evaluated, as shown in Table 4.

$$\dot{\gamma} = \frac{2r_a r_b \omega}{\left(r_b^2 - r_a^2\right)} \quad (1)$$

where r_a is the radius of the rotary propeller (here referred to that of the mixer blade), r_b is the radius of the container (here referring to the CPB mixer vessel), and ω represents the angular velocity of the blades. It is illustrated in Figure 4 that r_a = 60 mm and r_b = 55 mm. Similarly, the shear rate of the high-intensity mixing technology could be calculated according to the geometrical parameters of the mixer, as shown in Table 4. As the speed of the high-intensity mixing of CPB was generally controlled between 1000 and 2000 rpm, the mixing speed of the high-shearing-type mixer was not more than 600 rpm in the study according to Table 4.

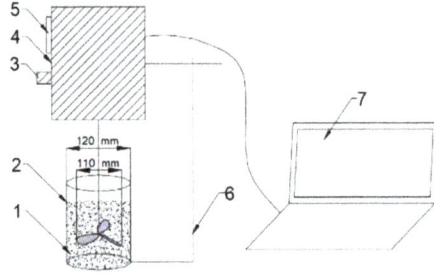

1-CPB sample; 2- Glass bottle; 3- Manual controller;
4- Mixer; 5- Speed display; 6- Holder; 7- Control computer.

Figure 4. High-shearing-type mixer.

Table 4. Estimated shear rate of the high-shearing-type mixer and high-intensity mixer.

High-Shearing-Type Mixer (rpm)	High-Intensity Mixer (rpm)	Estimated Shear Rate (s^{-1})
100	334.78	120.14
200	669.57	240.28
300	1004.35	360.42
400	1339.13	480.56
500	1673.91	600.70
600	2008.70	720.83

A sequence consisting of four steps was designed for the CPB mixing to better simulate the high-intensity mixing technology, which is illustrated in Figure 5. In Step 2, a relatively low speed (75 rpm) was set for the initial mixing (using the cement mortar mixer) to highlight the differences in the rheological behavior of CPBs under different mixing intensities at Step 4 (using the high-shearing-type mixer), specifically 75, 200, 300, 400, 500, and 600 rpm. After the initial three steps, the paste should have been sufficiently homogenized, and Step 4 was designed to measure how the high mixing speed affects the CPB samples.

Figure 5. Mixing treatment to prepare CPB. Mixing speeds ranged from 75, 200, 300, 400, 500, and 600 rpm for Step 4.

2.4. Experimental Methods

2.4.1. Inductively Coupled Plasma Mass Spectrometry

The rheological parameters of CPB (especially the viscosity) are closely related to the chemical environment. In order to test the changes in the chemical environment of CPB under different agitation conditions, inductively coupled plasma mass spectrometry (ICP) was used to gain insight into the pore solution chemistry, especially the concentration of calcium, sodium, and potassium ions, since they are representative of the ion concentration in the pore solution. Previous research has shown that the concentration changes of these ions are related to the hydration reaction and shrinking electric double layer, which affects the interaction force between particles.

To obtain the pore solution of CPB, a 0.50 μm filter paper and a funnel were used. Prepared CPB-D sample (100 g) was transferred to the filter paper in the funnel, and 100 g of deionized water was added.

The pore solution was obtained by vacuum filtration. The pore solution was diluted with 1:1 deionized water before the ICP test. Two bottles of about 5 mL samples were taken in the obtained pore solution for the ICP test. The ICP test uses SPECTRO FLAME S (Ocean Optics, Largo, FL, USA). The CPB-D prepared for each mixing speed was tested twice and the average of the two data was calculated.

2.4.2. Rheology

To monitor the rheological behavior of the CPB samples, an R/S four-paddle rotational rheometer (Brookfield RST, AMETEK Brookfieldb, Middleborough, MA, USA) was used. During the experiment, the four-paddle rotor was immersed into the slurry, spinning at a changing shearing rate. This process was under real-time monitoring and a shearing stress–rate curve was exported by software for further analysis. The temperature of the water bath equipped on this equipment remained at 20 °C.

The paste was poured into a round glass bottle, 95 mm in diameter and 115 mm in height, to a level of 90 mm. The bottle was then fastened after putting the mixing rotor (VT-40-20) of the Brookfield RST into the paste and the flow curve test was performed at a CSR (control shearing rate) mode by raising the rate from 0 to 120 s^{-1} in increments of 1 s^{-1} via a step-up approach. Although the CPB has thixotropic properties, the recovery time of the microstructure was much longer than the time required for the breakdown. The time required for the rheological test was very short, and the shear rate was less than the mixing process, and it can be considered that the microstructure breakdown by the mixing was not recovered. Therefore, from a general view, the paste of unclassified tailings borders on a Bingham fluid and thus Equation (2) could be used for data fitting [39]. The slope and intercept were obtained through the least squares regression. In order to ensure a steady shear stress condition, the curve in which the shear rate was increased from 20 to 120 s^{-1} was selected for fitting.

$$\tau = \tau_0 + \eta \dot{\gamma} \tag{2}$$

Equation (2) represents the Bingham model: τ represents the shear stress, Pa; τ_0 represents the yield stress, Pa; η represents the plastic viscosity, Pa·s; $\dot{\gamma}$ represents the shear rate, s^{-1}.

To verify the reproducibility of the experiment, two tests were conducted for each type of mixture and the average value was adopted. For example, 12 samples were prepared for CPB-A to be stirred at 6 different speeds in Step 4. A representative set of flow curves is shown to indicate the rheology of the material.

3. Results and Discussion

3.1. The Chemical Environment Changes of CPB

As stated in the introduction, high mixing intensity causes structural breakdown to dissolve more hydration products into the pore solution. To verify that this conjecture is correct, ICP was used to analyze the ion concentrations of the pore solution. Figure 6 shows ion concentrations in CPB-D prepared with different mixing speeds and the error bars in the graph represent the range of experimental results.

As shown in Figure 6, when the mixing speeds were between 300 and 400 rpm, the concentrations of ions tend to be in a trough. However, when the mixing speed exceeds 500 rpm, they show a significant increase, especially calcium ions. This change in ion concentration was due to the intensifying of the mixing speed since the cement-based material static curing for 30 min after preparation did not cause a significant change in ion elution [40]. This increase of the initial dissolution could be caused by the consecutive repetition of breakdown and restoration of the double electrode layers of particles.

As we know, the rheological properties of CPB are influenced by its microstructure [41]. However, when CPB is prepared at different mixing speeds, changes are likely to occur to its microstructural state because the concentrations of ions affect the microstructural rebuilding and breaking. Therefore, different mixing speeds change the chemical environment of the CPB, affect the aggregation kinetics of the particles, and may ultimately affect the rheological properties.

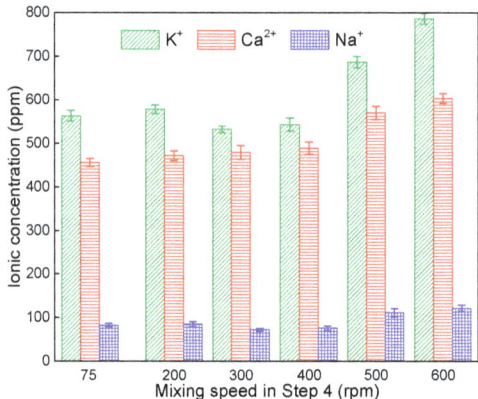

Figure 6. Results of the inductively coupled plasma mass spectrometry (ICP) test for CPB-D.

3.2. Rheological Properties

The experimental rheological curves are illustrated in Figure 7. Besides, from the flow curves, the Bingham parameters of yield stress and plastic viscosity were calculated using a fitted curve to the experimental data points. The rheograph in Figure 8 depicts the Bingham parameters as a function of the mixing speed in the high shear mixer, and the values in the graph represent the average values of the duplicate tests and the error bars in the graph represent the range of experimental results. In the sub-plots of Figure 7, each curve corresponds to a mixing speed of six different levels in Step 4 that was set for specimen preparation. Consistent with expectations, with an increase in the solids content, the CPB samples tended to possess a higher viscosity and yield stress but poorer flowability (except CPB-D). A higher solids content means the decrease in the spacing between particles and the increase in the interaction force, resulting in growing flow resistance in CPB. A synergistic effect of the mixing speed, solids content, and cement content used appeared to control the rheological behavior of CPB. It is known that the mixing intensity plays an important role in rheological properties of CPB, as shown in Figure 8, which, however, was often ignored or barely commented.

The experimental results indicate that the effect on the rheological properties of the paste was not obvious before the speed reaching 200 rpm, and once exceeding 200 rpm, the viscosity and yield stress reduced significantly and the flowability was enhanced in some CPB samples. Thus, the paste could be deemed having been "activated" or dispersed and the paste presented shear-thinning characteristics. It is true that the CPB in the field where it was prepared by a higher mixing speed, had better fluidity than in laboratory experiments where it was prepared by hand mixing.

However, a higher mixing intensity does not always correspond to a better flowability. Contrary to what is widely believed, it was indicated from the results that for mixing speeds exceeding 300–400 rpm, the viscosity and yield stress increased and the fluidity deteriorated. Such non-Newtonian flow behavior of paste is thought to be caused by changes in particle arrangements [42]. Compared with Newtonian fluids, such as water, the rheological behavior of CPB is complicated under different mixing speeds as tiny particles are suspended in the liquid. The results, as shown in Figure 8, also show that the effect of mixing speeds on the rheological properties was more sensitive in cemented paste backfill with the higher cement proportion (CPB-D) than the lower one (CPB-C).

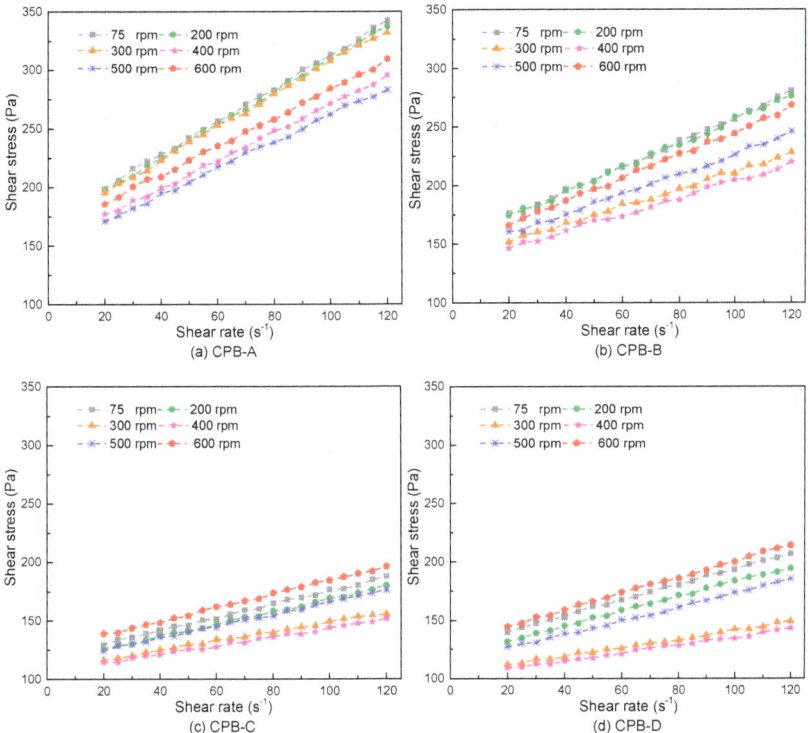

Figure 7. Influence of mixing speed on flow curve behaviors depending on different CPBs: (**a**) CPB-A; (**b**) CPB-B; (**c**) CPB-C; and (**d**) CPB-D.

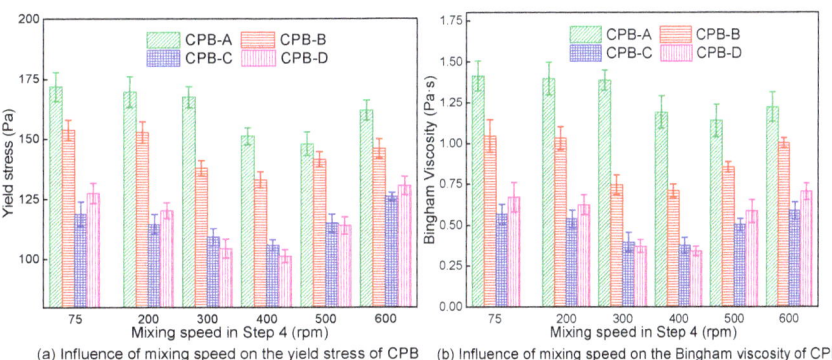

Figure 8. Influence of mixing speed on the yield stress (**a**) and Bingham viscosity (**b**) of CPB.

3.3. Thixotropic Breakdown

As shown in Figure 8, when the mixing speed gradually accelerated from 75 to 200 rpm, the rheological properties almost remained at the same level without significant changes but showed a slight downward trend in some samples. Although CPB is commonly recognized as a Bingham fluid, it could be found that a conversion trend to shear-thinning behavior appeared in some of the CPBs as soon as the mixing speed reached above a threshold that was connected with the solids content of the samples. For the CPBs examined in this research, the threshold speed occurred between 200 and 300 rpm for the majority of CPB, as shown in Figure 8.

According to the PFI-theory, a speed below the threshold is not enough to destroy the cohesion between particles, thus the rheological properties of the paste will not change significantly. Figure 9a depicts that a potential energy well was determined by the particle interaction forces (colloidal interactions for CPB) for each particle (i.e., an equilibrium position of minimum energy exists for each particle) [43]. The particle will not jump out of the well on condition that the mixing energy ΔE offered to the system is less than a certain value, as shown in Figure 9b. Once the shearing ceased, the elastic solid behavior occurred and the particle returns to its start position. However, when exceeding the threshold, the mixing energy was enough to disperse the particle flocculation structure (the mixing energy provided to the system was above the certain value). Therefore, the particle has the ability to get out of the well, as shown in Figure 9c, the microstructure of the paste undergoes thixotropic breakdown, and the rheological properties present shear-thinning behavior, thus the CPB system exhibits thixotropic behavior.

In the preparation of CPB, the CPB is expected to be made of homogeneous distributed particles and to have a good fluidity. The corresponding threshold (minimum mixing energy) is called "the first threshold". The reversibility of this mechanism has been described by Liu et al. [44]. The attractive forces and interparticle links between the particles formed again during the rest period, and the rheology characteristics of CPB will change over time, which, however, takes a longer time compared to that of the breakdown process [45,46]. The potential energy well tends to be deeper at rest with time because of the Brownian motion as well as a possible evolution of the colloidal interactions [47]; the particle needs a larger amount of energy $\Delta E'$ to get out of the well (increase of the shear stress in Figure 9d) [48].

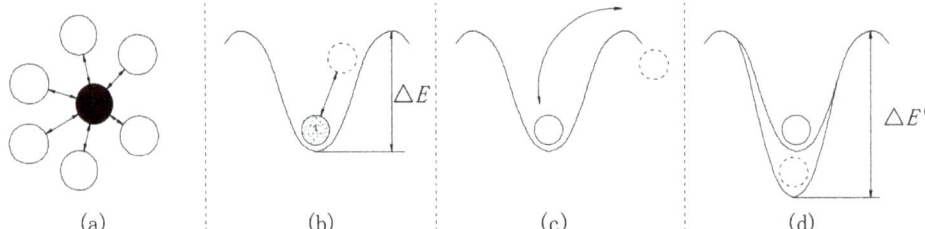

Figure 9. A simple illustration of the thixotropic behavior of CPB. (**a**) The particle interaction forces determine for each particle a potential energy well. (**b**) The particle does not leave the well. (**c**) The particle has the ability to get out of the well. (**d**) The potential energy well tends to be deeper at rest with time, the particle needs a larger amount of energy to get out of the well.

Further analyzing the experimental results of different groups (CPB-A, CPB-B, and CPB-C) in Figure 8, we can also find that the first threshold of mixing speed increases with the solids content. The increase in the solids content of CPB reduces the distance between the particles, that is to say, as the total potential between the particles increases, the energy required to separate the particles that adsorb each other also increases. Therefore, the solids content of CPB should be fully considered when selecting a cost-effective mixing speed.

The distance between two particles is an important factor affecting the interparticle forces, and the interparticle forces could affect the microstructure (aggregation or breakdown) of the CPB. When the particles are separated under the mixing action, the repulsive force of the particles is reduced, and more ions are adsorbed to the surface of the particles [49]. Therefore, as shown in Figure 6, the concentrations of ions in the pore solution decreased when the CPB undergoes thixotropic breakdown.

On the other hand, the water-cement (w/c) ratio is also a key factor affecting the rheological behavior of CPB. The solids content of CPB-D was slightly higher than that of CPB-C. The yield stress and viscosity of CPB-D were larger than those of CPB-C when the mixing speed was lower than the first threshold. It was consistent with the relationship between rheological parameters and solids content described in Section 3.2. However, we found that the rheological parameters (yield stress and viscosity) of CPB-D were less than CPB-C when the mixing speed exceeded the first threshold. Therefore, it can be assumed that the cement has a marked impact on the structural network development, which results in a more flocculated structure, and enhances the thixotropy of CPB.

3.4. Structural Breakdown

Therefore, it seems that a mixing speed above the first threshold within the range of 300 to 400 rpm was enough to disperse the CPB particles. There was no longer a significant decrease in viscosity of the paste with increasing mixing speed until reaching the second threshold of mixing intensity. For mixing speeds higher than the second threshold, the viscosity began to grow significantly in some CPBs and a shear-thickening phenomenon occurred.

It has some relation with the cement particles in CPB since a similar phenomenon also occurs in the high-speed mixing of cement slurry [50]. Since the rheological behavior of cement-based materials is affected by their changing microstructure [51] due to hydration, changes are more likely to happen to the microstructural state when the cement-based materials are prepared at varying mixing speeds [52]. The formation of agglomerates is not only due to hydration but also due to interparticle forces produced from the colloidal state of the finest cement grains [53]. As the distance between particles decreases, the thickness of electrical double layers of the particles becomes the key factor affecting the interparticle forces.

As we know, the thickness of electrical double layers of the particles is influenced by the ionic concentration [28,54], and the thickness could be represented by the Debye–Hückel length equation (Equation (3)) [55]:

$$\frac{1}{\kappa} = \sqrt{\frac{\varepsilon \varepsilon_0 RT}{2F^2 I}} \qquad (3)$$

where $1/\kappa$ is the thickness of electrical double layers, ε is the dielectric constant (relative permittivity) of the dispersion medium, ε_0 is the permittivity of the vacuum, T is the absolute temperature, R is the gas constant, I is the ionic strength, and F is the Faraday constant. A higher ionic concentration will result in a decrease of the thickness of electrical double layers, leading to stronger/increased agglomeration [56]. Additionally, the ionic concentration is known in the ICP test. Figure 10 shows the relationship between the ionic concentration (K^+) and viscosity of CPB-D. When the mixing speed exceeded 400 rpm, the ionic concentration (K^+) in the pore solution increased rapidly, and the changes of the two curves in the figure were consistent. It indicates that the rheological response of CPB was consistent with the change of the chemical environment under different mixing speeds.

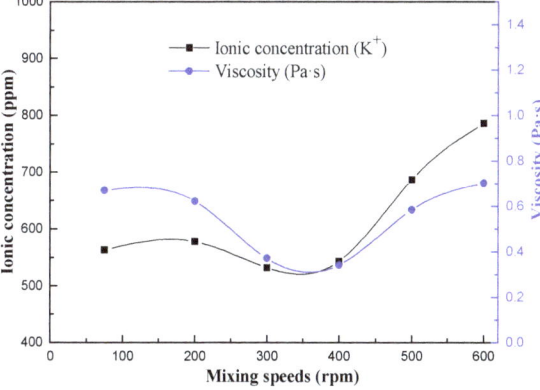

Figure 10. Relationship between ionic concentration (K^+) and viscosity of CPB-D.

This result shows that the chemical and physical qualities of the CPB system changed once the mixing speed differed. High-intensity mixing promoted the aggregation of particles. Figure 11 depicts how high-intensity mixing results in an increasing agglomeration of particles [57]. Under shear, the breakdown of particles directly leads to increased contact areas [58], and a structure of double electrode layers forms around the particle–water interface, as shown in Figure 11a. Subsequently, under the mixing action, the layer shrinks, and numerous ions and early hydration products dissolve into the water as shown in Figure 11b. According to the literature [59], the collapse of the layer reduces the electron repulsion between particles and the prevailing attraction is beneficial for aggregation of cement particles, as shown in Figure 11c. Therefore, it is believed that an increasing mixing speed during the sample preparation affects chemical reactions of CPB and thus transforms the agglomeration kinetics in CPB.

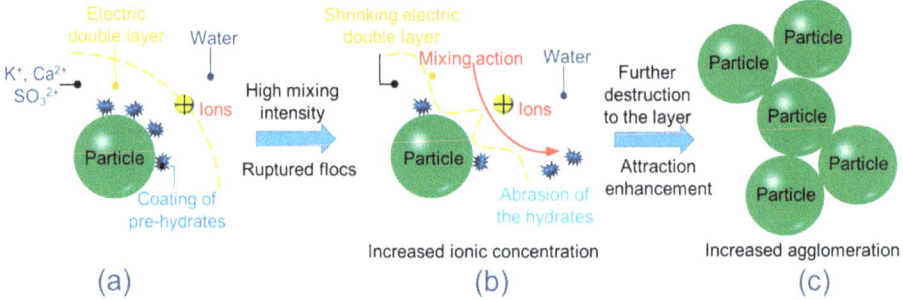

Figure 11. Schematic of the mechanism depicting how high-intensity mixing can lead to increased agglomeration. (**a**) Electric double layer of particle. (**b**) Shrinking electric double layer. (**c**) Particle aggregation.

Comparing the results of different groups of experiments of rheological properties, as shown in Figure 8, it can be found that the composition of CPB was an important factor affecting the shear-thickening phenomenon, and the second threshold of mixing speed increases as the solids content increases. CPBs of different compositions have different microstructures. When under shearing (mixing process), the microstructures of CPB respond to the shear-induced stresses with the interference of interparticle forces, leading to changes in the rheological properties. Thus, the higher sensitivity to mixing intensity (the mixing speed exceeded the second threshold) found in CPB-D as compared to CPB-C could be ascribed to the more cement inducing changes within the CPB that modified the fundamental microstructural response of the CPB systems.

The structural breakdown, which seems to be less recognized and less used, describes the damaging of certain linkages chemically formed among the particles, and also includes the collapse of electrical double layers. Although this mechanism looks like "shear-thickening" behavior, it does not belong to thixotropic behavior, because the structural breakdown is not recoverable. Unfortunately, this behavior is often confused with thixotropy. To obtain a CPB with a better performance, a structural breakdown should not occur in the CPB mixing process. Considering the factors such as energy saving and high productivity, an optimal speed was proposed, and in this study, it was around 300–400 rpm.

4. Conclusions

This study helps us to understand the mechanism of the effect of mixing speed on CPB rheology by determining two threshold values for the mixing speed corresponding to shear thinning and (pseudo) shear thickening. The results show that the rheological behavior of pastes was greatly affected by their shear history. Changes happened in the chemical and physical nature of cemented paste backfill once the preparation mixing speed altered.

(1) The results show that the two thresholds were related to the solids content and composition. For the majority of CPBs under examination in this study, the first threshold occurred between 200 and 300 rpm and the second threshold between 400 and 500 rpm, respectively. When the mixing speed for CPB preparation was between the two thresholds (the shear rate was between 360 and 600 s^{-1}), the paste tends to have lower viscosity and proper fluidity; this phenomenon is often referred to as thixotropic breakdown, which is a recoverable change. Considering the factors such as energy saving and high productivity, the authors proposed an optimal speed, and in this study, it was around 300–400 rpm.

(2) Far from general views, when the mixing speed for the sample preparation exceeds the second threshold, the double layer of cement particles collapses and numerous ions and early hydration products dissolve into the water, which promotes the agglomeration of particles and increases paste yield stress and Bingham viscosity. This trend was amplified in cemented paste backfill with increasing cement content.

(3) In previous studies, some researchers believed that CPB could be characterized by only two kinds of thixotropic behavior, such as shear-thickening and/or shear-thinning. The results presented in this paper suggest the observed shear-thickening was rather a pseudo shear-thickening provoked by irreversible chemical interactions between the particles. Therefore, the three concepts "structural breakdown", "thixotropic breakdown", and "thixotropic behavior" should not be confused, since a structural breakdown does not belong to thixotropic behavior and is unrecoverable.

Author Contributions: Conceptualization, L.Y. and H.W.; Data curation, H.L.; Formal analysis, X.Z.

Funding: This work received financial support from the National Key Technologies R&D Program for the 13th Five-Year Plan (2017YFC0602903) as well as the National Natural Science Foundation of China (51374304).

Acknowledgments: The authors thank Wuyan Li and Natsuko Sagawa for their help in refining the paper.

Conflicts of Interest: The authors declare that there are no conflicts of interest regarding the publication of this paper.

References

1. Pileggi, R.G.; Studart, A.R.; Pandolfelli, V.C.; Gallo, J. How mixing affects the rheology of refractory castables—Part 2. *Am. Ceram. Soc. Bull.* **2001**, *80*, 38–42.
2. Wu, A.; Ruan, Z.; Wang, Y.; Yin, S.; Wang, S.; Wang, Y.; Wang, J. Simulation of long-distance pipeline transportation properties of whole-tailings paste with high sliming. *J. Cent. South Univ.* **2018**, *25*, 141–150. [CrossRef]
3. Fall, M.; Adrien, D.; Célestin, J.; Pokharel, M.; Touré, M. Saturated hydraulic conductivity of cemented paste backfill. *Miner. Eng.* **2009**, *22*, 1307–1317. [CrossRef]
4. Cihangir, F.; Ercikdi, B.; Kesimal, A.; Ocak, S.; Akyol, Y. Effect of sodium-silicate activated slag at different silicate modulus on the strength and microstructural properties of full and coarse sulphidic tailings paste backfill. *Constr. Build. Mater.* **2018**, *185*, 555–566. [CrossRef]
5. Wang, Y.; Fall, M.; Wu, A. Initial temperature-dependence of strength development and self-desiccation in cemented paste backfill that contains sodium silicate. *Cem. Concr. Compos.* **2016**, *67*, 101–110. [CrossRef]
6. Mahlaba, J.; Kearsley, E.; Kruger, R.; Pretorius, P. Evaluation of workability and strength development of fly ash pastes prepared with industrial brines rich in SO$_4^{-}$ and Cl$^-$ to expand brine utilization. *Miner. Eng.* **2011**, *10*, 1077–1081. [CrossRef]
7. Yang, L.; Wang, H.; Wu, A.; Xing, P.; Gao, W. Thixotropy of unclassified pastes in the process of stirring and shearing. *J. Univ. Sci. Technol. Beijing* **2016**, *38*, 1343–1349.
8. Zhang, Q.; Wang, X. Performance of cemented coal gangue backfill. *J. Cent. South Univ.* **2007**, *14*, 216–219. [CrossRef]
9. Fall, M.; Célestin, J.; Pokharel, M.; Touré, M. A contribution to understanding the effect of temperatures on the mechanical properties of mine cemented tailings backfill: Experimental results. *Eng. Geol.* **2010**, *114*, 397–413. [CrossRef]

10. Wu, D.; Cai, S. Coupled effect of cement hydration and temperature on hydraulic behavior of cemented tailings backfill. *J. Cent. South Univ.* **2015**, *22*, 1956–1964. [CrossRef]
11. Baroud, G.; Samara, M.; Steffen, T. Influence of mixing method on the cement temperature-mixing time history and doughing time of three acrylic cements for vertebroplasty. *J. Biomed. Mater. Res. Part B Appl. Biomater.* **2004**, *68*, 112–119. [CrossRef] [PubMed]
12. Wendling, A.; Mar, D.; Wischmeier, N.; Anderson, D.; Mciff, T. Combination of modified mixing technique and low frequency ultrasound to control the elution profile of vancomycin-loaded acrylic bone cement. *Bone Jt. Res.* **2016**, *5*, 26–32. [CrossRef] [PubMed]
13. Du, K.; Li, X.; Yin, Z. A new manufacture method of backfill samples in lab—Illustrated with a case study. *J. Cent. South Univ.* **2013**, *20*, 1022–1028. [CrossRef]
14. Fall, M.; Benzaazoua, M.; Saa, E. Mix proportioning of underground cemented tailings backfill. *Tunn. Undergr. Space Technol. Inc. Trenchless Technol. Res.* **2008**, *23*, 80–90. [CrossRef]
15. Ferron, R.; Shah, S.; Fuente, E.; Negro, C. Aggregation and breakage kinetics of fresh cement paste. *Cem. Concr. Res.* **2013**, *50*. [CrossRef]
16. Toutou, Z.; Roussel, N. Multi scale experimental study of concrete rheology: From water scale to gravel scale. *Mater. Struct.* **2006**, *39*, 189–199. [CrossRef]
17. Jiang, H.; Fall, M. Yield stress and strength of saline cemented tailings materials in sub-zero environments: Slag-paste backfill. *Int. J. Miner. Process.* **2017**, *160*, 68–75. [CrossRef]
18. Ghirian, A.; Fall, M. *Paste Tailings Management*; Springer International Publishing: Berlin/Heidelberg, Germany, 2017; pp. 22–45.
19. Wallevik, O.; Feys, D.; Wallevik, J.; Khayat, K. Avoiding inaccurate interpretations of rheological measurements for cement-based materials. *Cem. Concr. Res.* **2015**, *78*, 100–109. [CrossRef]
20. Wallevik, J. Rheological properties of cement paste: Thixotropic behavior and structural breakdown. *Cem. Concr. Res.* **2009**, *39*, 14–29. [CrossRef]
21. Wallevik, J. Particle Flow Interaction Theory-Thixotropic Behavior and Structural Breakdown. In Proceedings of the Conference on Our World of Concrete and Structures, Singapore, 14–16 August 2011; pp. 1–6.
22. Hattori, K.; Izumi, K. Rheology of Fresh Cement and Concrete. In *Rheology of Fresh Cement and Concrete, Proceedings of the International Conference, London, UK*; CRC Press: Boca Raton, FL, USA, 1991; pp. 83–92.
23. Hattori, K.; Izumi, K. A rheological expression of coagulation rate theory. *J. Dispers. Sci. Technol.* **1982**, *3*, 129–193. [CrossRef]
24. Tattersall, G. The rheology of Portland cement pastes. *Br. J. Appl. Phys.* **1955**, *6*, 165–167. [CrossRef]
25. Ritchie, A. *The Rheology of Fresh Concrete*; Pitman Books Limited: Boston, MA, USA, 1983; pp. 73–95.
26. Wallevik, J. Rheology of Particle Suspensions: Fresh Concrete, Mortar and Cement Paste with Various Types of Lignosulfonates. Ph.D. Thesis, Norwegian University of Science and Technology, Trondheim, Norway, 2003.
27. Lapasin, R.; Papo, A.; Rajgelj, S. Flow behavior of fresh cement pastes. A comparison of different rheological instruments and techniques. *Cem. Concr. Res.* **1983**, *13*, 349–356. [CrossRef]
28. Hunter, R.J. *Foundations of Colloid Science*, 2nd ed.; Oxford University Press: New York, NY, USA, 2001; pp. 101–132, 131–144.
29. Williams, D.; Saak, A.; Jennings, H. The influence of mixing on the rheology of fresh cement paste. *Cem. Concr. Res.* **1999**, *29*, 1491–1496. [CrossRef]
30. Tattersall, G. Structural breakdown of cement pastes at constant rate of shear. *Nature* **1955**, *175*, 166. [CrossRef]
31. Ahari, R.; Erdem, T.; Ramyar, K. Thixotropy and structural breakdown properties of self-consolidating concrete containing various supplementary cementitious materials. *Cem. Concr. Compos.* **2015**, *59*, 26–37. [CrossRef]
32. Bullard, J.W.; Jennings, H.; Livingston, R.; Nonat, A.; Scherer, G.; Schweitzer, J.; Scrivener, K.; Thoma, J. Mechanisms of cement hydration. *Cem. Concr. Res.* **2011**, *41*, 1208–1223. [CrossRef]
33. Takahashi, K.; Bier, T.A. Westphal. Effects of mixing energy on technological properties and hydration kinetics of grouting mortars. *Cem. Concr. Res.* **2011**, *41*, 1167–1176. [CrossRef]
34. Cazacliu, B. In-mixer measurements for describing mixture evolution during concrete mixing. *Chem. Eng. Res. Des.* **2008**, *86*, 1423–1433. [CrossRef]
35. Cazacliu, B.; Roquet, N. Concrete mixing kinetics by means of power measurement. *Cem. Concr. Res.* **2009**, *39*, 182–194. [CrossRef]

36. He, Z.; Xie, K.; Zhang, C.; Xie, C. Activating mixing technology and its application in mine backfill. *GOLD* **2000**, *21*, 18–20.
37. Landriault, D.A. Backfill in underground mining. In *Underground mining methods: Engineering Fundamentals and International Case Studies*; Hustrulid, R.L., Bulloch, W., Eds.; Society for Mining, Metallurgy and Exploration-SME: Lilleton, CO, USA, 2001; pp. 601–614.
38. Barnes, A.; Merseyside, L.; Carnali, O. The vane-in-cup as a novel rheometer geometry for shear thinning and thixotropic materials. *J. Rheol.* **1990**, *34*, 841–866. [CrossRef]
39. Wu, D.; Fall, M.; Cai, S. Coupling temperature, cement hydration and rheological behaviour of fresh cemented paste backfill. *Miner. Eng.* **2013**, *42*, 76–87. [CrossRef]
40. Locher, F.W. *Zement-Grundlagen der Herstellung und Verwendung*; Vbt Verlag Bau+ Technik: Düsseldorf, Germany, 2000.
41. Kim, W.; Yang, S. Microstructures and Rheological Responses of Aqueous CTAB Solutions in the Presence of Benzyl Additives. *Langmuir* **2000**, *16*, 6084–6093. [CrossRef]
42. Dintzis, F.; Berhow, M.; Bagley, E.; Wu, Y.; Felker, F. Shear-thickening behavior and shear-induced structure in gently solubilized starches. *Cereal Chem.* **1996**, *73*, 638–643.
43. Roussel, N. A thixotropy model for fresh fluid concretes: Theory, validation and applications. *Cem. Concr. Res.* **2006**, *36*, 1797–1806. [CrossRef]
44. Liu, X.; Wu, A.; Wang, H.; Jiao, H.; Liu, S.; Wang, S. Experimental Studies on the Thixotropic Characteristics of Unclassified-tailings Paste Slurry. *J. Wuhan Univ. Technol.* **2014**, *38*, 539–543.
45. Rudman, M.; Blackburn, M.; Graham, J.; Pullum, L. Turbulent Pipe Flow of Shear-Thinning Fluids. *J. Non-Newton. Fluid Mech.* **2004**, *118*, 33–48. [CrossRef]
46. Larson, R. Constitutive equations for thixotropic fluids. *J. Rheol.* **2015**, *59*, 595–611. [CrossRef]
47. Barnes, H. Thixotropy—A review. *Non-Newton. Fluid Mech.* **1997**, *70*, 1–33. [CrossRef]
48. Barnes, H.; Nguyen, Q. Rotating vane rheometry—A review. *J. Non-Newton. Fluid Mech.* **2001**, *98*, S0257–S0377. [CrossRef]
49. Overbeek, J. Interparticle forces in colloid science. *Powder Technol.* **1984**, *37*, 195–208. [CrossRef]
50. Han, D.; Ferron, R. Influence of high mixing intensity on rheology, hydration, and microstructure of fresh state cement paste. *Cem. Concr. Res.* **2016**, *84*, 95–106. [CrossRef]
51. Nair, S.; Ferron, R. Set-on-demand concrete. *Cem. Concr. Res.* **2014**, *57*, 13–27. [CrossRef]
52. Struble, L.; Lei, W. Rheological changes associated with setting of cement paste. *Adv. Cem. Based Mater.* **1995**, *2*, 224–230. [CrossRef]
53. Erdem, K.; Khayat, K.; Yahia, A. Correlating, rheology of self-consolidating concrete to corresponding concrete-equivalent mortar. *ACI Mater. J.* **2010**, *106*, 154–160.
54. Cosgrove, T. *Colloid Science: Principles, Methods and Applications*; Blackwell Publishing: Ames, IA, USA, 2005; pp. 76–95.
55. Yang, M.; Neubauer, C.; Jennings, H. Interparticle potential and sedimentation behavior of cement suspensions-Review and results from paste. *Adv. Cem. Based Mater.* **1997**, *5*. [CrossRef]
56. Ferron, R. Formwork Pressure of Self-Consolidating Concrete: Influence of Flocculation Mechanisms, Structural Rebuilding, Thixotropy and Rheology. Ph.D. Thesis, Northwestern University, Evanston, IL, USA, 2008.
57. Takahashi, K.; Bier, T. Effects of mixing action on hydration kinetics and hardening properties of cement-based mortars. *Cem. Sci. Concr. Technol.* **2015**, *69*, 161–168. [CrossRef]
58. Han, D.; Ferron, R. Effect of mixing method on microstructure and rheology of cement paste. *Constr. Build. Mater.* **2015**, *93*, 278–288. [CrossRef]
59. Takahashi, K.; Bire, A. Mechanisms of Degradation in Rheological Properties Due to Pumping and Mixing. *Adv. Civ. Eng. Mater.* **2014**, *3*, 25–39. [CrossRef]

 © 2019 by the authors. Licensee MDPI, Basel, Switzerland. This article is an open access article distributed under the terms and conditions of the Creative Commons Attribution (CC BY) license (http://creativecommons.org/licenses/by/4.0/).

MDPI
St. Alban-Anlage 66
4052 Basel
Switzerland
Tel. +41 61 683 77 34
Fax +41 61 302 89 18
www.mdpi.com

Minerals Editorial Office
E-mail: minerals@mdpi.com
www.mdpi.com/journal/minerals

www.ingramcontent.com/pod-product-compliance
Lightning Source LLC
LaVergne TN
LVHW071943080526
838202LV00064B/6661